Fractional Order Processes

Fractional Order Processes

Simulation, Identification, and Control

by Seshu Kumar Damarla and
Madhusree Kundu

CRC Press
Taylor & Francis Group
Boca Raton London New York

CRC Press is an imprint of the
Taylor & Francis Group, an **informa** business

CRC Press
Taylor & Francis Group
6000 Broken Sound Parkway NW, Suite 300
Boca Raton, FL 33487-2742

CRC Press is an imprint of Taylor & Francis Group, an Informa business

Library of Congress Cataloging-in-Publication Data

Names: Damarla, Seshu Kumar., author. | Kundu, Madhusree, author.
Title: Fractional order processes : simulation, identification, and control / Seshu Kumar. Damarla and Madhusree Kundu.
Description: Boca Raton : Taylor & Francis, a CRC title, part of the Taylor & Francis imprint, a member of the Taylor & Francis Group, the academic division of T&F Informa, plc, 2018. | Includes bibliographical references and index.
Identifiers: LCCN 2018021873| ISBN 9781138586741 (hardback : acid-free paper) | ISBN 9780429504433 (ebook)
Subjects: LCSH: Fractional calculus. | Intelligent control systems–Mathematics. | Chaotic behavior in systems–Mathematical models.
Classification: LCC QA314 .D295 2018 | DDC 515/.83–dc23
LC record available at https://lccn.loc.gov/2018021873

Typeset in Palatino
by Integra Software Services Pvt. Ltd.

Visit the Taylor & Francis Web site at
http://www.taylorandfrancis.com

and the CRC Press Web site at
http://www.crcpress.com

Dedicated to the fond memory of my beloved father,
late Venugopalarao Damarla

Seshu Kumar Damarla

Dedicated to my students

Madhusree Kundu

Contents

Preface

This book discusses significant applications of triangular functions in simulation, identification, and control of fractional-order processes. Processes exhibiting fractional-order dynamics are called fractional-order processes. Fractional Calculus (FC) is an active branch of mathematical analysis that deals with the theory of differentiation and integration of arbitrary order. It is also known as Generalized Integral and Differential Calculus, and Calculus of Arbitrary Order. The concept of the fractional-order derivative was first discussed by Leibniz and L'Hospital almost three hundred years ago (i.e., at the end of the seventeenth century), a time when the foundations of the integer order calculus were developed by Isaac Newton and Gottfried Wilhelm Leibniz. Leibniz introduced the symbol $d^nf(x)/dx^n$, $n \in N$, to denote the nth derivative of a function $f(x)$. In Leibnizs' letter to Guillaume de l'Hospital, dated 30 September 1695 (which is considered as the date of birth of fractional calculus), he raised the question about the possibility of generalizing the operation of classical differentiation to noninteger orders. This question aroused l'Hospital's inquisitiveness and he replied to Leibniz with another question: "What does $d^nf(x)/dx^n$ if $n = 1/2$ mean?" Leibniz replied, "It will lead to a paradox from which one day useful consequence will be drawn." L'Hosptial's curiosity about the meaning of the semiderivative (i.e., $d^nf(x)/dx^n$, $n = 1/2$, a fraction or rational number) gave rise to the name of this subject (FC), and its name has remained the same, even though n can be any real number (rational or irrational). Although the same name is used throughout this book due to historical reasons, it should be understood to be noninteger (arbitrary real number) calculus, to be exact. As a matter of fact, even complex numbers can be allowed. Since the inception of FC, many great mathematicians (pure and applied), such as N. H. Abel, M. Caputo, L. Euler, A. K. Grunwald, J. Fourier, J. Hadamard, G. H. Hardy, O. Heaviside, H. J. Holmgren, P. S. Laplace, G. W. Leibniz, A. V. Letnikov, J. Liouville, B. Riemann, M. Riesz, and H. Weyl have contributed to this field. FC remained unknown to many applied mathematicians, scientists, and engineers because several definitions formulated for fractional-order derivative worked only in some situations. The mathematical theory of the subject seemed very different from that of integer order calculus. FC was considered to be an abstract area involving only mathematical manipulation of little or no use, and was thought to have no applications.

Almost three decades ago, the mathematics and applied mathematics fraternity realized the potential of FC and started developing essential mathematical theory to establish it. Since then, FC has emerged as an important and efficient tool for the study of dynamical systems where classical calculus reveals strong limitations. The books and monographs of Oldham and Spanier (1974), Oustaloup (1991, 1994, 1995), Miller and Ross (1993), Samko, Kilbas, and Marichev (1993), Kiryakova (1994), Carpinteri and Mainardi (1997), Podlubny (1999), and Hilfer (2000) have been instrumental in introducing FC to the pure and applied mathematics community. FC has been applied in diffusion processes, modeling of the mechanical properties of materials, signal processing, advection and dispersion of solutes in natural porous or fractured media, image processing, modeling of the behavior of viscoelastic and viscoplastic materials under external influences, pharmacokinetics, bioengineering, description of mechanical systems subject to damping, relaxation, and reaction kinetics of polymers, ultraslow processes, connections to the theory of random walks, finance, control theory, and psychology. FC has substantial applications within the various fields of mathematics itself. One of the major advantages of FC is that it includes the integer order calculus as a special case (i.e., a superset of integer order calculus). Therefore, FC can accomplish what its counterpart cannot achieve, especially capturing the memory and heredity of a process. FC is a useful and efficient tool to reveal many phenomena in nature because nature has memory. We believe that FC will be the only type of calculus in the future.

There are too many books available on the subject of FC. However, these books separate out topics in a way that can be confusing to students and less experienced researchers. There are some books addressing the "pure mathematical" side of the problems without taking into consideration those questions that arise in the applications mentioned earlier, and some that present the engineer's point of view without rigorous mathematical justification. Whereas a unified approach to address a variety of topics starting from theory to application along with source codes being available is very much desirable for less experienced researchers, it will save their time to be invested toward further advancements. All of our struggles at one point of time as beginners and our efforts to make a way out gave us an invaluable experience, one that impelled us to write this book.

An important goal of this book is to employ triangular orthogonal functions and triangular strip operational matrices to devise new numerical methods for simulation, identification, and control of fractional-order processes. The use of orthogonal functions as bases of expansion for squared integrable real-valued functions is a standard method in mathematical analysis and computational techniques. Numerous sets of orthogonal bases functions are available in mathematics. The existing sets of orthogonal functions can be categorized into two classes. The first class includes the classical sets of continuous functions such as sine-cosine functions, Legendre, Laguerre, Chebyshev, Jacobi, and Hermite orthogonal functions, and so on, which are continuous over their intervals of definition and consequently are well suited to approximate continuous functions. The second class consists of piecewise constant functions with inherent discontinuities. Walsh, block pulse, and Haar functions fall under the second category. The triangular orthogonal function sets, which are the foundation of most of numerical methods formulated in this book, are a complementary pair of piecewise linear polynomial function sets evolved from a simple dissection of block pulse function (BPF) set. The reason for choosing orthogonal functions as basis of the numerical methods presented in this book is that they can reduce the calculus of continuous dynamical systems to an attractive algebra, that is, they can convert integral, integro-differential, differential, differential-algebraic, and partial differential equations into a set of algebraic equations. The triangular functions are only selected among the existing enormous orthogonal functions because it is much easier to work with them than with others.

This book is organized into 10 chapters. The objectives, original contributions, and key findings of each chapter are summarized in the following paragraphs.

Chapter 1 is the backbone of the book, as it contains all of the mathematical postulations used in the subsequent chapters of the book. The special mathematical functions that play a vital role in establishing the proper definition of operators of fractional calculus are briefly reviewed. The most widely used definitions and properties of fractional-order integrals and fractional-order derivatives are provided. To analyze the behavior of linear lumped fractional-order systems, the Laplace transforms of fractional-order operators are derived, and then fractional-order systems are categorized along with a discussion about their stability. Different types of fractional-order controllers as well as advantages and disadvantages of the classical types are discussed. Triangular orthogonal functions are presented along with approximation formulae for estimating functions and fractional-order integrals. Finally, triangular strip operational matrices, which are the basis of the proposed robust controller tuning technique in Chapter 8, are derived from the classical finite difference formula and the Grunwald-Letnikov fractional-order derivative. Source codes developed in MATLAB® for triangular function operational matrices and triangular strip operational matrices are provided.

Weakly singular integral equations are extremely difficult to solve. In Chapter 2, an effective numerical method is developed using triangular orthogonal functions to solve weakly singular (WS) Fredholm, WS Volterra, and WS Volterra-Fredholm integral equations. Mathematical theory regarding the existence of unique solutions to weakly singular Volterra-Fredholm integral equations is provided. The convergence of the approximate solution to the actual solution is studied theoretically and numerically. The proposed method is tested on a set of test problems and then applied to physical process models described by weakly singular integral equations.

Abel's integral equation, one of the very first integral equations, was seriously investigated by Niels Henrik Abel in 1823 and by Liouville in 1832 as a fractional power of the operator of antiderivation. This equation is encountered in the inversion of seismic travel times, stereology of spherical particles, spectroscopy of gas discharges (more generally, "tomography" of cylindrically or spherically symmetric objects like, e.g., globular clusters of stars), and determination of the refractive index of optical fibers and electrochemistry. In Chapter 3, a novel numerical method using triangular orthogonal functions is developed to solve Abel's integral equation (fractional-order integral equation) of the first and second kind. It is proved that Abel's integral equations considered in this chapter have a unique solution in the given interval. The convergence analysis is carried out theoretically and numerically to prove that the proposed numerical algorithm can offer accurate approximate solutions that are very close to the true solutions of Abel's integral equations under consideration, provided that a relatively small step size is employed. A wide variety of Abel's integral equations is solved to demonstrate the applicability, accuracy, and stability of the proposed numerical algorithm. Encouraged by this success on the test problems, the proposed numerical method is applied to solve problems in electrochemistry, which are modeled by Abel's integral equations. The obtained results confirm the practical appropriateness of the numerical algorithm for applications of Abel's integral equations.

Integro-differential equations of fractional order find their applications in heat transfer, thermodynamics, electrical conduction of polymers, and many more. It is well known that most of physical process models involving fractional-order integro-differential equations do not have exact solutions. This fact has been the driving force for numerous researchers to conduct research toward the development of efficient numerical methods to simulate such physical process models. The objective of Chapter 4 is to propose a novel numerical method based on triangular orthogonal functions for the numerical solutions of fractional-order integro-differential equations such as Fredholm integro-differential equations of fractional order, Volterra integro-differential equations of fractional order, and Fredholm-Volterra integro-differential equations of fractional order. It is theoretically shown that there exists a unique solution to the general form of the system of fractional-order integro-differential equations considered in this chapter. Convergence analysis is conducted to prove that in the limit of step size tends to zero, the proposed numerical method ensures the convergence of the approximate solution to the exact solution of fractional-order integro-differential equations considered. Numerical examples as well as physical process models involving fractional-order integro-differential equations are solved to demonstrate the effectiveness of the proposed numerical method.

The development of a single numerical method that is able to solve different forms of fractional-order differential equations and fractional-order differential-algebraic equations is the prime objective of Chapter 5. Before construction of the numerical method, it is shown that the general form of system of fractional-order differential equations encompassing the aforesaid different forms has a unique solution in the given interval.

Convergence analysis is carried out to show that the approximate solution obtained by the proposed method can approach the original solution as the step size decreases to zero. The proposed method is applied to physical process models such as the Bagley-Torvik equation, the two-point Bagley-Torvik equation, the plant-herbivore model, the computer virus model, the chemical Akzo Nobel problem, Robertson's system describing the kinetics of autocatalytic reaction, and the high irradiance response of photo morphogenesis. In addition to the proposed method, the most popular semi-analytical techniques such as the Adomian decomposition method (ADM), the homotopy analysis method (HAM), and the fractional differential transform method with Adomian polynomials (FDTM) are implemented as well on physical process models involving a stiff system of differential equations or stiff differential-algebraic equations. It is astonishing to note that ADM, HAM, and FDTM fail to simulate those process models even in the neighborhood of the initial time point 0, although they have successfully simulated many other physical process models. By contrast, the proposed method is able to produce valid approximate solution not only in the vicinity of the initial time point 0 but also in the desired time interval, which can be quite a bit larger than [0, 1].

In Chapter 6, the triangular functions–based numerical method is formulated to simulate fractional diffusion-wave equation. It is theoretically proved that the proposed method converges the approximate solution to the original solution of fractional diffusion-wave equation in the limit of step size tends to zero.

Because the geometric and physical interpretation of fractional calculus is not as distinct as integer calculus, it is difficult to model real systems as fractional-order systems directly based on mechanistic analysis. Therefore, system identification is a practical way to model a fractional-order system, using experimental or simulated process data. Chapter 7 introduces an arbitrary order (note that the order can be integer or noninteger) system identification method based on the triangular orthogonal functions to estimate parameters including arbitrary differential orders and initial conditions of a model under consideration from experimental or simulated data. Five identification problems encompassing integer- and noninteger-order linear and nonlinear systems are given to validate the proposed method. It is proved that the proposed method works well for various kinds of input excitation signal such as step signal, pseudo random binary signal, square wave signal, Sawtooth wave signal, and pulse signal. The obtained results are compared with the results of some of the existing numerical methods, and it is found that the proposed method demonstrates superior performance over those methods.

To tackle plant uncertainty issues, many controller design methods are developed. The convenient one among these methods is designing a robust fractional $PI^{\lambda}D^{\mu}$ controller. Chapter 8 proposes a simple tuning technique aimed to produce a robust noninteger order PID controller exhibiting iso-damping property during the reparameterization of a plant. The required robustness property is achieved by allowing the fractional PID control system to imitate the dynamics of a reference system with Bode's ideal transfer function in its forward path. The objective of designing robust controller by tracking the dynamics of reference control system is defined mathematically as an H_∞-optimal control problem. Fractional differential systems are transformed into algebraic equations by the use of triangular strip operational matrices. The H_∞-optimal control problem is then changed to an ∞-norm minimization of a parameter $(K_C, K_I, K_d, \lambda, \mu)$ varying square matrix. Global optimization techniques; Luus-Jaakola direct searche, and particle swarm optimization are employed to find the optimum values of fractional PID controller parameters. The proposed method of control system design is implemented in heating furnace temperature control, automatic voltage regulator systems, and some integer and fractional-order process models. Fractional PI^λ, fractional PD^μ, $PI^\lambda D^\mu D^{\mu 2}$, fractional PID with fractional-order filter, and the series form of fractional PID controller are designed as optimal controllers using the triangular strip operational matrix–based control design method. The performance of the proposed fractional-order controller tuning technique is found to be better than the performance of some fractional-order controller tuning methodologies reported in the literature. Triangular strip operational matrices proposed from the perspective of mathematics (for the solution of fractional differential and partial differential equation) finds its elegant application in the proposed method of control system design.

Chapter 9 presents a new numerical method to find an equivalent finite dimensional integer order system for an infinite dimensional fractional-order system. The approximated rational integer order system owns characteristics close enough to that of irrational fractional-order system that can be used in place of the original fractional-order system. In comparison to Crone, Carlson, Charef, Matsuda, and continued fraction expansion approximation methods, the proposed method offers lower order rational approximation that precisely mimics the dynamics of the original irrational fractional-order system. One significant attribute of the proposed method, which none of the abovementioned methods possess, is that the order of rational approximation can be controlled while obtaining accurate approximation to the fractional-order system.

Optimal control problems arise naturally in various areas of science, engineering, and mathematics. Considerable work has been done in the area of integer optimal control problems (IOCPs), whose dynamics are described by conventional integer differential equations. Recently, it has been demonstrated that fractional differential equations are more accurate than integer differential equations to describe the dynamic behavior of many real-world processes. Fractional optimal control problems (FOCPs) are a subclass of classical optimal control problems whose dynamics are described by fractional differential equations. It is well known that the analytical solutions of FOCPs generally do not exist except for special cases, and therefore, numerical methods to obtain an approximate solution have become the preferred approach for solving FOCPs. An effective numerical method is devised in Chapter 10 to solve FOCPs. The proposed method is implemented on a set of examples comprising linear time invariant optimal control problems, time-variant fractional-order optimal control problems, nonlinear fractional-order optimal control problems, and two-dimensional integer order optimal control problems. Comparisons between the obtained results and that of some of the existing numerical methods shows the superiority of the proposed method over most of the others.

This book may be treated as an interdisciplinary reference book, and it also may be used as a textbook for courses related to numerical methods for fractional-order systems, fractional-order process identification, fractional-order controller design, and fractional-order optimal control. The book is suitable for engineering and basic science researchers and scientists irrespective of their disciplines. The broader scope of the book makes it suitable for the budding researcher. Final year mechanical, electrical, chemical, mathematics, physics, and biomedical graduate students will find support when consulting this book for research-oriented courses. The reader is expected to be acquainted with classical calculus (differential and integral calculus, elementary theory of integral, differential, integro-differential, and partial differential equations), process identification and control, and optimal control. After reading Chapter 1, the reader will be able to understand the concepts presented in the subsequent chapters. The extensive literature survey on each concept addressed in the book is deliberately avoided. Every chapter begins with the necessary basic information and adequate background for the reader to grasp the concepts discussed in the chapter. Source codes developed in MATLAB are provided in each chapter, allowing the interested reader to take advantage of these codes to broaden and enhance the scope of the book itself as well as develop new results. We hope that readers will find this book useful and valuable in the advancement of their knowledge and their field. We look forward to receiving comments and suggestions from researchers, pure and applied mathematicians, scientists, and engineers.

India, April 2018

Seshu Kumar Damarla
Madhusree Kundu

Acknowledgments

We are grateful to the reviewers: Dr. Sillas Hadjiloucas (Department of Bio-engineering, University of Reading), Mr. Amr Elsonbaty (Engineering Mathematics and Physics Department, Faculty of Engineering, Mansoura University, Egypt), Dr. YangQuan Chen (University of California Merced), and three anonymous reviewers for their valuable suggestions that helped to enhance the quality of the book.

We would like to thank the CRC press team, in particular, Dr. Gagandeep Singh, Senior Editor (Acquisitions) for engineering/environmental sciences, and CRC Press, for their patience, encouragement, invaluable suggestions, and keeping trust in us during the process of writing this book.

We are thankful to HOD, Chemical Engineering, NIT Rourkela, and Director, NIT Rourkela for their kind cooperation during preparation of the book. Our sincere thanks go to those who directly or indirectly helped us during preparation of the book.

Last but not least, we thank our families for their unconditional love and support.

About the Authors

Seshu Kumar Damarla was born in the year 1985 in Chirala, Prakasam, Andhra Pradesh, India. He did his B.Tech (Chemical Engineering) from Bapatla Engineering College, Bapatla, Andhra Pradesh, India (2008), and M.Tech (Chemical Engineering) from NIT Rourkela, Odisha, India (2011). Mr. Damarla submitted his Ph.D dissertation (Developing Numerical Methods for Simulation, Identification and Control of Fractional Order Process) to NIT Rourkela, Odisha, India (2017). Mr. Damarla served as an Assistant Professor for a short duration (from 5th August 2011 to 31st December 2011) in Department of Chemical Engineering at Maulana Azad National Institute of Technology Bhopal, Madhya Pradesh, India, and for a year (from July 2017 to July 2018) in Department of Chemical Engineering at C.V. Raman College of Engineering, Bhubaneswar, Odisha, India. Mr. Damarla has published a couple of research articles in the internationally refereed journals to his credit and also published in the proceedings of national and international conferences. Mr. Damarla co-authored a reference textbook *Chemometric Monitoring: Product Quality Assessment, Process Fault Detection, and Applications* (CRC Press). Mr. Damarla has been a referee for *Acta Biotheoretica, Journal of King Saud Science, and Applied and Computational Mathematics*. Mr. Damarla is a member of International Association of Engineers (IAENG), Fractional Calculus and Application Group, and Allahabad Mathematical Society.

Madhusree Kundu started her academic pursuits with a graduation in chemistry, with honors (University of Calcutta) followed by graduation and post-graduation in chemical engineering from the Rajabazar Science College, University of Calcutta, (1990–1992). Dr. Kundu gained experience as a process engineer at Simon Carves (I) Ltd., Kolkata (1993–1998). In the next phase of her scholarly pursuit, Dr. Kundu earned her Ph.D from the Indian Institute of Technology, Kharagpur (1999–2004), and started her academic profession as the faculty of the Chemical Engineering Group, BITS Pilani, Rajasthan (2004–2006). She joined the NIT Rourkela in 2007 and is continuing there as Professor in the Department of Chemical Engineering. Apart from teaching, she has focused her research activities in chemometrics along with fractional order process modeling and control, solution thermodynamics, and fluid-phase equilibria. Dr. Kundu has authored several research articles in International refereed journals and has a few book chapters, and a reference text book (*Chemometric Monitoring: Product Quality Assessment, Process Fault Detection, and Applications*, CRC Press) to her credit.

1

Mathematical Postulations

1.1 Special Functions

In this section, the important functions that are essential for the formulation of fractional-order integral and fractional-order derivative, and that will be used in the following chapters, are provided.

1.1.1 Gamma Function [1]

The gamma function is the generalization of the factorial function to non-integer numbers. It plays a vital role in defining fractional-order integrals and fractional-order derivatives.

The complete gamma function is defined by the Euler limit as

$$\Gamma(x) = \lim_{N \to \infty} \left[\frac{N!N^x}{x(x+1)(x+2)\ldots\ldots(x+N)} \right] \tag{1.1}$$

and by the integral transform as

$$\Gamma(x) = \int_0^\infty e^{-y} y^{x-1} dy, \; x > 0. \tag{1.2}$$

The definition in Equation (1.2) is more practicable than that in Equation (1.1), even though it is restricted to positive values of x.

Following are some useful properties of the gamma function.

The recurrence relationship given here is the most important property of the gamma function, and it can be obtained by applying the integration by parts to Equation (1.2):

$$\Gamma(x+1) = x\Gamma(x). \tag{1.3}$$

For $x \le 0$, the gamma function in Equation (1.2) becomes infinite. But the ratios of the gamma functions of negative integers are, however, finite, as shown in the next equation;

$$\frac{\Gamma(-P)}{\Gamma(-Q)} = (-1)^{Q-P}\frac{Q!}{P!}, \text{ where } P, Q \text{ are arbitrary positive integers.} \tag{1.4}$$

The reflection of the gamma function $\Gamma(x)$ is

$$\Gamma(-x) = \frac{-\Pi\text{cosec}(\Pi x)}{\Gamma(x+1)} \tag{1.5}$$

and the duplication is

$$\Gamma(2x) = \frac{4^x\Gamma(x)\Gamma\left(x+\frac{1}{2}\right)}{2\sqrt{\Pi}}. \tag{1.6}$$

The incomplete gamma function is

$$\gamma^*(c,x) = \frac{c^{-x}}{\Gamma(x)}\int_0^c y^{x-1}exp(-y)dy = exp(-x)\sum_{j=0}^{\infty}\frac{x^j}{\Gamma(j+c+1)}, \tag{1.7}$$

where $\gamma^*(c,x)$ is a finite single-valued analytic function of x and c.

The recursive relation of $\gamma^*(c,x)$ is

$$\gamma^*(c-1,x) = x\gamma^*(c,x) + \frac{exp(-x)}{\Gamma(c)}. \tag{1.8}$$

1.1.2 Beta Function [1]

The complete beta function, which is also known as Euler's integral of the second kind, is defined by the beta integral as

$$B(p,q) = \int_0^1 y^{p-1}(1-y)^{q-1}dy,\ p>0,\ q>0. \tag{1.9}$$

The following relationship between the complete beta function and the complete gamma function can be used if either p or q is nonpositive; if this is the case, the integral in Equation (1.9) diverges:

$$B(p,q) = \frac{\Gamma(p)\Gamma(q)}{\Gamma(p+q)}, \ \forall p,q. \tag{1.10}$$

The incomplete beta function of argument x is defined by

$$B_x(p,q) = \int_0^x y^{p-1}(1-y)^{q-1}dy. \tag{1.11}$$

1.1.3 Mittag-Leffler Function

The following one-parameter Mittag-Leffler function, introduced by Mittag-Leffler in [2], is an essential function used in modelling physical processes with the help of the fractional calculus concepts:

$$E_\alpha(z) = \sum_{k=0}^{\infty} \frac{z^k}{\Gamma(\alpha k+1)}. \tag{1.12}$$

The classical exponential function can be acquired from Equation (1.12) if $\alpha = 1$. The two-parameter Mittag-Leffler function, which is equally important as Equation (1.12) in fractional calculus, is given in the next equation [3]:

$$E_{\alpha,\beta}(z) = \sum_{k=0}^{\infty} \frac{z^k}{\Gamma(\alpha k+\beta)}, \ \alpha,\beta > 0. \tag{1.13}$$

1.1.4 Hypergeometric Function

The generalized hypergeometric function, which embraces several analytical functions as specific or limiting cases, is defined as

$${}_pF_q(a_1,\ldots\ldots,a_p,b_1,\ldots\ldots,b_q;z) = \frac{\Gamma(b_1)\ldots\ldots\Gamma(b_q)}{\Gamma(a_1)\ldots\ldots\Gamma(a_p)} \sum_{k=0}^{\infty} \frac{\Gamma(a_1+k)\ldots\ldots\Gamma(a_p+k)}{\Gamma(b_1+k)\ldots\ldots\Gamma(b_q+k)} \frac{z^k}{k!}, \tag{1.14}$$

where b_i's are not nonpositive integers.

The series in Equation (1.14) converges for all z if $p \leq q$, and for $|z| < 1$ if $p = q + 1$. It diverges for all nonzero z if $p > q + 1$.

For the special case of $p = 2$ and $q = 1$, the generalized hypergeometric function in Equation (1.14) produces the well-known Gauss hypergeometric function

$$ {}_2F_1(a, b, c; z) = \frac{\Gamma(c)}{\Gamma(a)\Gamma(b)} \sum_{k=0}^{\infty} \frac{\Gamma(a+k)\Gamma(b+k)}{\Gamma(c+k)} \frac{z^k}{k!}, \tag{1.15} $$

which frequently arises in physical problems.

The relation among the hypergeometric function, the incomplete gamma function, and the incomplete beta function is

$$ \gamma^*(v, z) = \frac{1}{\Gamma(v+1)} {}_1F_1(v, v+1; z) = \frac{1}{\Gamma(v+1)} e^{-z} {}_1F_1(1, v+1; z), \tag{1.16} $$

$$ B_\tau(x, y) = x^{-1}\tau^x {}_2F_1(x, 1-y, x+1; \tau) = x^{-1}\tau^x (1-\tau)^y {}_2F_1(x+y, 1, x+1; \tau). \tag{1.17} $$

1.1.5 Error Function and Complementary Error Function

The error function, which comes across in integrating normal distribution and is an entire function, is defined as

$$ erf(z) = \frac{2}{\sqrt{\Pi}} \int_0^z e^{-t^2} dt. \tag{1.18} $$

When the argument z becomes infinite, then $erf(\infty) = 1$. The series form of the error function is

$$ erf(z) = \frac{2}{\sqrt{\Pi}} \sum_{n=0}^{\infty} \frac{(-1)^n z^{2n+1}}{(2n+1)n!} = \frac{2}{\sqrt{\Pi}} \left(z - \frac{z^3}{3} + \frac{z^5}{10} - \frac{z^7}{42} + \frac{z^9}{216} + \ldots\ldots \right). \tag{1.19} $$

The error function can be expressed by means of the hypergeometric function:

$$ erf(x) = 2\Pi^{-1/2} x e^{-x^2} {}_1F_1\left(1, \frac{3}{2}; x^2\right) = 2\Pi^{-1/2} x {}_1F_1\left(\frac{1}{2}, \frac{3}{2}; -x^2\right). \tag{1.20} $$

The complementary error function is an entire function defined by

$$erfc(z) = 1 - erf(z) = 1 - \frac{2}{\sqrt{\Pi}}\int_0^z e^{-t^2}dt = \frac{2}{\sqrt{\Pi}}\int_z^\infty e^{-t^2}dt \tag{1.21}$$

and its series asymptotic expansion is

$$erfc(z) = \frac{e^{-z^2}}{z\sqrt{\Pi}}\left(1+\sum_{n=1}^{\infty}(-1)^n\frac{1\cdot 3\cdot 5\cdot \ldots (2n-1)}{(2z^2)^n}\right) = \frac{e^{-z^2}}{z\sqrt{\Pi}}\left(1+\sum_{n=1}^{\infty}(-1)^n\frac{(2n)!}{n!(2z)^{2n}}\right). \tag{1.22}$$

1.1.6 Bessel Functions

Of all higher transcendental functions, the Bessel functions are possibly most omnipresent. They appear often in physical phenomena such as electromagnetic waves in a cylindrical waveguide, pressure amplitudes of inviscid rotational flows, heat conduction in a cylindrical object, diffusion problems on a lattice, dynamics of floating bodies, signal processing, and so on.

The Bessel functions of the first kind are the solutions to the Bessel differential equations

$$x^2\frac{d^2y}{dx^2} + x\frac{dy}{dx} + (x^2 - n^2)y = 0, \tag{1.23}$$

which are finite at the origin $x = 0$.

The Bessel functions of the first kind, which are also called cylindrical functions or cylindrical harmonics, are defined by infinite series expansion as

$$J_n(x) = \sum_{m=0}^{\infty}\frac{(-1)^m}{m!\Gamma(m+n+1)}\left(\frac{x}{2}\right)^{2m+n}, \quad n \text{ is the order of the Bessel function.} \tag{1.24}$$

The Bessel functions of the second kind defined here, occasionally called Weber functions or Neumann functions, are the solutions to Equation (1.23) when it becomes singular at $x = 0$:

$$Y_n(x) = \frac{J_n(x)\,cos(n\Pi) - J_{-n}(x)}{sin(n\Pi)}. \tag{1.25}$$

1.2 Definitions and Properties of Fractional-Order Operators

In this section, the notion of generalizing classical calculus to derive definitions of fractional-order integral and fractional-order derivative is presented [4].

1.2.1 Grunwald-Letnikov (GL) Fractional-Order Derivative

Extrapolating the applicability of classical backward difference formula to derivative of non-integer order gives rise to the formation of the Grunwald-Letnikov (GL) fractional-order derivative. Perhaps, GL is the first definition proposed for differentiation of noninteger order.

Let us consider a real function $f(t)$ $(t\in[0,b])$. The function is said to be in the space $C_\mu, \mu\in R$ if there exists a real number $p(>\mu)$, such that $f(t)=t^p f_1(t)$, where $f_1(t)\in C[a,\infty)$, and it is said to be in the space C_μ^n if and only if $f^{(n)}\in C_\mu$, $n\in N$.

We now express the nth order derivative (n is an integer) of the casual function $f(t)$ (i.e., $f(t)=0,\ t<0$) in terms of backward difference formula:

$$\frac{d^n f(t)}{dt^n} \cong \frac{\nabla^n f(t)}{h^n} = h^{-n}\sum_{i=0}^{N}\left((-1)^i\binom{n}{i}f(t-ih)\right), \tag{1.26}$$

where $\binom{n}{i}=\frac{n!}{i!(n-i)!}$, $h=(b-a)/N$, N is the total number of equidistant nodes in the interval $[0,b]$.

Equation (1.26) is the discretized form of nth order derivative of the function, $f(t)$. Because $\binom{n}{i}$ becomes zero for all values of i greater than n, the upper limit of summation in this definition can be increased to infinity.

Rewrite Equation (1.26) as

$${}_{0}^{GL}D_t^n f(t) \cong h^{-n}\sum_{i=0}^{\infty}\left((-1)^i\binom{n}{i}f(t-ih)\right),\ D=\frac{d}{dt}. \tag{1.27}$$

The following definition for the Grunwald-Letnikov fractional-order derivative can be obtained by putting α in place of n in Equation (1.27):

$${}_{0}^{GL}D_t^\alpha f(t) \cong h^{-\alpha}\sum_{i=0}^{\infty}\left((-1)^i\binom{\alpha}{i}f(t-ih)\right). \tag{1.28}$$

When the sign of α is negative, Equation (1.28) turns out to be a fractional-order integral.

The integral transform definition of GL fractional-order derivative is [5]

$$ {}_{0}^{GL}D_{t}^{\alpha}f(t) = \sum_{k=0}^{n-1} \frac{f^{(k)}(0)t^{-\alpha+k}}{\Gamma(-\alpha+k+1)} + \frac{1}{\Gamma(n-\alpha)} \int_{0}^{t} (t-\tau)^{n-\alpha-1} f^{(n)}(\tau)d\tau, \quad (1.29) $$

where $n-1 \leq \alpha < n, n \in Z^{+}, t > 0$.

The reason why the fractional-order derivative possesses nonlocal property is that the term $\binom{\alpha}{i}$ in Equation (1.28) will never become zero; that is, determining the fractional-order derivative of any function requires its entire history. Therefore, it needs infinite memory and thus is more suitable to explain long memory processes mathematically. It is worth mentioning here that classical calculus is a particular case of the fractional calculus.

The GL fractional-order derivative in Equation (1.29) is the left fractional-order derivative, because the lower terminal of the fractional integral is fixed at the left end of the interval $[0, b]$ and the upper terminal moves in the interval. If the upper terminal of the fractional integral in Equation (1.29) is fixed at the right end of the interval $[0, b]$ and the lower terminal is moving, then the GL fractional derivative is called the right fractional derivative. Let us suppose that the independent variable t is time and the function $f(t)$ describes the dynamic behavior of a process. If $\tau < t$ (t is the current instant), then the past of this process can be described by the state $f(\tau)$. If $\tau > t$, then the state $f(\tau)$ belongs to the future of the process. Therefore, the left fractional derivative is only considered throughout this book as it requires the past information of the process to describe it mathematically.

1.2.2 Riemann-Liouville (RL) Fractional-Order Integral

Cauchy's formula for repeated integration, which reduces n-fold integration of function $f(t)$ to single integral, is:

$$ f^{-n}(t) = {}_{0}J_{t}^{n}f(t) = \frac{1}{n-1} \int_{0}^{t} (t-x)^{n-1} f(x)dx, \quad (1.30) $$

where n is a positive integer.

Equation (1.30) can be written as

$$ {}_0J_t^n f(t) = \frac{1}{\Gamma(n)} \int_0^t (t-x)^{n-1} f(x)dx, \tag{1.31} $$

where $\Gamma(n)$ is a well-known Euler's Gamma function: $\Gamma(n) = \int_0^\infty e^{-x}x^{n-1}dx$.

Equation (1.31) permits us to replace n with α to obtain a fractional-order integral:

$$ {}_0J_t^\alpha f(t) = \frac{1}{\Gamma(\alpha)} \int_0^t (t-x)^{\alpha-1} f(x)dx. \tag{1.32} $$

1.2.3 Riemann-Liouville Fractional-Order Derivative

The left Riemann-Liouville (RL) fractional-order derivative of function $f(t)$ is defined as

$$ {}_0^{RL}D_t^\alpha f(t) = D^n J^{n-\alpha} f(t) = \frac{1}{\Gamma(m-\alpha)} \frac{d^n}{dt^n} \int_0^t (t-\tau)^{n-\alpha-1} f(\tau)d\tau, t > 0, \tag{1.33} $$

where α is a noninteger that satisfies the relation $n-1 < \alpha \le n, n \in Z^+$.

The right Riemann-Liouville fractional-order derivative is

$$ {}_0^{RL}D_t^\alpha f(t) = D^n J^{n-\alpha} f(t) = \frac{1}{\Gamma(n-\alpha)} \frac{d^n}{dt^n} \int_t^b (t-\tau)^{n-\alpha-1} f(\tau)d\tau, t < b. \tag{1.34} $$

1.2.4 Caputo Fractional-Order Derivative

As we shall show in the next subsection, Riemann-Liouville fractional differential equations lack widespread physical applications because of the need for fractional-order initial conditions. To enable fractional calculus concepts to be applied in different applied branches of science and technology, Caputo modified Equation (1.34) as shown in the following definition [6].

The left Caputo fractional-order derivative is

$$ {}_0^{C}D_t^\alpha f(t) = J^{n-\alpha} f^n(t) = \frac{1}{\Gamma(n-\alpha)} \int_0^t (t-\tau)^{n-\alpha-1} f^n(\tau)d\tau, \ t > 0 \tag{1.35} $$

and the right Caputo fractional derivative is

$$ {}_{0}^{C}D_{t}^{\alpha}f(t) = J^{n-\alpha}f^{n}(t) = \frac{1}{\Gamma(n-\alpha)}\int_{t}^{b}(t-\tau)^{n-\alpha-1}f^{n}(\tau)d\tau,\ t<b. \qquad (1.36) $$

When there are homogenous initial conditions, Riemann-Liouville and Caputo fractional-order derivatives are equivalent. The definitions of operators of fractional calculus presented so far will become the operators of classical calculus when the fractional-order is equal to an integer. Like operators of classical calculus, fractional-order operators also have a physical interpretation. Fractional-order integrals can be understood as the area under a shape changing curve, whereas fractional-order derivatives imply the integer order derivative of area under the shape-changing curve [7].

1.2.5 Properties of GL, RL, and Caputo Fractional-Order Derivatives

Some useful properties of fractional-order operators that we shall use in the following chapters are provided here [4, 5].

For $f(t)\in C_{\mu}, \mu > -1$ and $n-1 \le \alpha < n, p-1 \le \beta < p, p, n, q \in Z^{+}, \alpha, \beta \in R^{+}$:

- ${}_{0}J_{t}^{\alpha}{}_{0}J_{t}^{\beta}f(t) = {}_{0}J_{t}^{\beta}{}_{0}J_{t}^{\alpha}f(t) = {}_{0}J_{t}^{\alpha+\beta}f(t)$ (semigroup and commutative property).
- $\lim_{\alpha\to n}\left({}_{0}J_{t}^{\alpha}f(t)\right) = {}_{0}J_{t}^{n}f(t)$ (consistency property with the integer order integral).
- ${}_{0}^{GL}D_{t}^{\alpha}c = (ct^{-\alpha})/\Gamma(1-\alpha)$, ${}_{0}^{RL}D_{t}^{\alpha}c = (ct^{-\alpha})/\Gamma(1-\alpha)$, ${}_{0}^{C}D_{t}^{\alpha}c = 0$, c is a constant.
- ${}_{0}^{C}D_{t}^{\alpha}f(t) = {}_{0}^{RL}D_{t}^{\alpha}\left(f(t) - \sum_{k=0}^{n-1}\frac{t^{k}}{k!}f^{(k)}(0)\right)$.
- ${}_{0}^{C}D_{t}^{\alpha}{}_{0}J_{t}^{\alpha}f(t) = {}_{0}^{RL}D_{t}^{\alpha}{}_{0}J_{t}^{\alpha}f(t) = f(t)$ holds for $n = 1$.
- ${}_{0}J_{t}^{\alpha}{}_{0}^{C}D_{t}^{\alpha}f(t) = f(t) - \sum_{k=0}^{n-1}\frac{t^{k}}{k!}f^{(k)}(0)$.
- ${}_{0}^{C}D_{t}^{\alpha}\left({}_{0}D_{t}^{q}f(t)\right) = {}_{0}D_{t}^{q}\left({}_{0}^{C}D_{t}^{\alpha}f(t)\right) = {}_{0}^{C}D_{t}^{\alpha+q}f(t)$, $f^{(s)}(0) = 0,\ s = n, n+1, \ldots, q$.
- ${}_{0}^{RL}D_{t}^{\alpha}\left(D_{0,t}^{q}f(t)\right) = {}_{0}D_{t}^{q}\left({}_{0}^{RL}D_{t}^{\alpha}f(t)\right) = {}_{0}^{RL}D_{t}^{\alpha+q}f(t)$, $f^{(s)}(0) = 0,\ s = 0, 1, \ldots, q$.

- $\begin{cases} {}_0D_t^{\alpha}\left({}_0D_t^{\beta}f(t)\right) = {}_0D_t^{\beta}\left({}_0D_t^{\alpha}f(t)\right) \neq {}_0D_t^{\alpha+\beta}f(t) \\ {}_0D_t^{\alpha}\left({}_0D_t^{\beta}f(t)\right) \neq {}_0D_t^{\beta}\left({}_0D_t^{\alpha}f(t)\right) = {}_0D_t^{\alpha+\beta}f(t) \end{cases}$,

${}_0D_t^{\gamma} \in [{}_0^{RL}D_t^{\gamma}, {}_0^{C}D_t^{\gamma}], \gamma \in [\alpha, \beta]$.

1.3 Laplace Transforms of Fractional-Order Operators

Before obtaining Laplace transforms of fractional-order integrals and fractional-order derivatives, let us briefly review the basics of the Laplace transform.

Let us suppose the function, $f(t)$. The Laplace transform of $f(t)$ is defined by

$$L\left(f(t)\right) = F(s) = \int_0^{\infty} e^{-st}f(t)dt. \tag{1.37}$$

This definition is valid if and only if the function $f(t)$ is of exponential order β, that is $e^{-\beta t}|f(t)| \leq M$, $\forall t > \delta$, where M, δ are positive constants.

The original function $f(t)$ can be retrieved from $F(s)$ by using the following inverse Laplace transform:

$$f(t) = L^{-1}\left(F(s)\right) = \frac{1}{2\Pi i} \lim_{a \to \infty} \int_{c-ia}^{c+ia} e^{st}F(s)ds, \; c = \mathrm{Re}(s) > c_0, \tag{1.38}$$

where c_0 lies in the right half plane of the absolute convergence of Equation (1.37).

Let us assume two functions $f_1(t)$ and $f_2(t)$ ($f_1(t) = f_2(t) = 0$, $t < 0$). The convolution of $f_1(t)$ and $f_2(t)$ creates a function $f_3(t)$ according to the following equation:

$$f_3(t) = f_1(t) * f_2(t) = \int_0^t f_1(t-\tau)f_2(\tau)d\tau = \int_0^t f_1(\tau)f_2(t-\tau)d\tau. \tag{1.39}$$

The mathematical operation in Equation (1.39) is valuable in mathematical physics and probability theory.

The Laplace transform of Equation (1.39) is

$$L\big(f_3(t)\big) = L\big(f_1(t) * f_2(t)\big) = L\big(f_1(t)\big)L\big(f_2(t)\big) = F_1(s)F_2(s). \tag{1.40}$$

Equation (1.39) will be used in the derivation of the Laplace transform of the Riemann-Liouville fractional-order integral. For the Laplace transform of the fractional-order derivative, we need the following equation:

$$L\left(D^n f(t)\right) = s^n F(s) - \sum_{k=0}^{n-1} s^{n-k-1} f^{(k)}(0) = s^n F(s) - \sum_{k=0}^{n-1} s^k f^{(n-k-1)}(0). \tag{1.41}$$

Let us recall the Riemann-Liouville fractional-order integral:

$${}_0J_t^\alpha f(t) = \frac{1}{\Gamma(\alpha)} \int_0^t (t-\tau)^{\alpha-1} f(\tau)d\tau = \frac{t^{\alpha-1} * f(t)}{\Gamma(\alpha)}. \tag{1.42}$$

Following Equation (1.39),

$$L\Big({}_0J_t^\alpha f(t)\Big) = \frac{1}{\Gamma(\alpha)} L\Big(t^{\alpha-1} * f(t)\Big) = \frac{1}{\Gamma(\alpha)} L(t^{\alpha-1}) L\Big(f(t)\Big) = \frac{F(s)}{s^\alpha}. \tag{1.43}$$

By using Equations (1.40) and (1.42), we can get the Laplace transform of the Riemann-Liouville fractional-order derivative:

$$\begin{aligned}
L\left({}^{RL}_{\ 0}D_t^\alpha f(t)\right) &= L\left({}_0D_t^n\, {}_0J_t^{n-\alpha} f(t)\right) = s^n L\left({}_0J_t^{n-\alpha} f(t)\right) - \sum_{k=0}^{n-1} s^k \left[{}_0D_t^{n-k-1}\, {}_0J_t^{n-\alpha} f(t)\right]_{t=0}, \\
&= s^n \frac{F(s)}{s^{n-\alpha}} - \sum_{k=0}^{n-1} s^k \left[{}_0J_t^{-(n-k-1)}\, {}_0J_t^{n-\alpha} f(t)\right]_{t=0}, \\
&= s^\alpha F(s) - \sum_{k=0}^{n-1} s^k \left[{}_0J_t^{-(\alpha-k-1)} f(t)\right]_{t=0}, \\
&= s^\alpha F(s) - \sum_{k=0}^{n-1} s^k \left[{}^{RL}_{\ 0}D_t^{\alpha-k-1}(f(t))\right]_{t=0}, n-1 \le \alpha < n,\ n \in Z^+.
\end{aligned} \tag{1.44}$$

Similarly, the Laplace transform of the Caputo fractional-order derivative is

$$\begin{aligned} L\left({}_0^C D_t^\alpha f(t)\right) &= L\left({}_0 J_t^{n-\alpha}{}_0 D_t^n f(t)\right) = \frac{L\left({}_0 D_t^n f(t)\right)}{s^{n-\alpha}}, \\ &= s^{n-\alpha}\left(s^n F(s) - \sum_{k=0}^{n-1} s^{n-k-1}\left[f^{(k)}(t)\right]_{t=0}\right), \\ &= s^\alpha F(s) - \sum_{k=0}^{n-1} s^{\alpha-k-1}\left[f^{(k)}(t)\right]_{t=0}, \quad n-1 \leq \alpha < n,\ n \in Z^+. \end{aligned} \tag{1.45}$$

Equation (1.44) reveals that, in most scenarios, the RL fractional-order derivative does not work in physical process models, as it requires evaluation of the fractional-order derivative of $f(t)$ at the lower terminal $t = 0$. There are very few situations in which these fractional initial conditions have a clear physical meaning [8]. From an application point of view, the Caputo form is more appropriate as it takes the values of function and its $(m - 1)$ integer order derivatives at the lower terminal $t = 0$.

The Laplace transform of the GL fractional-order derivative can be obtained as shown here by using Equations (1.29) and (1.44):

$$\begin{aligned} L\left({}_0^{GL} D_t^\alpha f(t)\right) &= \sum_{k=0}^{n-1} \frac{\left[f^{(k)}(t)\right]_{t=0}(-\alpha+k)!}{\Gamma(-\alpha+k+1)s^{-\alpha+k+1}} + s^\alpha F(s) - \sum_{k=0}^{n-1} s^{\alpha-k-1}\left[f^{(k)}(t)\right]_{t=0}, \\ &= s^\alpha F(s),\ \alpha \in [0,1]. \end{aligned} \tag{1.46}$$

1.4 Fractional-Order Systems [9]

Developing first principle models utilizing fractional-order integrals and/or fractional-order derivatives for a linear lumped parameter SISO system results in the following fractional-order system:

$$\begin{aligned} a_n {}_0^{GL} D_t^{\alpha_n} y(t) + a_{n-1} {}_0^{GL} D_t^{\alpha_{n-1}} y(t) + \cdots\cdots + a_0 {}_0^{GL} D_t^{\alpha_0} y(t) &= b_n {}_0^{GL} D_t^{\beta_n} u(t) + A, \\ \sum_{k=0}^{n}\left(a_k {}_0^{GL} D_t^{\alpha_k} y(t)\right) &= \sum_{k=0}^{m}\left(b_k {}_0^{GL} D_t^{\beta_k} u(t)\right), \end{aligned} \tag{1.47}$$

where $A = b_{n-1} {}_0^{GL} D_t^{\beta_{n-1}} u(t) + \cdots\cdots + b_0 {}_0^{GL} D_t^{\beta_0} u(t)$, *$a_k$, and b_k are constants, α_k and β_k are real numbers, and $y(t)$ and $u(t)$ are the output and the input, respectively, of the system.*

Assuming homogeneous initial conditions, a Laplace transform operation performed on Equation (1.47) gives the SISO linear time invariant fractional-order transfer function:

$$G(s) = \frac{Y(s)}{U(s)} = \frac{b_m s^{\beta_m} + b_{m-1} s^{\beta_{m-1}} + \ldots\ldots + b_0 s^{\beta_0}}{a_n s^{\alpha_n} + a_{n-1} s^{\alpha_{n-1}} + \ldots\ldots + a_0 s^{\alpha_0}}. \quad (1.48)$$

If the fractional orders on both sides of Equation (1.47) are integer multiples of the base order α, that is, $\alpha_k = \beta_k = k\alpha$, then the fractional-order system is said to be a commensurate-order system:

$$G(s) = \frac{Y(s)}{U(s)} = \frac{\sum_{k=0}^{m} b_k (s^{\alpha})^k}{\sum_{k=0}^{n} a_k (s^{\alpha})^k}; \quad (1.49)$$

if not, then it is a noncommensurate order system.

The commensurate-order system in Equation (1.49) can be considered as a pseudo-rational function, $H(\lambda)$, of the variable $\lambda = s^{\alpha}$:

$$H(\lambda) = \frac{\sum_{k=0}^{m} b_k \lambda^k}{\sum_{k=0}^{n} a_k \lambda^k}. \quad (1.50)$$

If the order of commensurate-order systems is $^1/_q$ ($q \in Z^+$), then such a system is called a rational commensurate-order system.

Similar to integer systems, the stability criteria for commensurate-order fractional systems can be defined as

$$|arg(\lambda_i)| > \frac{\alpha \Pi}{2}, \ |arg(\lambda_i)| > \frac{\Pi}{2q}, \quad (1.51)$$

where λ_i's are poles of the commensurate-order system $H(\lambda)$.

For commensurate as well as noncommensurate order systems, the condition for bounded input-bounded output (BIBO) stability is

$$\lim_{s \to \infty} |G(s)| < M, \ M \text{ is a finite value.} \quad (1.52)$$

1.5 Fractional-Order PI$^\lambda$, PD$^\mu$, and PI$^\lambda$D$^\mu$ Controller [9]

The output of the traditional PID controller is

$$u(t) = K_C + K_{I0}J_t^1 e(t) + K_{d0}D_t^1 e(t). \quad (1.53)$$

The transfer function form of the PID controller can be obtained by assuming zero initial conditions

$$G_C(s) = K_C + \frac{K_I}{s} + K_d s. \quad (1.54)$$

The classical control actions—proportional, integral, and derivative—have positive as well as negative effects over the controlled system behavior. The proportional action increases the speed of the response and decreases the steady state error and the relative stability. The integral action eliminates the steady state error but decreases the relative stability. The derivative action increases the relative stability but makes the controlled system sensitive to high-frequency noisy signals. If the more general control actions of the form $s^p, \frac{1}{s^q}, p, q \in R^+$ are considered, the more acceptable tradeoffs between positive and negative effects can be accomplished. It will be noticed that the classical PID controller can meet only three performance specifications; however, if the order of the integrator and the differentiator can be an arbitrary order (integer and noninteger) and can be variable, the respective controller will bear five unknowns, that is, the controller gains (K_C, K_I, K_d) and the order of the integrator and differentiator (p, q), and can achieve five control objectives, thus leading to the formulation of the generalized (fractional-order) PID controller. The ability of fulfilling more control objectives cannot be gained unless efficient tuning techniques are available.

The following fractional-order integro-differential Equation expresses the output of the fractional-order PID controller:

$$u(t) = K_C e(t) + K_{I0}J_t^\lambda e(t) + K_d\, {}_0^{GL}D_t^\mu e(t), \quad (1.55)$$

where J$^\lambda$ is the fractional integral of order λ, and D$^\mu$ is the Grunwald-Letnikov fractional derivative of order μ.

Supposing null initial conditions and applying a Laplace Transform to the above equation, the transfer function of the fractional-order PID controller can be expressed by

$$G_C(s) = \frac{U(s)}{E(s)} = K_C + \frac{K_I}{s^\lambda} + K_d s^\mu,\ \lambda, \mu \in R. \quad (1.56)$$

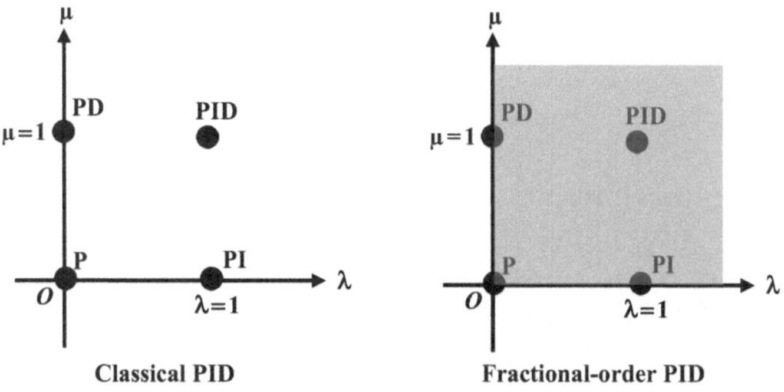

FIGURE 1.1
Classical PID controller versus fractional-order PID controller

If $\lambda = 0$ and $\mu = 0$, the fractional-order PID controller turns to be the traditional proportional controller. For $\lambda = 1$ and $\mu = 0$, Equation (1.56) changes to the classical PI controller. The classical proportional derivative controller can be obtained from Equation (1.56) when $\lambda = 0$ and $\mu = 1$. It is clear from Figure 1.1 that the classical P, PI, PD, and PID are the particular cases of the fractional-order PID controller. There are only four control configurations possible with PID controller, whereas numerous control structures can be obtained from the fractional-order PID controller by selecting the values of λ and μ in the Euclidean plane.

1.6 Triangular Orthogonal Functions

The orthogonal triangular function (TF) sets developed by Deb et al. [10] are a complementary pair of piecewise linear polynomial function sets evolved from a simple dissection of block pulse function(BPF) set [10, 11]. The authors of [10, 11] derived a complementary pair of operational matrices for first-order integration in the TF domain and demonstrated that the TF domain technique for dynamical systems analysis is computationally more effective than the BPF domain technique. In this section, first, we review block pulse functions in brief and then we introduce the method of dissecting the block pulse function set to formulate a complementary pair of orthogonal triangular function sets.

1.6.1 Review of Block Pulse Functions

Let us consider a square integral function $f(t)$ of Lebesgue measure, which is continuous in the interval $[0, T]$. Divide the interval into m subintervals of constant width $h = T/m$ as $[t_i, t_{i+1}]$, $i = 0, 1, \ldots\ldots m - 1$.

Let $\psi_m(t)$ be a set of block pulse functions containing m component functions in the interval $[0, T)$

$$\psi_m(t) = [\psi_0(t), \psi_1(t), \psi_2(t), \ldots\ldots, \psi_{m-1}(t)]^T_{1\times m}, \tag{1.57}$$

where $[\ldots]^T$ signifies transpose.

The ith component of the BPF vector $\psi_m(t)$ is defined as

$$\psi_i(t) = \begin{cases} 1, & ih \leq t < (i+1)\mathrm{h}, \\ 0, & \text{otherwise}, \end{cases}, \; i = 0, 1, 2, \ldots\ldots\ldots, m-1. \tag{1.58}$$

The square integrable function $f(t)$ can be approximated by BPFs as

$$f(t) = \sum_{i=0}^{m-1} f_i \psi(t) = [f_0, f_1, \ldots\ldots, f_{m-1}]\psi_m(t) = F^T \psi_m(t), \tag{1.59}$$

where the constant coefficients f_i's are defined as $f_i = \frac{1}{h} \int\limits_{ih}^{(i+1)h} f(t)dt$.

The BPF estimate for the first-order integration of $f(t)$ can be derived as [12]

$$Jf(t) = \int\limits_0^t f(\tau)d\tau \cong \int\limits_0^t F^T \psi_m(\tau)d\tau = F^T \int\limits_0^t \psi_m(\tau)d\tau = F^T P \psi_m(t), \tag{1.60}$$

where P is the operational matrix for the first-order integration in the BPF domain: $P = \frac{h}{2}\begin{bmatrix} 1 & 2 & 2 & \cdots & \cdots & 2 \\ 0 & 1 & 2 & 2 & \cdots & 2 \\ 0 & 0 & 1 & 2 & \cdots & 2 \\ 0 & 0 & 0 & 1 & \vdots & \vdots \\ \vdots & \cdots & \cdots & \vdots & \vdots & 2 \\ 0 & 0 & 0 & \cdots & 0 & 1 \end{bmatrix}_{m\times m}$.

The Riemann-Liouville fractional-order integral of $f(t)$ can be expressed by BPFs as [12]

$${}_0J_t^\alpha f(t) \cong F^T F_\alpha \psi_m(t), \tag{1.61}$$

where

$$F_\alpha = \begin{bmatrix} 1 & \xi_1 & \xi_2 & \cdots & \cdots & \xi_{m-1} \\ 0 & 1 & \xi_1 & \xi_2 & \cdots & \xi_{m-2} \\ 0 & 0 & 1 & \xi_1 & \cdots & \xi_{m-3} \\ \vdots & \vdots & \vdots & \vdots & \vdots & \xi_{m-4} \\ 0 & 0 & 0 & 0 & \vdots & \vdots \\ 0 & 0 & 0 & 0 & 0 & 1 \end{bmatrix}, \xi_k = (k+1)^{\alpha+1} - 2k^{\alpha+1} + (k-1)^{\alpha+1}, k = 1, 2, \ldots\ldots m-1.$$

1.6.2 Complementary Pair of Triangular Orthogonal Function Sets

Let us divide the first component of the BPF vector $\psi_m(t)$ into a complementary pair of linear polynomial functions as shown in Figure 1.2.

$$\psi_0(t) = T1_0(t) + T2_0(t), \tag{1.62}$$

where $T1_0(t) = \left(1 - \frac{t}{h}\right)$ and $T2_0(t) = \left(\frac{t}{h}\right)$,
and the second component $\psi_1(t)$,

$$\psi_1(t) = T1_1(t) + T2_1(t), \tag{1.63}$$

where $T1_1(t) = 1 - \left(\frac{t-h}{h}\right)$ and $T2_1(t) = \left(\frac{t-h}{h}\right)$.

In the same fashion, we can divide the remaining components of $\psi_m(t)$ into respective complementary pairs of linear polynomial functions. Thus, for the whole set of BPFs, we now have two sets of linear polynomial functions, namely,$T1_m(t)$ and $T2_m(t)$ each contains m component functions in the interval $[0, T]$.

$$\psi_m(t) = T1_m(t) + T2_m(t), \tag{1.64}$$

where $T1_m(t) = [T1_0(t), T1_1(t), \ldots\ldots, T1_{m-1}(t)]^T$ and $T2_m(t) = [T2_0(t), T2_1(t), \ldots\ldots, T2_{m-1}(t)]^T$.

The triangular function vectors; $T1_m(t)$ and $T2_m(t)$ together form the entire set of block pulse function set; hence, $T1_m(t)$ and $T2_m(t)$ complement each other as far as BPF set is considered. We recognize from Figure 1.2 that the shapes of $T1_i$'s and $T2_i$'s are left-handed and right-handed triangles, respectively. So, these two sets are named as the left-handed triangular function vector (LHTF) and the right-handed triangular function vector (RHTF), respectively.

Now we define the ith component of the LHTF vector $T1_m(t)$ as

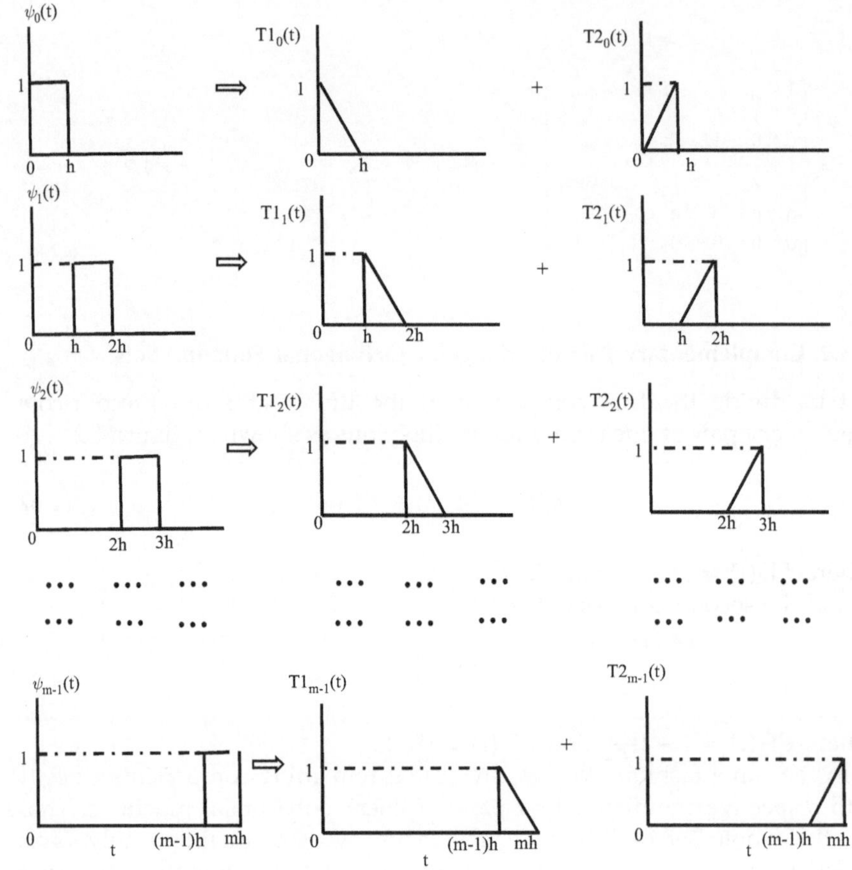

FIGURE 1.2
Generation of TFs from BPFs

$$T1_i(t) = \begin{cases} 1 - \left(\dfrac{t - ih}{h}\right), & ih \le t < (i+1)h, \\ 0, & \text{otherwise}, \end{cases}, \quad i = 0, 1, 2, \ldots\ldots, m-1 \tag{1.65}$$

and the ith component of the RHTF vector $T2_m(t)$ as

$$T2_i(t) = \begin{cases} \left(\dfrac{t - ih}{h}\right), & ih \le t < (i+1)h, \\ 0, & \text{otherwise}, \end{cases}, \quad i = 0, 1, 2, \ldots\ldots, m-1. \tag{1.66}$$

Like BPFs, TFs can also be employed for the approximation of the square integrable function $f(t)$ in the interval $[0, T]$.

$$f(t) = \lim_{m \to \infty} \left(\sum_{i=0}^{m-1} \left(c_i T1_i(t) + d_i T2_i(t) \right) \right). \tag{1.67}$$

Truncating the TFs series expansion in Equation (1.67) to m finite terms gives the practical approximation to the function, $f(t) \in L^2([0, T])$,

$$f(t) \approx \sum_{i=0}^{m-1} \left(c_i T1_i(t) + d_i T2_i(t) \right) = C^T T1_m(t) + D^T T2_m(t), \tag{1.68}$$

where $C^T = [c_0 \quad c_1 \quad \ldots \quad c_{m-1}]$, $D^T = [d_0 \quad d_1 \quad \ldots \quad d_{m-1}]$, $c_i = f(ih)$, $d_i = f((i+1)h)$.

The error analysis of estimating the function, $f(t)$, in the TFs domain is thoroughly studied in [7].

The expressions for the coefficients; c_i's and d_i's emphasis that the function evaluations at the equidistant nodes, t_i, $i = 0, 1, 2, \ldots\ldots, m-1$, are enough to find their numerical values. Whereas the coefficients in the BPF series representation in Equation (1.59) demand the integration of $f(t)$. Thus, the function approximation in TF domain is computationally more effective compared to that in BPF domain.

Because the members of $T1_m(t)$ and $T2_m(t)$ are mutually disjoint, their product can be expressed in the TF domain itself:

$$T1(t)T1^T(t) = \begin{bmatrix} T1_0(t) & 0 & 0 & 0 & 0 \\ 0 & T1_1(t) & 0 & \ldots & 0 \\ 0 & 0 & T1_2(t) & \vdots & 0 \\ \vdots & \vdots & \vdots & \vdots & 0 \\ 0 & 0 & 0 & 0 & T1_{(m-1)}(t) \end{bmatrix} = diag\left(T1(t)\right), \tag{1.69}$$

$$T2(t)T2^T(t) = \begin{bmatrix} T2_0(t) & 0 & 0 & 0 & 0 \\ 0 & T2_1(t) & 0 & \ldots & 0 \\ 0 & 0 & T2_2(t) & \vdots & 0 \\ \vdots & \vdots & \vdots & \vdots & 0 \\ 0 & 0 & 0 & 0 & T2_{(m-1)}(t) \end{bmatrix} = diag\left(T2(t)\right), \tag{1.70}$$

$$T1(t)T2^T(t) = \begin{bmatrix} 0 & 0 & 0 & 0 & 0 \\ 0 & 0 & 0 & \dots & 0 \\ 0 & 0 & 0 & \vdots & 0 \\ \vdots & \vdots & \vdots & \vdots & 0 \\ 0 & 0 & 0 & 0 & 0 \end{bmatrix} = O_{m\times m}, \tag{1.71}$$

where O is a null matrix.

Each member of the LHTF vector, $T1_m(t)$, and the RHTF vector, $T2_m(t)$, possesses orthogonal property, hence, TFs are called orthogonal TFs.

$$\int_0^T T1_i(t)T1_j(t)dt = \begin{cases} \frac{h}{3}, & \text{if} i == j, \\ \frac{h}{6}, & \text{if} i \neq j, \end{cases} , \int_0^T T2_i(t)T2_j(t)dt = \begin{cases} \frac{h}{3}, & \text{if} i == j, \\ \frac{h}{6}, & \text{if } i \neq j, \end{cases} \quad \forall i, j \in [0,\ m-1] \tag{1.72}$$

Similar to the time function approximation by TFs as explained in Equation (1.68), any function, for example, $G\big(f(t)\big)$, which can be linear or nonlinear, can be expanded into TFs:

$$G\big(f(t)\big) = [c_0^n \quad c_1^n \quad c_2^n \quad \dots \quad c_{m-1}^n]T1_m(t) + [d_0^n \quad d_1^n \quad d_2^n \quad \dots \quad d_{m-1}^n]T2_m(t), \tag{1.73}$$

where $c_i = G\big(f(t_i)\big)$, $d_i = c_{i-1}$.

The product of square integrable functions $h_1(t) \cdot h_2(t) \dots\dots h_n(t)$ can be approximated in the TF domain as

$$h_1(t) \cdot h_2(t) \dots\dots h_n(t) \cong [c_0 \quad c_1 \quad \dots \quad c_{m-1}]T1_m(t) + [d_0 \quad d_1 \quad \dots \quad d_{m-1}]T2_m, \tag{1.74}$$

where $c_i = h_1(jh) \cdot h_2(jh) \dots\dots h_n(jh)$, $d_i = c_{i+1}$.

1.6.3 Expansion of Two Variable Function via TFs

Let us consider the function in two variables, $f(x,t) \in L^2(J_1 \times J_2)$, $J_1 = [0,T]$, $J_2 = [0,T]$. We now describe $f(x,t)$ by using the TFs series. The two intervals are split into an equal number (m) of subintervals using the constant step size, h, as described here:

$$x_i \in [ih, (i+1)h),\ t_i \in [ih, (i+1)h),\ i \in [0, m-1]. \tag{1.75}$$

By using Equation (1.68),

$$f(x,t) \approx T1^T(x) \underbrace{\begin{bmatrix} f(0,t) \\ f(h,t) \\ \vdots \\ \vdots \\ f\Big((m-2)h,t\Big) \\ f\Big((m-1)h,t\Big) \end{bmatrix}}_{C_1} + T2^T(x) \underbrace{\begin{bmatrix} f(h,t) \\ f(2h,t) \\ \vdots \\ \vdots \\ f\Big((m-1)h,t\Big) \\ f(mh,t) \end{bmatrix}}_{C_2}. \tag{1.76}$$

Each element of the coefficient column vectors C_1 and C_2 in Equation (1.76) can further be expanded by using TFs with respect to the independent variable t.

$$f(x,t) \approx T1^T(x) \underbrace{\begin{bmatrix} F_{11}^T T1(t) + F_{21}^T T2(t) \\ F_{12}^T T1(t) + F_{22}^T T2(t) \\ \vdots \\ \vdots \\ F_{1(m-1)}^T T1(t) + F_{2(m-1)}^T T2(t) \\ F_{1m}^T T1(t) + F_{2m}^T T2(t) \end{bmatrix}}_{C_1} + T2^T(x) \underbrace{\begin{bmatrix} F_{31}^T T1(t) + F_{41}^T T2(t) \\ F_{32}^T T1(t) + F_{42}^T T2(t) \\ \vdots \\ \vdots \\ F_{3(m-1)}^T T1(t) + F_{4(m-1)}^T T2(t) \\ F_{3m}^T T1(t) + F_{4m}^T T2(t) \end{bmatrix}}_{C_2}, \tag{1.77}$$

where

$$\begin{aligned} F_{1(i+1)}^T &= \Big[f(ih,0) \quad f(ih,h) \quad f(ih,2h) \quad \ldots \quad \ldots \quad f\Big(ih,(m-1)h\Big)\Big], i \in [0, m-1], \\ F_{2(j+1)}^T &= \Big[f(jh,h) \quad f(jh,2h) \quad f(jh,3h) \quad \ldots \quad f\Big(jh,(m-1)h\Big) \quad f(jh,mh)\Big], j \in [0, m-1], \\ F_{3k}^T &= \Big[f(kh,0) \quad f(kh,h) \quad f(kh,2h) \quad \ldots \quad f\Big(kh,(m-2)h\Big) \quad f\Big(kh,(m-1)h\Big)\Big], k \in [1, m], \\ F_{4l}^T &= \Big[f(lh,h) \quad f(lh,2h) \quad f(lh,3h) \quad \ldots \quad f\Big(lh,(m-1)h\Big) \quad f(lh,mh)\Big], l \in [1, m]. \end{aligned}$$

Equation (1.77) can be written as

$$f(x,t) \approx T1^T(x)\big(F_1 T1(t) + F_2 T2(t)\big) + T2^T(x)\big(F_3 T1(t) + F_4 T2(t)\big), \tag{1.78}$$

where F_1, F_2, F_3 and F_4 are square matrices of size $m \times m$, and are defined as

$$F_1 = \begin{bmatrix} F_{11} \\ F_{12} \\ F_{13} \\ \vdots \\ F_{1m} \end{bmatrix}_{m\times m}, F_2 = \begin{bmatrix} F_{21} \\ F_{22} \\ F_{23} \\ \vdots \\ F_{2m} \end{bmatrix}_{m\times m}, F_3 = \begin{bmatrix} F_{31} \\ F_{32} \\ F_{33} \\ \vdots \\ F_{3m} \end{bmatrix}_{m\times m}, F_4 = \begin{bmatrix} F_{41} \\ F_{42} \\ F_{43} \\ \vdots \\ F_{4m} \end{bmatrix}_{m\times m}.$$

1.6.4 The TF Estimate of the First-Order Integral of Function $f(t)$

One-fold integration of square integrable function $f(t)$ is

$$_0J_t^1 f(t) = \int_0^t f(t)dt. \tag{1.79}$$

Substituting the TF estimate of $f(t)$ in Equation (1.68) leads to

$$\begin{aligned}\int_0^t f(t)dt &\cong \int_0^t (C^T T1_m(t) + D^T T2_m(t))dt = C^T \int_0^t T1_m(t)dt + +D^T \int_0^t T2_m(t)dt, \\ &= C^{\mathrm{T}} \left[\int_0^t T1_0(t)dt, \quad \ldots \quad \ldots\ldots, \quad \int_0^t T1_{m-1}(t)dt \right]^T \\ &\quad + D^T \left[\int_0^t T2_0(t)dt, \quad \ldots \quad \ldots\ldots, \quad \int_0^t T2_{m-1}(t)dt \right]^T.\end{aligned} \tag{1.80}$$

Integration of the function $f(t)$ is now changed to integration of LHTF set and RHTF set. Because the function $f(t)$ is square integrable, its TF estimate is also square integrable.

The graph of $T1_i(t)$ versus t and $T2_i(t)$ versus t depicted in Figure 1.2 can be expressed mathematically as

$$T1_i(t) = u(t - ih) - \frac{t - ih}{h} u(t - ih) + \frac{t - (i + 1)h}{h} u(t - (i + 1)h), \tag{1.81}$$

$$T2_i(t) = \frac{(t - ih)}{h} u(t - ih) - \frac{t - (i + 1)h}{h} u(t - (i + 1)h) - u(t - (i + 1)h). \tag{1.82}$$

We now integrate each component of LHTF set $T1_m(t)$ using Equation (1.81) and express the result in terms of LHTF set $T1_m(t)$ and RHTF set $T2_m(t)$.

$$\begin{aligned}\int_0^t T1_0(t)dt &= \int_0^t \left(1-\frac{t}{h}\right)u(t)dt + \int_h^t \left(\frac{t-h}{h}\right)u(t-h)dt,\\ &= \left(t-\frac{t^2}{2h}\right)u(t) - \left\{\left(t-\frac{t^2}{2h}\right)-\frac{h}{2}\right\}u(t-h),\\ &= \frac{h}{2}T2_0(t) + \frac{h}{2}[(T1_1(t)+T2_1(t)) + (T1_2(t)+T2_2(t)) + \cdots + (T1_{m-1}(t)+T2_{m-1}(t))],\\ &= \frac{h}{2}[0\ \ 1\ \ 1\ \ \ldots\ \ 1\ \ 1]T1_m(t) + \frac{h}{2}[1\ \ 1\ \ 1\ \ \ldots\ \ 1\ \ 1]T2_m(t).\end{aligned} \tag{1.83}$$

$$\int_0^t T1_1(t)dt = \frac{h}{2}[0\ \ 0\ \ 1\ \ \ldots\ \ 1\ \ 1]T1_m(t) + \frac{h}{2}[0\ \ 1\ \ 1\ \ \ldots\ \ 1\ \ 1]T2_m(t). \tag{1.84}$$

$$\vdots$$

$$\int_0^t T1_{m-1}(t)dt = \frac{h}{2}[0\ \ 0\ \ 0\ \ \ldots\ \ 0\ \ 0]T1_m(t) + \frac{h}{2}[0\ \ 0\ \ 0\ \ \ldots\ \ 0\ \ 1]T2_m(t). \tag{1.85}$$

Therefore, the first-order integration of LHTF set $T1_m(t)$ is

$$\int_0^t T1_m(t)dt \cong P_1 T1_m(t) + P_2 T2_m(t), \tag{1.86}$$

where $P_1 = \frac{h}{2}\begin{bmatrix} 0 & 1 & 1 & \ldots & 1\\ 0 & 0 & 1 & 1 & 1\\ 0 & 0 & 0 & \vdots & \vdots\\ \vdots & \vdots & 0 & \vdots & 1\\ 0 & 0 & \ldots & \ldots & 0\end{bmatrix}_{m\times m}$, $P_2 = \frac{h}{2}\begin{bmatrix} 1 & 1 & 1 & \ldots & 1\\ 0 & 1 & 1 & \vdots & 1\\ 0 & 0 & 1 & \vdots & 1\\ \vdots & \vdots & 0 & \vdots & 1\\ 0 & 0 & \ldots & 0 & 1\end{bmatrix}_{m\times m}$.

Following the same procedure, the first-order integration of RHTF set $T2_m(t)$ using Equation (1.82) is

$$\int_0^t T2_m(t)dt \cong P_1T1_m(t) + P_2T2_m(t) = \int_0^t T1_m(t)dt. \tag{1.87}$$

Equation (1.79) becomes

$$\int_0^t f(t)dt \cong (C^T + D^T)\left(P_1T1_m(t) + P_2T2_m(t)\right). \tag{1.88}$$

Here P_1 and P_2 are complement to each other as far as P is considered. This complementary pair is acting as a first-order integral in the TF domain.

1.6.5 The TF Estimate of Riemann-Liouville Fractional-Order Integral of $f(t)$

The Riemann-Liouville fractional-order integral of the function $f(t)$ is

$${}_0J_t^{\alpha}f(t) = \frac{1}{\Gamma(\alpha)}\int_0^t (t-\tau)^{\alpha-1}f(\tau)d\tau. \tag{1.89}$$

Using integral convolution property, we rewrite the above equation as

$${}_0J_t^{\alpha}f(t) = \frac{1}{\Gamma(\alpha)}\int_0^t (t-\tau)^{\alpha-1}f(\tau)d\tau = \frac{1}{\Gamma(\alpha)}\left(t^{\alpha-1} * f(t)\right). \tag{1.90}$$

Replacing $f(t)$ with its TF estimate,

$$\begin{aligned}\frac{1}{\Gamma(\alpha)}\left(t^{\alpha-1} * f(t)\right) &= C^T\left(\frac{t^{\alpha-1} * T1_m(t)}{\Gamma(\alpha)}\right) + D^T\left(\frac{t^{\alpha-1} * T2_m(t)}{\Gamma(\alpha)}\right)\\ &= C^T(J^{\alpha}T1_m(t)) + D^T(J^{\alpha}T2_m(t)),\\ &= C^T[J^{\alpha}T1_0(t),\ \ J^{\alpha}T1_1(t),\ \ \ldots\ldots,\ \ J^{\alpha}T1_{m-1}(t)]^T\\ &\quad + D^T[J^{\alpha}T2_0(t),\ \ J^{\alpha}T2_1(t),\ \ \ldots\ldots,\ \ J^{\alpha}T2_{m-1}(t)]^T\end{aligned} \tag{1.91}$$

Similar to Equation (1.83), we compute the $\alpha-$ order Riemann-Liouville fractional integral of $T1_0(t)$ and express the result by means of complementary pair of TF sets.

$$
\begin{aligned}
{}_0J_t^{\alpha}T1_0(t) &= \frac{1}{\Gamma(\alpha)}\int_0^t \left((t-\tau)^{\alpha-1}\left(1-\frac{\tau}{h}\right)\right)d\tau + \frac{1}{\Gamma(\alpha)}\int_h^t \left((t-\tau)^{\alpha-1}\left(\frac{\tau-h}{h}\right)\right)d\tau, \\
&= \frac{1}{\Gamma(\alpha)}\left(\frac{\tau^{\alpha}}{\alpha}-\frac{t\tau^{\alpha}}{h\alpha}+\frac{\tau^{\alpha+1}}{h(\alpha+1)}\right)_0^t - \frac{1}{\Gamma(\alpha)}\left(\frac{\tau^{\alpha+1}}{h(\alpha+1)}-\left(\frac{t\tau^{\alpha}}{h\alpha}-\frac{\tau^{\alpha}}{\alpha}\right)\right)_h^t, \\
&= \frac{h^{\alpha}}{\Gamma(\alpha+2)}[0 \quad \varsigma_1 \quad \varsigma_2 \quad \cdots \quad \cdots \quad \varsigma_{m-1}]T1_m(t) \\
&\quad + \frac{h^{\alpha}}{\Gamma(\alpha+2)}[\varsigma_1 \quad \varsigma_2 \quad \cdots \quad \cdots \quad \cdots \quad \varsigma_m]T2_m(t),
\end{aligned} \tag{1.92}
$$

where $\varsigma_j = \left(j^{\alpha}(1+\alpha-j)+(j-1)^{(\alpha+1)}\right)$, $j = 1,2,\ldots,m-1$.

$$
{}_0J_t^{\alpha}T1_1(t) = \frac{h^{\alpha}}{\Gamma(\alpha+2)}[0 \quad 0 \quad \varsigma_1 \quad \varsigma_2 \quad \cdots \quad \varsigma_{m-2}]T1_m(t) + \frac{h^{\alpha}}{\Gamma(\alpha+2)}[0 \quad \varsigma_1 \quad \varsigma_2 \quad \cdots \quad \cdots \quad \varsigma_{m-1}]T2_m(t). \tag{1.93}
$$

$$\vdots$$

$$
{}_0J_t^{\alpha}T1_{m-1}(t) = \frac{h^{\alpha}}{\Gamma(\alpha+2)}[0 \quad 0 \quad 0 \quad 0 \quad \cdots \quad 0]T1_m(t) + \frac{h^{\alpha}}{\Gamma(\alpha+2)}[0 \quad 0 \quad 0 \quad 0 \quad \cdots \quad \varsigma_1]T2_m(t). \tag{1.94}
$$

Therefore, the Riemann-Liouville fractional integral of order α of LHTF set $T1_m(t)$ is

$$
\frac{1}{\Gamma(\alpha)}\int_0^t (t-\tau)^{\alpha-1}T1_m(\tau)d\tau = P_1^{\alpha}T1_m(t) + P_2^{\alpha}T2_m(t), \tag{1.95}
$$

where

$$P_1^\alpha = \frac{h^\alpha}{\Gamma(\alpha+2)}\begin{bmatrix} 0 & \varsigma_1 & \varsigma_2 & \varsigma_3 & \cdots & \varsigma_{m-1} \\ 0 & 0 & \varsigma_1 & \varsigma_2 & \cdots & \varsigma_{m-2} \\ 0 & 0 & 0 & \varsigma_1 & \cdots & \varsigma_{m-3} \\ \vdots & 0 & \vdots & 0 & \vdots & \vdots \\ 0 & \vdots & \vdots & \vdots & \vdots & \varsigma_1 \\ 0 & \cdots & \cdots & 0 & 0 & 0 \end{bmatrix}, P_2^\alpha = \frac{h^\alpha}{\Gamma(\alpha+2)}\begin{bmatrix} \varsigma_1 & \varsigma_2 & \varsigma_3 & \varsigma_4 & \cdots & \varsigma_m \\ 0 & \varsigma_1 & \varsigma_2 & \varsigma_3 & \cdots & \varsigma_{m-1} \\ 0 & 0 & \varsigma_1 & \varsigma_2 & \cdots & \varsigma_{m-2} \\ \vdots & 0 & \vdots & \varsigma_1 & \vdots & \vdots \\ 0 & \vdots & \vdots & \vdots & \vdots & \varsigma_2 \\ 0 & \cdots & \cdots & 0 & 0 & \varsigma_1 \end{bmatrix}.$$

Following the same procedure as we applied for LHTF set, the Riemann-Liouville fractional-order integral of RHTF set $T2_m(t)$ using Equation (1.82) is derived as.

$$_0J_t^\alpha T2_m(t) = \frac{1}{\Gamma(\alpha)}\int_0^t (t-\tau)^{\alpha-1} T2_m(\tau)d\tau = P_3^\alpha T1_m(t) + P_4^\alpha T2_m(t), \tag{1.96}$$

where

$$P_3^\alpha = \frac{h^\alpha}{\Gamma(\alpha+2)}\begin{bmatrix} 0 & \xi_1 & \xi_2 & \xi_3 & \cdots & \xi_{m-1} \\ 0 & 0 & \xi_1 & \xi_2 & \cdots & \xi_{m-2} \\ 0 & 0 & 0 & \xi_1 & \cdots & \xi_{m-3} \\ \vdots & 0 & \vdots & 0 & \vdots & \vdots \\ 0 & \vdots & \vdots & \vdots & \vdots & \xi_1 \\ 0 & \cdots & \cdots & 0 & 0 & 0 \end{bmatrix}, P_4^\alpha = \frac{h^\alpha}{\Gamma(\alpha+2)}\begin{bmatrix} \xi_1 & \xi_2 & \xi_3 & \xi_4 & \cdots & \xi_m \\ 0 & \xi_1 & \xi_2 & \xi_3 & \cdots & \xi_{m-1} \\ 0 & 0 & \xi_1 & \xi_2 & \cdots & \xi_{m-2} \\ \vdots & 0 & \vdots & \xi_1 & \vdots & \vdots \\ 0 & \vdots & \vdots & \vdots & \vdots & \xi_2 \\ 0 & \cdots & \cdots & 0 & 0 & \xi_1 \end{bmatrix},$$

$$\xi_j = j^{\alpha+1} - (j+\alpha)(j-1)^\alpha.$$

From Equations (1.89), (1.95) and (1.96),

$$_0\tilde{J}_t^\alpha f(t) = \frac{1}{\Gamma(\alpha)}\int_0^t (t-\tau)^{\alpha-1}\tilde{f}(\tau)d\tau \cong (C^T P_1^\alpha + D^T P_3^\alpha)T1_m(t) + (C^T P_2^\alpha + D^T P_4^\alpha)T2_m(t). \tag{1.97}$$

For the special case of $\alpha = 1$,

$$P_1^\alpha = P_3^\alpha = P_1,\ P_2^\alpha = P_4^\alpha = P_2. \tag{1.98}$$

So the TF estimate of fractional-order integral will be reduced to the TF estimate of the first-order integral when $\alpha = 1$.

1.6.6 Error Analysis

Let us denote the TF estimate of function $f(t)$ as

$$\widetilde{f}(t) = C^T T1_m(t) + D^T T2_m(t). \tag{1.99}$$

We replace $f(t)$ with $\widetilde{f}(t)$ in Equation (1.89) and we call the resulting integral the mth approximate of the $\alpha-$ order Riemann-Liouville fractional integral of $f(t)$.

$${}_0\widetilde{J}_t^\alpha f(t) = \frac{1}{\Gamma(\alpha)}\int_0^t (t-\tau)^{\alpha-1}\widetilde{f}(\tau)d\tau = \frac{1}{\Gamma(\alpha)}\int_0^t (t-\tau)^{\alpha-1}\left(C^T T1_m(\tau) + D^T T2_m(\tau)\right)d\tau. \tag{1.100}$$

The absolute error between the exact fractional integral ${}_0J_t^\alpha f(t)$ and the mth approximate ${}_0\widetilde{J}_t^\alpha f(t)$ is

$$\varepsilon_m = \left|{}_0J_t^\alpha f(t) - {}_0\widetilde{J}_t^\alpha f(t)\right|. \tag{1.101}$$

Theorem 1.6.1: If the function $f(t)$ is represented by a complementary pair of LHTF and RHTF sets, then

(i) $|f(t) - \widetilde{f}(t)| \le \frac{T^2}{2m^2}|f''(ih)| + O\left(\frac{1}{m^3}\right), t \in [ih, (i+1)h), i = 0, 1, 2, \ldots, m-1.$

(ii) $\varepsilon_m \le \frac{MT^{2+\alpha}}{2m^2\Gamma(\alpha+1)} + O\left(\frac{1}{m^3}\right), t \in [ih, (i+1)h), |f''(ih)| \le M, \forall i, M$ is finite positive value.

Proof:

(i) From Equations (1.65) to (1.66), we can approximate $f(t)$ in the ith interval as

$$\begin{aligned}\widetilde{f}(t) &= f(ih)T1_i(t) + f\left((i+1)h\right)T2_i(t) = f(ih)\left(1 - \left(\frac{t-ih}{h}\right)\right) + f\left((i+1)h\right)\left(\frac{t-ih}{h}\right),\\ &= f(ih) + f\left((i+1)h\right)\left(\frac{t-ih}{h}\right) - f(ih)\left(\frac{t-ih}{h}\right) = f(ih) + \left(\frac{f\left((i+1)h\right) - f(ih)}{h}\right)(t-ih),\\ &= f(ih) + f'(ih)(t-ih),\ h \to 0.\end{aligned} \tag{1.102}$$

Expanding the exact function $f(t)$ by Taylor series with the center ih as

$$f(t) = f(ih) + (t-ih)f'(ih) + \frac{(t-ih)^2}{2}f''(ih) + \sum_{k=3}^{\infty}\frac{(t-ih)^k}{k!}f^{(k)}(ih). \tag{1.103}$$

From Equations (1.102) and (1.103), the absolute error between the function and its TF estimate can be determined as

$$|f(t) - \tilde{f}(t)| = \frac{(t - ih)^2}{2}|f''(ih)| + O(t - ih)^3. \tag{1.104}$$

Because $(t - ih) < h$ and $mh = T$, the above equation becomes

$$|f(t) - \tilde{f}(t)| = \frac{T^2}{2m^2}|f''(ih)| + O\left(\frac{1}{m^3}\right). \tag{1.105}$$

(ii) The absolute error between the exact fractional integral ${}_0J_t^{\alpha}f(t)$ and the mth approximate ${}_0\tilde{J}_t^{\alpha}f(t)$ is

$$\begin{aligned}
\varepsilon_m &= \left|{}_0J_t^{\alpha}f(t) - {}_0\tilde{J}_t^{\alpha}f(t)\right| = \frac{1}{\Gamma(\alpha)}\int_0^t (t-\tau)^{\alpha-1}\left|f(\tau) - \tilde{f}(\tau)\right|d\tau,\\
&= \frac{1}{\Gamma(\alpha)}\left[\sum_{r=0}^{i-1}\int_{rh}^{(r+1)h}(t-\tau)^{\alpha-1}\left|f(\tau) - \tilde{f}(\tau)\right|d\tau + \int_{ih}^{t}(t-\tau)^{\alpha-1}\left|f(\tau) - \tilde{f}(\tau)\right|d\tau\right],\\
&\le \frac{1}{\Gamma(\alpha)}\left[\sum_{r=0}^{i-1}\int_{rh}^{(r+1)h}(t-\tau)^{\alpha-1}\left(\frac{T^2}{2m^2}|f''(ih)| + O\left(\frac{1}{m^3}\right)\right)d\tau + \int_{ih}^{t}(t-\tau)^{\alpha-1}\left(\frac{T^2}{2m^2}|f''(ih)| + O\left(\frac{1}{m^3}\right)\right)d\tau\right],\\
&\le \frac{1}{\Gamma(\alpha)}\left(\frac{T^2}{2m^2}|f''(ih)| + O\left(\frac{1}{m^3}\right)\right)\left[\sum_{r=0}^{i-1}\int_{rh}^{(r+1)h}(t-\tau)^{\alpha-1}d\tau + \int_{ih}^{t}(t-\tau)^{\alpha-1}d\tau\right],\\
&\le \frac{t^{\alpha}}{\Gamma(\alpha+1)}\frac{T^2}{2m^2}|f''(ih)| + O\left(\frac{1}{m^3}\right).
\end{aligned} \tag{1.106}$$

We now consider the following assumption.

$$Max|f''(ih)| \le M,\ \forall i,\ i = 0, 1, 2, \ldots, m - 1, \tag{1.107}$$

where M is finite positive value.

From Equations (1.106) and (1.107), the absolute error between ${}_0J_t^{\alpha}f(t)$ and ${}_0\tilde{J}_t^{\alpha}f(t)$ can be estimated as

$$\varepsilon_m \le \frac{MT^{2+\alpha}}{2m^2\Gamma(\alpha+1)} + O\left(\frac{1}{m^3}\right). \tag{1.108}$$

To confirm whether the maximal absolute error caused by TFs will be smaller than the theoretical upper bound derived in Equation (1.108), we consider the function $f(t) = t$ in the interval $[0, 1]$ divided into five equal subintervals ($m = 5$).

TABLE 1.1

Absolute errors using TFs and BPFs.

t	$\lvert {}_0J_t^\alpha f(t) - {}_0\tilde{J}_t^\alpha f(t)\rvert$	$\lvert {}_0J_t^\alpha f(t) - {}_0^{BPF}J_t^\alpha f(t)\rvert$
0	0	0.0336417669602688
0.2	0	0.0615115192345456
0.4	0	0.0796544626999182
0.6	0	0.0943263867874177
0.8	0	0.106992253349687
1	0	0.118304589557845

The exact fractional integral of function $f(t)$ is

$$ {}_0J_t^\alpha t = [0\ 0.067283533920\ 0.19030657238\ 0.34961549778\ 0.53826827136\ 0.75225277806]. \quad (1.109) $$

Using Equation (1.97) with $\alpha = 0.5$, $T = 1$ and $h = 0.2$, the TF estimate of $J^\alpha t$ is obtained as

$$ \begin{aligned} {}_0\tilde{J}_t^\alpha t =& [0\ 0.067283533920\ 0.19030657238\ 0.34961549778\ 0.53826827136]T1_m(t) \\ &+ [\ 0.067283533920\ 0.19030657238\ 0.34961549778\ 0.53826827136 \\ &0.75225277806]T2_m(t). \end{aligned} \quad (1.110) $$

In the BPF domain, $J^\alpha t$ is approximated as

$$ \begin{aligned} {}_0^{BPF}J_t^\alpha t \ &= [0.0336417669\ 0.128795053\ 0.2699610350\ 0.4439418845 \\ &0.6452605247\ 0.870557367621]\psi_m(t). \end{aligned} \quad (1.111) $$

Table 1.1 presents the absolute errors given by TF domain analysis and BPF domain analysis. The piecewise linear nature of TFs made them capable of estimating the fractional integral accurately even with small value of m. Therefore, the TF estimate of Riemann-Liouville fractional integral is effective.

1.6.7 MATLAB® Code for Generalized Triangular Function Operational Matrices

Programs 1.1 and 1.2 create the generalized triangular function operational matrices for integral of arbitrary order.

Program 1.1

```
function [P1alph,P2alph]=TOF1(t0,T,m,alpha)
% t0 is left end point of the interval [a, b], T is the right
end point of the interval, m is the number of subintervals,
alpha is the order of integration.
h=(T-t0)/(m+0);t=[t0:h:T];P1alph=zeros(length(t)-1,
length(t)-1);
P2alph=zeros(length(t)-1,length(t)-1);geta1=zeros(1,
length(t));
for j=0:1:length(t)-1
   if j>0
   geta2(j+0)=(j^alpha)*(1+alpha-j)+(j-1)^(alpha+1);
   end
   if j>0
    geta1(j-0)=(alpha-j)*((j+1)^alpha)+j^(alpha+1);
   end
end
geta3=geta2(2:end);
for ii=1:1:length(t)-1
   for kk=1:1:length(t)-1
      if ii==kk
         P2alph(ii,ii)=geta2(1);
      end
      if kk>ii
         P1alph(ii,kk)=geta2(kk-ii);P2alph(ii,kk)=geta3
           (kk-ii);
      end
   end
end
P1alph=((h^alpha)/gamma(alpha+2))*P1alph;
P2alph=((h^alpha)/gamma(alpha+2))*P2alph;
if alpha==0
    P1alph=1;P2alph=1;
end
end
```

Program 1.2

```
function [P3alph,P4alph]=TOF12(t0,T,m,alpha)
h=(T-t0)/(m);t=[t0:h:T];P1alph=zeros(length(t)-1,length
(t)-1);
P2alph=zeros(length(t)-1,length(t)-1);geta1=zeros(1,
length(t));
for j=0:1:length(t)+0
   if j>0
   geta2(j+0)=(j^(alpha+1)-(j+alpha)*(j-1)^(alpha));
```

```
    end
    if j>0
     geta1(j+1)=((j+2)^(alpha+1))-(2+alpha+j)*(j+1)
       ^alpha;
    end
end
geta1(1)=1-(2^alpha)*(1-alpha);geta3=geta2(2:end);
for ii=1:1:length(t)-1
    for kk=1:1:length(t)-1
         if ii==kk
            P2alph(ii,ii)=geta2(1);
        end
        if kk>ii
            P1alph(ii,kk)=geta2(kk-ii);
            P2alph(ii,kk)=geta3(kk-ii);
        end
    end
end
P1alph=((h^alpha)/gamma(alpha+2))*P1alph;
P2alph=((h^alpha)/gamma(alpha+2))*P2alph;
if alpha==0
   P1alph=1;P2alph=1;
end
P3alph= P1alph;P4alph= P2alph;
end
```

1.7 Triangular Strip Operational Matrices for Classical and Fractional Derivatives

In this section, we build triangular strip operational matrices from the discretized forms of classical derivative and fractional-order derivative [13].

1.7.1 Operational Matrix for Classical Derivative

Let us first derive triangular strip operational matrix for n^{th} order derivative (n is a positive integer) of casual function $f(t)$, which is continuous in the interval $[0, T]$. We start with the first-order derivative of $f(t)$. Generate N equidistant nodes in the interval $[0, T]$ with step size h, that is, $t_i = t_0 + ih$, $i = 0, 1, 2, \ldots\ldots, N$.

To find out the numerical solution of $f'(t)$, the following first-order backward difference formula can be used.

$$f'(t_i) \cong \frac{1}{h}\nabla f(t_i) = \frac{1}{h}(f_i - f_{i-1}),\ i = 0, 1, 2, ..., N. \tag{1.112}$$

From the above formula, N expressions can be written as

$$\begin{aligned} \frac{1}{h}\nabla f(t_0) &= \frac{1}{h}(f_0 - f_{-1}) \text{ at node } 1, \\ \frac{1}{h}\nabla f(t_1) &= \frac{1}{h}(f_1 - f_0) \text{ at node } 2, \\ &\vdots \\ \frac{1}{h}\nabla f(t_N) &= \frac{1}{h}(f_N - f_{N-1}) \text{ at node } N. \end{aligned} \tag{1.113}$$

The system of equations in Equation (1.113) can be shown in the vector-matrix form below.

$$\underbrace{\begin{bmatrix} h^{-1}f(t_0) \\ h^{-1}\nabla\left(f(t_1)\right) \\ h^{-1}\nabla\left(f(t_2)\right) \\ \vdots \\ h^{-1}\nabla\left(f(t_{N-1})\right) \\ h^{-1}\nabla\left(f(t_N)\right) \end{bmatrix}}_{Y} = \frac{1}{h}\underbrace{\begin{bmatrix} 1 & 0 & 0 & 0 & \dots & 0 \\ -1 & 1 & 0 & \dots & \dots & 0 \\ 0 & -1 & 1 & 0 & \dots & \vdots \\ \vdots & 0 & \vdots & \vdots & 0 & \vdots \\ 0 & 0 & 0 & \vdots & 1 & 0 \\ 0 & 0 & \dots & 0 & -1 & 1 \end{bmatrix}}_{B_N^1}\underbrace{\begin{bmatrix} f_0 \\ f_1 \\ f_2 \\ \vdots \\ f_{N-1} \\ f_N \end{bmatrix}}_{F_N}, \tag{1.114}$$

where F_N is the vector of function values obtained at equidistant nodes, B_N^1 is the lower triangular strip operational matrix, and Y is the vector of approximated values of $f'(t)$.

Likewise, we can get the following triangular strip operational matrix for n^{th} order derivative of $f(t)$.

$$B_N^n = \frac{1}{h}\begin{bmatrix} w_0 & 0 & 0 & 0 & \dots & \dots & 0 & 0 \\ w_1 & w_0 & 0 & 0 & \dots & \dots & 0 & 0 \\ w_2 & w_1 & w_0 & 0 & \dots & \dots & 0 & 0 \\ \vdots & w_2 & w_1 & w_0 & 0 & \dots & \vdots & \vdots \\ w_n & \vdots & \vdots & \vdots & \vdots & 0 & 0 & 0 \\ 0 & \vdots & \vdots & \vdots & \vdots & \vdots & 0 & 0 \\ \vdots & 0 & \vdots & \vdots & \vdots & \vdots & \vdots & 0 \\ 0 & \dots & 0 & w_n & \dots & w_2 & w_1 & w_0 \end{bmatrix}, \tag{1.115}$$

where $w_j = (-1)^j \binom{n}{j}, j = 0, 1, 2, \ldots\ldots, N$.

1.7.2 Operational Matrix for Fractional-Order Derivative

Let us recall the definition of fractional-order derivative in the Grunwald-Letnikov sense:

$$ {}_0^{GL}D_t^{\alpha} f(t) \cong \frac{1}{h^{\alpha}} \nabla^{\alpha} f(t_k) = h^{-\alpha} \sum_{i=0}^{k} \left((-1)^i \binom{\alpha}{i} f(t - ih) \right), \tag{1.116} $$

where $k = 0, 1, 2, \ldots, N$.

There are such $N+1$ equations. Each equation computes the approximate value of $f^{\alpha}(t_k)$ at node t_k and these equations are simultaneously written in the vector-matrix form given in the next equation.

$$ \underbrace{\begin{bmatrix} h^{-\alpha}\nabla^{\alpha} f(t_0) \\ h^{-\alpha}\nabla^{\alpha} f(t_1) \\ h^{-\alpha}\nabla^{\alpha} f(t_2) \\ \vdots \\ h^{-\alpha}\nabla^{\alpha} f(t_{N-1}) \\ h^{-\alpha}\nabla^{\alpha} f(t_N) \end{bmatrix}}_{Y} = \frac{1}{h^{\alpha}} \underbrace{\begin{bmatrix} w_0^{\alpha} & 0 & 0 & 0 & \cdots & 0 \\ w_1^{\alpha} & w_0^{\alpha} & 0 & 0 & \cdots & 0 \\ w_2^{\alpha} & w_1^{\alpha} & w_0^{\alpha} & 0 & \cdots & 0 \\ \vdots & \vdots & \vdots & \vdots & \cdots & \cdots \\ w_{N-1}^{\alpha} & \vdots & w_2^{\alpha} & w_1^{\alpha} & w_0^{\alpha} & 0 \\ w_N^{\alpha} & w_{N-1}^{\alpha} & \vdots & w_2^{\alpha} & w_1^{\alpha} & w_0^{\alpha} \end{bmatrix}}_{B_N^{\alpha}} \underbrace{\begin{bmatrix} f_0 \\ f_1 \\ f_2 \\ \vdots \\ f_{N-1} \\ f_N \end{bmatrix}}_{F_N}, \tag{1.117} $$

where $w_k^{\alpha} = (-1)^k \binom{\alpha}{k}, k = 0, 1, 2, \ldots\ldots, \text{N}$.

Here, F_N is the vector of function values obtained at equidistant nodes, B_N^{α} is the lower triangular strip operational matrix for fractional derivatives, and Y is the vector of approximated values of ${}_0^{GL}D_t^{\alpha} f(t)$.

1.7.3 MATLAB Code for Triangular Strip Operational Matrices

The following program generates triangular strip operational matrix for derivative of arbitrary order.

Program 1.3

```
function B = ban(alpha,N,h)
%  alpha - order of differentiation (real, not necessarily
integer)
%  N  - size of the resulting matrix B (N x N)
%  h  - step of discretization; default is h=1
B = zeros(N,N);
```

```
if nargin <= 1 || nargin > 3
  error('BAN: Wrong number of input parameters')
else
  bc=fliplr(bcrecur(alpha,N-1));
  for k=1:N
    B(k,1:k)=bc((N-k+1):N);
  end
endif nargin == 3
  B=1*B;
end
end
% bcrecur.m is called in the main program 'ban.m'
function y=bcrecur(a, n)
y=cumprod([1, 1 - ((a+1) ./ (1:n))]);
end
```

References

[1] K. B. Oldham, J. Spanier (1974). *The Fractional Calculus: Theory and Applications of Differentiation and Integration to Arbitrary Order*. New York: Dover Publications.

[2] M. G. Mittag-Leffler (1903). Sur la nouvelle fonction Eα(x). *Comptes Rendus Acad. Sci. Paris.*, vol. 137, pp. 554–558.

[3] R. Gorenflo, A. A. Kilbas, S. V. Rogosin (1998). On the generalized Mittag-Leffler type functions. *Integral Transform. Spec. Funct.*, vol. 7, pp. 215–224.

[4] I. Podlubny (1999). *Fractional Differential Equations*. New York: Academic Press.

[5] C. Li, W. Deng (2007). Remarks on fractional derivatives. *Appl. Math. Comput.*, vol. 187, pp. 777–784.

[6] M. Caputo (1967). Linear models of dissipation whose Q is almost frequency – Part II. *J. Roy. Austral. Soc.*, vol. 13, pp. 529–539.

[7] S. Das (2011). *Functional Fractional Calculus*. Berlin: Springer-Verlag.

[8] N. Heymans, I. Podlubny (2005). Physical interpretation of initial conditions for fractional differential equations with Riemann-Liouville fractional derivatives. *Rheol. Acta*, vol. 45, pp. 765–771.

[9] C. A. Monje, Y. Q. Chen, B. M. Vinagre, D. Xue, V. Feliu (2010). *Fractional-Order Systems and Controls: Fundamentals and Applications*. London: Springer-Verlag.

[10] A. Deb, A. Dasgupta, G. Sarkar (2006). A new set of orthogonal functions and its application to the analysis of dynamic systems. *J. Franklin Inst.*, vol. 343, pp. 1–26.

[11] A. Deb, G. Sarkar, A. Sengupta (2007). *Triangular Orthogonal Functions for the Analysis of Continuous Time Systems*. Gurgaeon, India: Elsevier.

[12] Y. Li, N. Sun (2011). Numerical solution of fractional differential equations using the generalized block pulse operational matrix. *Comput. Math. Appl.*, vol. 62, pp. 1046–1054.

[13] I. Podlubny (2000). Matrix approach to discrete fractional calculus. *Fract. Calc. Appl. Anal.*, vol. 3, pp. 359–386.

2

Numerical Method for Simulation of Physical Processes Represented by Weakly Singular Fredholm, Volterra, and Volterra-Fredholm Integral Equations

An equation in which the unknown, generally a function of one or more independent variables, appears under an integral sign is called an integral equation (IE). Integral equations have progressively been paid attention, since the emergence of Abel's integral equations in 1823, predominantly for two obvious reasons. The first is that there are some cases in which the integral equations are the natural mathematical models representing physical processes. The second is that the integral equations provide an expedient and practically useful alternative to differential equations.

The standard integral equation is

$$y(x) = f(x) + \lambda \int_{g(x)}^{h(x)} k(x,t)y(t)dt, \tag{2.1}$$

where $g(x)$ and $h(x)$ are the limits of integration, λ is a constant parameter, $k(x,t)$ is a function of two variables; x and t, usually called the kernel or the nucleus of the integral equation, $y(x)$ is the unknown function that will be determined. The functions $f(x)$ and $k(x,t)$ are given in advance. It may be noticed that the limits of integration, $g(x)$ and $h(x)$, may be both constants, variables, or mixed. The rather general definition of an integral equation in Equation (2.1) allows for many different specific forms. Of many various kinds, two distinct ways depending on the limits of integration are the Fredholm integral equations, which were studied by a Swedish mathematician; *Ivar Fredholm* [1, 2], and the Volterra integral equations, which were introduced by an Italian mathematician; *Vito Volterra* [3].

For the Fredholm integral equation, the limits of integration are fixed at constant values; a and b as shown below.

$$y(x) = f(x) + \lambda \int_a^b k(x,t)y(t)dt. \quad (2.2)$$

For the Volterra integral equation, the lower limit of integration is a constant while the upper limit changes within the interval of integration as given in the next equation.

$$y(x) = f(x) + \lambda \int_a^x k(x,t)y(t)dt. \quad (2.3)$$

Equation (2.1) is said to be singular if one or both limits of integration are infinite, or if the kernel $k(x,t)$ is singular within the domain of definition. The singular integral equations appear in numerous forms such as Abel's equation, Fredholm singular integral equation, Volterra singular integral equation (its particular case is Abel's equation), Cauchy type integral equation, Carleman equation, integral equation of the logarithmic kernel, Wiener-Hopf integral equation, and so on [4]. The weakly singular integral equation is a special case of singular integral equations, which has weak singular kernel, and which can be transformed into a regular integral equation by converting the weak singular kernel to a kernel that is bounded in the domain of definition or interval of integration. Such a weak singular kernel assumes the form specified in the following equation:

$$k(x,t) = \frac{k_1(x,t)}{|x-t|^{\alpha}}, \text{ where } k_1(x,t) \text{ is a bounded function, } 0 < \alpha < 1. \quad (2.4)$$

The weakly singular integral equation can be defined as

$$y(x) = f(x) + \lambda \int_{g(x)}^{h(x)} \frac{k_1(x,t)}{|x-t|^{\alpha}} y(t)dt. \quad (2.5)$$

If the Fredholm or Volterra IE contains weak singular kernel, then it is said to be weakly singular Fredholm or Volterra IE.

The weakly singular linear Fredholm IE is defined as

$$y(x) = f(x) + \lambda \int_a^b \frac{k_1(x,t)}{|x-t|^{\alpha}} y(t)dt \quad (2.6)$$

and the weakly singular linear Volterra IE as

$$y(x) = f(x) + \lambda \int_a^x \frac{k_1(x,t)}{|x-t|^\alpha} y(t)dt. \tag{2.7}$$

If the unknown function $y(x)$ appears only under the integral sign of Equation (2.6) or Equation (2.7), then the integral equation is called the first kind weakly singular Fredholm or Volterra IE, respectively. If the unknown function appears both inside and outside of the integral sign such as in Equation (2.6) or Equation (2.7), then the integral equation is called the second kind weakly singular Fredholm or Volterra IE, respectively. If $f(x)$ is identically zero in the first kind (or second) weakly singular Fredholm (or Volterra) IE, then the resulting IE is called homogeneous first kind (or second kind) weakly singular Fredholm (or Volterra) IE, otherwise inhomogeneous.

The more general form of weakly singular Fredholm and Volterra integral IEs are provided in the following equations.

$$\lambda_0 y(x) = \lambda_1 f(x) + \lambda_2 \int_a^b \frac{k_1(x,t)}{|x-t|^\alpha} H(t, y(t))dt. \tag{2.8}$$

$$\lambda_0 y(x) = \lambda_1 f(x) + \lambda_2 \int_a^x \frac{k_1(x,t)}{|x-t|^\alpha} H(t, y(t))dt. \tag{2.9}$$

In Equation (2.8) or Equation (2.9), if λ_0, λ_1 and λ_2 are non-zero, then IE is called the second kind weakly singular Fredholm-Hammerstein or Volterra-Hammerstein IE, respectviely. If $H(x, y(x)) = y(x)$, then Equation (2.8) (or Equation (2.9)) reduces to Equation (2.6) (or Equation (2.7)). The other forms of weakly singular Fredholm or Volterra IEs can be regained from Equation (2.8) or Equation (2.9) by considering the appropriate values for λ_0, λ_1 and λ_2, and the suitable forms for $f(x)$ and $H(x, y(x))$.

The weakly singular integral equations, particularly weakly singular Volterra IEs, are encountered in heat and mass transfer, fluid mechanics, electrochemistry, and so on. Their exact or closed-form solutions play an important role in proper understanding of qualitative features of many phenomena in those applied fields. A very small number of such physical process models may have closed form solutions obtained either analytically or numerically, but a larger number either do not have exact solutions at all or have closed form solutions, which are extremely tough to derive. So the only alternative to gain insights into such physical processes, which are better modeled by singular or nonsingular integral equations, is to find approximate solutions. As the mathematical models become more realistic, the singular or nonsingular integral equations become unavoidable. The numerical and semianalytical methods for getting accurate approximate solutions to singular or nonsingular integral equation models have been the subject of a great deal of research for over half a century, which has resulted in the rapid growth of literature on their numerical solution. The numerical solution of weakly singular Fredholm, Volterra, and Fredholm-Volterra integral equations describing physical processes are the prime objectives of this chapter. The stated goals are accomplished by devising an efficient and simple numerical method, which is based on the triangular orthogonal functions. This chapter is a further contribution to what is becoming a subject of increasing concern to scientist, engineers, and mathematicians.

2.1 Existence and Uniqueness of Solution

The succeeding general form of weakly singular Fredholm-Volterra-Hammerstein integral equation of second kind, embracing various forms of weakly singular integral equations presented in the preceding section, is considered:

$$\lambda_0 y(x) = \lambda_1 f(x) + \lambda_2 \int_a^b \left(\frac{k_1(x,t)}{(x-t)^{\beta_1}} G(t, y(t)) \right) dt + \lambda_3 \int_a^x \left(\frac{k_2(x,t)}{(x-t)^{\beta_2}} H(t, y(t)) \right) dt, \tag{2.10}$$

where $P-1 \le \beta_1, \beta_2 < P$ ($P \in Q^+$, Q^+ is the collection of positive integers), $x \in [a,b]$, $t \in [a,b] (a < b, b \in \Re^+)$, $y(x) : [a,b] \to [a,\infty)$, $f(x) : [a,b] \to [a,\infty)$, $k_1(x,t) : ([a,b] \times [a,b]) \to [a,\infty)$, $k_2(x,t) : ([a,b] \times [a,b]) \to [a,\infty)$, $G(x, y(x)) : ([a,b] \times [a,\infty)) \to [a,\infty)$, $H(x, y(x)) : ([a,b] \times [a,\infty)) \to [a,\infty)$, $\lambda_0, \lambda_1, \lambda_2, \lambda_3$ are arbitrary constants.

In the aforementioned equation, $f(x)$, $k_1(x,t)$ and $k_2(x,t)$ are known functions, $G(x, y(x))$ and $H(x, y(x))$ are nonlinear functions of independent variable x and unknown function, $y(x)$, which is to be found out.

Without loss of generality, the following assumptions are made.

$$\lambda_i = 1, \; i \in [0,3]. \tag{2.11}$$

Equation (2.10) can be written as the fixed point equation $Ay(x) = y(x)$, where A is defined as

$$Ay(x) = f(x) + \int_a^b \left(\frac{k_1(x,t)}{(x-t)^{\beta_1}} G(t, y(t)) \right) dt + \int_a^x \left(\frac{k_2(x,t)}{(x-t)^{\beta_2}} H(t, y(t)) \right) dt. \tag{2.12}$$

Banach space is defined as $C : [a, b] \times [a, \infty)$ with $d : [a, \infty) \times [a, \infty) \to [a, \infty)$, $d\left(\tilde{y}(x), y(x)\right) = \| \tilde{y}(x) - y(x) \|_\infty$.

It is supposed that the functions $G(x, y(x))$ and $H(x, y(x))$ satisfy, with respect to the second argument, the condition of Lipschitz continuity

$$d(G(x, \tilde{y}(x)) - G(x, y(x))) \le L_1 d(\tilde{y}(x), y(x)), \tag{2.13}$$

where L_1 is a Lipschitz constant, $L_1 \in (0, 1)$,

$$d(H(x, \tilde{y}(x)) - H(x, y(x))) \le L_2 d(\tilde{y}(x), y(x)), \tag{2.14}$$

where L_2 is a Lipschitz constant, $L_2 \in (0, 1)$.

Let $A : C[a, b] \to C[a, b]$ be such that

$$Ay(x) = f(x) + \int_a^b \left(\frac{k_1(x,t)}{(x-t)^{\beta_1}} G(t, y(t)) \right) dt + \int_a^x \left(\frac{k_2(x,t)}{(x-t)^{\beta_2}} H(t, y(t)) \right) dt. \tag{2.15}$$

Therefore, for $x > a$,

$$d\left(A\tilde{y}(x) - Ay(x)\right) = \left\| \int_a^b \left(\frac{k_1(x,t)}{(x-t)^{\beta_1}} (G(t, \tilde{y}(t)) - G(t, y(t))) \right) dt + \int_a^x \left(\frac{k_2(x,t)}{(x-t)^{\beta_2}} (H(t, \tilde{y}(t)) - H(t, y(t))) \right) dt \right\|, \tag{2.16}$$

$$d\left(A\tilde{y}(x) - Ay(x)\right) \le \int_a^b \left(\frac{\| k_1(x,t) \|_\infty}{|x-t|^{\beta_1}} d\left(G(t,\tilde{y}(t)) - G(t,y(t))\right) \right) dt + \int_a^x \left(\frac{\| k_2(x,t) \|_\infty}{|x-t|^{\beta_2}} d\left(H(t,\tilde{y}(t)) - H(t,y(t))\right) \right) dt. \tag{2.17}$$

The following is assumed.

$$M_1 = \max_{x,\, t \in [a,\, b]} |k_1(x,t)|,\ M_2 = \max_{x,\, t \in [a,\, b]} |k_2(x,t)|. \tag{2.18}$$

From Equations (2.13), (2.14), (2.17) and (2.18),

$$\begin{aligned} d\left(A\tilde{y}(x) - Ay(x)\right) &\le \int_a^b \left(\frac{M_1}{|x-t|^{\beta_1}} L_1 d\left(\tilde{y}(t) - y(t)\right) \right) dt + \int_a^x \left(\frac{M_2}{|x-t|^{\beta_2}} L_2 d\left(\tilde{y}(t) - y(t)\right) \right) dt, \\ &\le M_1 L_1 d\left(\tilde{y}(x) - y(x)\right) \underbrace{\left\| \int_a^b \left(\frac{1}{(x-t)^{\beta_1}} \right) dt \right\|_\infty}_{IE1} + M_2 L_2 d\left(\tilde{y}(x) - y(x)\right) \underbrace{\left\| \int_a^x \left(\frac{1}{(x-t)^{\beta_2}} \right) dt \right\|_\infty}_{IE2}. \end{aligned} \tag{2.19}$$

Taking weakly singular integral *IE1*,

$$\begin{aligned} \left\| \int_a^b \left(\frac{1}{(x-t)^{\beta_1}} \right) dt \right\|_\infty &= \Gamma(\alpha_1) \left\| \frac{1}{\Gamma(\alpha_1)} \int_a^x \left((x-t)^{\alpha_1 - 1} t^0 \right) dt + \frac{1}{\Gamma(\alpha_1)} \int_x^b \left((x-t)^{\alpha_1 - 1} t^0 \right) dt \right\|_\infty \\ &\quad (\textit{where}\ \alpha_1 = 1 - \beta_1), \\ &= \Gamma(\alpha_1) \left(\left(\frac{\|t\|_\infty{}^{\alpha_1}}{\Gamma(\alpha_1 + 1)} \right)_a^x + \left(\frac{\|t\|_\infty{}^{\alpha_1}}{\Gamma(\alpha_1 + 1)} \right)_x^b \right), \\ &= \frac{|b|^{\alpha_1} - |a|^{\alpha_1}}{\alpha_1}. \end{aligned} \tag{2.20}$$

Likewise, the weakly singular integral *IE2* can be solved as

$$
\begin{aligned}
\| \int_a^x \left(\frac{1}{(x-t)^{\beta_2}} \right) dt \|_\infty &= \Gamma(\alpha_2) \| \frac{1}{\Gamma(\alpha_2)} \int_a^x (x-t)^{\alpha_2 - 1} t^0 dt \|_\infty, \\
&= \Gamma(\alpha_2) \left(\frac{\| t \|_\infty{}^{\alpha_2}}{\Gamma(\alpha_2 + 1)} \right)_a^x (\textit{where } \alpha_2 = 1 - \beta_2), \\
&= \Gamma(\alpha_2) \left(\frac{\| x \|_\infty{}^{\alpha_2} - \| a \|_\infty{}^{\alpha_2}}{\Gamma(\alpha_2 + 1)} \right), \\
&= \frac{|b|^{\alpha_2} - |a|^{\alpha_2}}{\alpha_2} (\textit{where } \alpha_2 = 1 - \beta_2), \\
&= \Gamma(\alpha_2) \left(\frac{\| x \|_\infty{}^{\alpha_2} - \| a \|_\infty{}^{\alpha_2}}{\Gamma(\alpha_2 + 1)} \right), \\
&= \frac{|b|^{\alpha_2} - |a|^{\alpha_2}}{\alpha_2} (\text{since } \| x \|_\infty = b).
\end{aligned}
\tag{2.21}
$$

As a and b are positive scalar integers,

$$
\| \int_a^b \left(\frac{1}{(x-t)^{\beta_1}} \right) dt \|_\infty = \frac{b^{\alpha_1} - a^{\alpha_1}}{\alpha_1}, \tag{2.22}
$$

$$
\| \int_a^x \left(\frac{1}{(x-t)^{\beta_2}} \right) dt \|_\infty = \frac{b^{\alpha_2} - a^{\alpha_2}}{\alpha_2}. \tag{2.23}
$$

By utilizing Equations (2.22) and (2.23) in Equation (2.19), we get the following.

$$
d(A\tilde{y}(x) - Ay(x)) \leq \Omega d(\tilde{y}(x) - y(x)), \tag{2.24}
$$

where $\Omega = \left(\frac{M_1 L_1}{\alpha_1} (b^{\alpha_1} - a^{\alpha_1}) + \frac{M_2 L_2}{\alpha_2} (b^{\alpha_2} - a^{\alpha_2}) \right)$.

In accordance with contraction mapping theorem, Equation (2.10) bears a unique solution in $C[a, b]$ if $\Omega < 1$.

2.2 The Proposed Numerical Method

The procedure of numerically solving the weakly singular Fredholm-Volterra-Hammerstein integral equation (Equation (2.10)) with the help of triangular orthogonal functions is thoroughly discussed in this section.

Equation (2.10) can be rewritten in the form

$$\lambda_0 y(x) = \lambda_1 f(x) + \lambda_2 \Gamma(\alpha_1) \left(\underbrace{\frac{1}{\Gamma(\alpha_1)} \int_a^x \left((x-t)^{\alpha_1 - 1} k_1(x,t) G(t,y(t)) \right) dt}_{IE3} + \underbrace{\frac{1}{\Gamma(\alpha_1)} \int_x^b \left((x-t)^{\alpha_1 - 1} k_1(x,t) G(t,y(t)) \right) dt}_{IE4} \right) + \lambda_3 \Gamma(\alpha_2) \left(\underbrace{\frac{1}{\Gamma(\alpha_2)} \int_a^x \left((x-t)^{\alpha_2 - 1} k_2(x,t) H(t,y(t)) \right) dt}_{IE5} \right), \tag{2.25}$$

where $\alpha_1 = 1 - \beta_1$, $\alpha_2 = 1 - \beta_2$.

The idea is to convert the above weakly singular integral equation into a system of nonlinear algebraic equations by replacing the original functions as well as the singular integrals with their approximations obtained in the TF domain.

The following can be obtained by using Equations (1.68) and (1.78).

$$y(x) \cong C^T T1_m(x) + D^T T2_m(x), \tag{2.26}$$

$$f(x) \cong C_0^T T1_m(x) + D_0^T T2_m(x), \tag{2.27}$$

$$G\left(t, y(t)\right) \cong C_G^T T1_m(t) + D_G^T T2_m(t), \tag{2.28}$$

$$H\left(t, y(t)\right) \cong C_H^T T1_m(t) + D_H^T T2_m(t), \tag{2.29}$$

$$k_i(x,t) \approx T1_m^T(x) \left(F_1^i T1_m(t) + F_2^i T2_m(t) \right) + T2_m^T(x) \left(F_3^i T1_m(t) + F_4^i T2_m(t) \right), \; i \in [1,2]. \tag{2.30}$$

Multiplying $k_1(x,t)$ by $G(t, y(t))$,

$$\begin{aligned}k_1(x,t)G(t,y(t)) &= (T1_m^T(x)(F_1^1T1_m(t)+F_2^1T2_m(t))+T2_m^T(x)(F_3^1T1_m(t)+F_4^1T2_m(t)))\\ &\quad \times(T1_m^T(t)C_G+T2_m^T(t)D_G),\\ &= T1_m^T(x)F_1^1T1_m(t)T1_m^T(t)C_G+T1_m^T(x)F_2^1T2_m(t)T1_m^T(t)C_G+T1_m^T(x)F_1^1T1_m(t)T2_m^T(t)D_G\\ &\quad +T1_m^T(x)F_2^1T2_m(t)T2_m^T(t)D_G+T2_m^T(x)F_3^1T1_m(t)T1_m^T(t)C_G+T2_m^T(x)F_4^1T2_m(t)T1_m^T(t)C_G\\ &\quad +T2_m^T(x)F_3^1T1_m(t)T2_m^T(t)D_G+T2_m^T(x)F_4^1T2_m(t)T2_m^T(t)D_G.\end{aligned} \tag{2.31}$$

Employing Equations (1.69), (1.70) and (1.71) in Equation (2.31),

$$\begin{aligned}k_1(x,t)G(t,y(t)) &= T1_m^T(x)F_1^1diag(T1_m(t))C_G+T1_m^T(x)F_2^1OC_G+T1_m^T(x)F_1^1OD_G\\ &\quad +T1_m^T(x)F_2^1diag(T2_m(t))D_G+T2_m^T(x)F_3^1diag(T1_m(t))C_G\\ &\quad +T2_m^T(x)F_4^1OC_G+T2_m^T(x)F_3^1OD_G+T2_m^T(x)F_4^1diag(T2_m(t))D_G.\end{aligned} \tag{2.32}$$

Equation (2.32) can also be written as

$$\begin{aligned}k_1(x,t)G(t,y(t)) &= T1_m^T(x)F_1^1diag(C_G)T1_m(t)+T1_m^T(x)F_2^1diag(D_G)T2_m(t)\\ &\quad +T2_m^T(x)F_3^1diag(C_G)T1_m(t)+T2_m^T(x)F_4^1diag(D_G)T2_m(t).\end{aligned} \tag{2.33}$$

In the same fashion, the final expression for the product of $k_2(x,t)$ and $H(t,y(t))$ is achieved as

$$\begin{aligned}k_2(x,t)H(t,y(t)) &= T1_m^T(x)F_1^2diag(C_H)T1_m(t)+T1_m^T(x)F_2^2diag(D_H)T2_m(t)\\ &\quad +T2_m^T(x)F_3^2diag(C_H)T1_m(t)+T2_m^T(x)F_4^2diag(D_H)T2_m(t).\end{aligned} \tag{2.34}$$

The weakly singular integral *IE3* can be reduced to a system of nonlinear algebraic equations as follows.

Taking the fractional integral of $k_1(x,t)G(t,y(t))$ of order α_1, with respect to t,

$$\begin{aligned}\frac{1}{\Gamma(\alpha_1)}\int_a^x((x-t)^{\alpha_1-1}k_1(x,t)G(t,y(t)))dt &= A_{1x}^{\alpha_1}(T1_m(t))+A_{2x}^{\alpha_1}\left(T2_m(t)\right),\\ &= A_1B_1+A_2B_2,\end{aligned} \tag{2.35}$$

where $B_1=P_1^{\alpha_1}T1_m(x)+P_2^{\alpha_1}T2_m(x)$, $B_2=P_3^{\alpha_1}T1_m(x)+P_4^{\alpha_1}T2_m(x)$,

$$A_1 = \left(T1_m^T(x)F_1^1 diag(C_G) + T2_m^T(x)F_3^1 diag(C_G)\right),$$

$$A_2 = \left(T1_m^T(x)F_2^1 diag(D_G) + T2_m^T(x)F_4^1 diag(D_G)\right).$$

Employing the disjointness properties of TFs,

$$\begin{aligned} &\frac{1}{\Gamma(\alpha_1)}\int_a^x \left((x-t)^{\alpha_1-1} k_1(x,t)G\left(t,y(t)\right)\right)dt \\ &= T1_m^T(x)\left(F_1^1 diag(C_G)P_1^{\alpha_1} + F_2^1 diag(D_G)P_3^{\alpha_1}\right)T1_m(x) \\ &+T2_m^T(x)\left(F_3^1 diag(C_G)P_2^{\alpha_1} + F_4^1 diag(D_G)P_4^{\alpha_1}\right)T2_m(x). \end{aligned} \tag{2.36}$$

Equation (2.36) can be written as

$$\frac{1}{\Gamma(\alpha_1)}\int_a^x \left((x-t)^{\alpha_1-1} k_1(x,t)G(t,y(t))\right)dt = X_1 T1_m(x) + X_2 T2_m(x), \tag{2.37}$$

where $X_1 = diag\left(F_1^1 diag(C_G)P_1^{\alpha_1}\right) + diag\left(F_2^1 diag(D_G)P_3^{\alpha_1}\right),$

$$X_2 = diag\left(F_3^1 diag(C_G)P_2^{\alpha_1}\right) + diag\left(F_4^1 diag(D_G)P_4^{\alpha_1}\right).$$

Similarly, the weakly singular integrals (*IE4* and *IE5*) can be transformed into the system of algebraic equations as derived below:

$$\frac{1}{\Gamma(\alpha_1)}\int_x^b \left((x-t)^{\alpha_1-1} k_1(x,t)G\left(t,y(t)\right)\right)dt = X_3 T1_m(x) + X_4 T2_m(x), \tag{2.38}$$

where $X_3 = diag\left(F_1^1 diag(C_G)\widetilde{P}_1^{\alpha_1}\right) + diag\left(F_2^1 diag(D_G)\widetilde{P}_3^{\alpha_1}\right),$

$$X_4 = diag\left(F_3^1 diag(C_G)\widetilde{P}_2^{\alpha_1}\right) + diag\left(F_4^1 diag(D_G)\widetilde{P}_4^{\alpha_1}\right).$$

$$\frac{1}{\Gamma(\alpha_2)}\int_a^x \left((x-t)^{\alpha_2-1}k_2(x,t)H\left(t,y(t)\right)\right)dt = X_5T1_m(x) + X_6T2_m(x), \quad (2.39)$$

where $X_5 = diag\left(F_1^2 diag(C_H)P_1^{\alpha_2}\right) + diag\left(F_2^2 diag(D_H)P_3^{\alpha_2}\right)$,

$$X_6 = diag\left(F_3^2 diag(C_H)P_2^{\alpha_2}\right) + diag\left(F_4^2 diag(D_H)P_4^{\alpha_2}\right).$$

Utilizing Equations (2.26), (2.27), (2.37) (2.38) and (2.39) in Equation (2.25),

$$\begin{aligned}\lambda_0\left(C^T T1_m(x) + D^T T2_m(x)\right) &= \lambda_1\left(C_0^T T1_m(x) + D_0^T T2_m(x)\right)\\ &+\lambda_2\Gamma(\alpha_1)\left(X_1T1_m(x) + X_2T2_m(x)\right)\\ &+\lambda_2\Gamma(\alpha_1)\left(X_3T1_m(x) + X_4T2_m(x)\right)\\ &+\lambda_3\Gamma(\alpha_2)\left(X_5T1_m(x) + X_6T2_m(x)\right),\end{aligned} \quad (2.40)$$

Equating the coefficients of LHTF and RHTF,

$$\lambda_0 C^T = \lambda_1 C_0^T + \lambda_2\Gamma(\alpha_1)X_1 + \lambda_2\Gamma(\alpha_1)X_3 + \lambda_3\Gamma(\alpha_2)X_5, \quad (2.41)$$

$$\lambda_0 D^T = \lambda_1 D_0^T + \lambda_2\Gamma(\alpha_1)X_2 + \lambda_2\Gamma(\alpha_1)X_4 + \lambda_3\Gamma(\alpha_2)X_6. \quad (2.42)$$

Solving the system of simultaneous nonlinear algebraic equations in Equations (2.41) and (2.42) yields the approximate solution to the weakly singular Fredholm-Volterra-Hammerstein integral equation.

2.3 Convergence Analysis

It is theoretically shown here that the TF solution of Equation (2.10) approaches the exact solution when a sufficiently small step size (large number of subintervals) is considered.

Let $y(t)$ and $\widetilde{y}(t)$ be the exact and TF solution, respectively, of Equation (2.10). The error between the exact and TF approximate solution is defined in the following equation:

$$\varepsilon_j = |y(t) - \widetilde{y}(t)|,\ t\in[jh,(j+1)h],\ j\in[0,m-1]. \quad (2.43)$$

The TF approximation for $y(t)$ in the jth subinterval is

$$\tilde{y}(t) = c_j T1_j(t) + d_j T2_j(t) = y(jh)\left(1 - \frac{t - jh}{h}\right) + y((j+1)h)\left(\frac{t - jh}{h}\right). \quad (2.44)$$

The next expression can be obtained by making use of backward difference formula in the above equation:

$$\tilde{y}(t) = y(jh) + \frac{y\big((j+1)h\big) - y(jh)}{h}(t - jh) = y(jh) + \left(\frac{dy(t)}{dt}\right)_{t=jh}(t - jh). \quad (2.45)$$

The Taylor series expansion of the exact solution around the point jh is

$$\begin{aligned} y(t) = y(jh) &+ \left(\frac{dy(t)}{dt}\right)_{t=jh}(t - jh) + \left(\frac{d^2y(t)}{dt^2}\right)_{t=jh}\frac{(t - jh)^2}{2} \\ &+ \sum_{k=3}^{\infty}\left(\left(\frac{d^ky(t)}{dt^k}\right)_{t=jh}\frac{(t - jh)^k}{k!}\right). \end{aligned} \quad (2.46)$$

The above series is truncated to the first three terms.

$$y(t) = y(jh) + \left(\frac{dy(t)}{dt}\right)_{t=jh}(t - jh) + \left(\frac{d^2y(t)}{dt^2}\right)_{t=jh}\frac{(t - jh)^2}{2}. \quad (2.47)$$

Equations (2.45) and (2.47) are used in Equation (2.43).

$$\varepsilon_j = \left|\left(\frac{d^2y(t)}{dt^2}\right)_{t=jh}\frac{(t - jh)^2}{2}\right| = \left|\left(\frac{d^2y(t)}{dt^2}\right)_{t=jh}\right|\left|\frac{(t - jh)^2}{2}\right|, \ t \in [jh, (j+1)h]. \quad (2.48)$$

Since $t \in [jh, (j+1)h]$ and t takes only the discrete values i.e. $0, h, 2h, \ldots, jh, \ldots, mh$, the term $(t - jh)$ is always less than or equal to h, and the assumption in Equation (2.49) can be made.

$$M_j = \left|\left(\frac{d^2y(t)}{dt^2}\right)_{t=jh}\right|, \ t \in [jh, (j+1)h]. \quad (2.49)$$

Therefore, Equation (2.48) becomes

$$\varepsilon_j = M_j \frac{h^2}{2}. \tag{2.50}$$

E is defined as the vector of errors attained in each subinterval.

$$E = \left[\varepsilon_0 \quad \varepsilon_1 \quad \varepsilon_2 \quad \ldots \quad \varepsilon_j \quad \ldots \quad \varepsilon_{m-1} \right], \tag{2.51}$$

where ε_j can be determined from Equation (2.50).

The Euclidean norm of the vector E can be computed as

$$\begin{aligned} \| E \|_2 &= \left(\sum_{j=0}^{m} |E(j+1)|^2 \right)^{1/2} = \left(\sum_{j=0}^{m} |\varepsilon_j|^2 \right)^{1/2}, \\ &= \left(\sum_{j=0}^{m} \left| M_j \frac{h^2}{2} \right|^2 \right)^{1/2} = \left(\sum_{j=0}^{m} M_j^2 \frac{h^4}{4} \right)^{1/2}. \end{aligned} \tag{2.52}$$

In the limit of step size tends to zero, the Euclidean norm of the vector E ultimately becomes zero as proved below:

$$\lim_{h \to 0} \| E \|_2 = \lim_{h \to 0} \sqrt{\left(\sum_{j=0}^{m} M_j^2 \frac{h^4}{4} \right)} = 0. \tag{2.53}$$

The maximum norm or infinity norm of the vector E can be determined as

$$\begin{aligned} \| E \|_\infty &= \max \left(|E(1)|, |E(2)|, |E(3)|, \ldots\ldots, |E(j)|, \ldots\ldots, |E(m+1)| \right), \\ &= \max \left(\left| M_0 \frac{h^2}{2} \right|, \left| M_1 \frac{h^2}{2} \right|, \left| M_2 \frac{h^2}{2} \right|, \ldots\ldots, \left| M_j \frac{h^2}{2} \right|, \ldots\ldots, \left| M_m \frac{h^2}{2} \right| \right). \end{aligned} \tag{2.54}$$

As shown in Equation (2.55), the infinity norm of the vector E approaches zero as the step size decreases to zero:

$$\lim_{h \to 0} \| E \|_\infty = \max \left(\lim_{h \to 0} \left| M_0 \frac{h^2}{2} \right|, \lim_{h \to 0} \left| M_1 \frac{h^2}{2} \right|, \ldots \lim_{h \to 0} \left| M_j \frac{h^2}{2} \right| \ldots, \lim_{h \to 0} \left| M_m \frac{h^2}{2} \right| \right) = 0. \tag{2.55}$$

2.4 Numerical Experiments

The proposed numerical method is tested in this section to corroborate whether it can be applied to solve various weakly singular integral equations, and it can compete with the existing numerical methods deployed for the same purpose. In addition to examining validity and performance, investigation is also carried out to make sure that the method produces stable numerical solution even in the case of corrupted source function. The performance indices such as infinity norm (L1) and Euclidean norm (L2), which are defined below, are used to quantify the performance of the proposed numerical algorithm.

The absolute error between the exact and TF approximate solution is defined as

$$\varepsilon(t) = |y(t) - \tilde{y}(t)|, \; t \in [a, b]. \tag{2.56}$$

The infinity norm (L1) of $\varepsilon(t)$ is defined as

$$\| \varepsilon(t) \|_{\infty} = \max_{j \in [0, m]} | \varepsilon(t_j) | \tag{2.57}$$

and the Euclidean norm (L2) is

$$\| \varepsilon(t) \|_{2} = \left(\sum_{j=0}^{m} | \varepsilon(t_j) |^2 \right)^{1/2}. \tag{2.58}$$

The computational time (CT) or CPU usage is calculated in seconds.

In Subsection 2.4.1, every test problem is solved using the BPFs-based numerical method as well as the proposed TFs-based numerical method. The mathematical theory for the BPFs-based numerical method is not provided in this chapter.

2.4.1 Investigation of Validity and Accuracy

Example 2.1: The weakly singular (WS) Fredholm-Hammerstein integral equation (IE) of 2nd kind

The weakly singular Fredholm-Hammerstein integral equation of the second kind is

$$y(t) - \int_0^1 |t - x|^{-0.5} \left(y(x) \right)^2 dx = f(t), \tag{2.59}$$

where
$f(t) = (t(1-t))^{0.5} + \frac{16}{15}t^{2.5} + 2t^2(1-t)^{0.5} + \frac{4}{3}t(1-t)^{3/2} + \frac{2}{5}(1-t)^{2.5} - \frac{4}{3}t^{3/2} - 2t(1-t)^{0.5} - \frac{2}{3}(1-t)^{3/2}$.

The given problem has the exact solution $y(t) = \sqrt{t(1-t)}$.

The given problem is solved numerically by using the BPFs-based numerical method, and the proposed TFs-based numerical method. Table 2.1 compares four numerical solutions (TF solution, Legendre Wavelets (LW) solution [5], Haar Wavelets (HW) [6] and BPF solution) with the actual solution. The TF and BPF solution are more accurate than LW and HW solution. When the TF and BPF solution are compared, the performance of the TFs-based numerical method is superior to that of the BPFs-based numerical method. As the BPFs are the zeroth degree polynomial, they cannot provide accurate approximations to higher order (i.e., order greater than 0) polynomial functions; consequently, the BPF solution is always less accurate than the TF solution. The expression given in Equation (2.59) for estimating the time function in the BPF domain involves the first-order integration, causing the BPFs-based numerical method requires more CPU usage or higher computational time; thus, it can never be faster than the proposed TFs-based numerical method. The maximum error obtained by piecewise collocation method in [7] is 7.1e-04, which is greater than

TABLE 2.1

Numerical solution of Example 2.1 acquired via various methods

t	Proposed method	LW in Ref. [5]	HW in Ref. [6]	BPFs-based method	ES
0	7.787809e-05	0.0507080452	0.1167357717	0	0
0.1	0.2999990565	0.3012573003	0.3117776466	0.2999991591	0.3000000000
0.2	0.3999991191	0.4015798695	0.4021691763	0.3999988108	0.4000000000
0.3	0.4582567481	0.4584563412	0.4567744935	0.4582561131	0.4582575694
0.4	0.4898971893	0.4900950434	0.4878027709	0.4898962668	0.4898979485
0.5	0.4999993076	0.4999923223	0.4996742893	0.4999981198	0.5000000000
0.6	0.4898973297	0.4898950393	0.4878028614	0.4898958889	0.4898979485
0.7	0.4582570339	0.4582563414	0.4567742823	0.4582553448	0.4582575694
0.8	0.3999995629	0.4015293214	0.4021683379	0.3999976217	0.4000000000
0.9	0.2999996912	0.3013050185	0.3117732826	0.2999974773	0.3000000000
1	5.1647897e-16	0.0507080452	0.1167400343	0.0006647347	0
			Performance Indices		
CT	29.972104	-	-	922.882420	N/A
L1	7.787809e-05	0.0507080452	0.1167400343	6.647347e-04	N/A
L2	1.083541e-04	0.0717691354	0.1659984670	6.660618e-04	N/A

The step size of 0.002 is used for TF and BPF solution. ES is exact solution.

that is achieved via the proposed method. Among these five approximate solutions, the TF solution is found to be the best.

Example 2.2: WS linear Fredholm IE of 2nd kind

The following linear weakly singular Fredholm integral equation of second kind is considered:

$$y(t) - \int_0^1 |t-x|^{-0.5} y(x)dx = t - 2\sqrt{1-t} - \frac{4}{3}t^{3/2} + \frac{4}{3}(1-t)^{3/2}, \qquad (2.60)$$

with the exact solution $y(t) = t$.

In Table 2.2, the numerical solution attained in [5] and [8] are compared with the TF and BPF solutions. The proposed numerical method shows better performance than the modified collocation method (MCM) [8] and BPFs-based numerical method, and gives the actual solution as the Legendre wavelets–based numerical method yielded [5].

Example 2.3: WS Fredholm-Hammerstein IE of 1st kind

The first kind Fredholm-Hammerstein integral equation with weak singular kernel is

$$\frac{9}{40}\left[(1-t)^{2/3} + 2t(1-t)^{2/3} + 4t^{5/3} - 3t^2(1-t)^{2/3} - 3t^{8/3}\right] = \int_0^1 \left(\frac{\left(x(s)\right)^2}{|t-s|^{1/3}} ds\right). \qquad (2.61)$$

TABLE 2.2

Numerical solution of Example 2.2 gained via various methods

t	Proposed method	LW in Ref. [5]	MCM in Ref. [8]	BPFs-based method	ES
0	0	0	0.000179	-5.81509e-05	0
0.25	0.25	0.25	0.251425	0.24993285	0.25
0.5	0.5	0.5	0.498799	0.49991776	0.5
0.75	0.75	0.75	0.749872	0.74988369	0.75
1	1	1	0.998124	1.04768370	1
		Performances Indices			
CT	101.248277	-	-	1932.33419	N/A
L1	0	0	0.0018759999	0.04768370	N/A
L2	0	0	0.0026534556	0.047926481	N/A

The step size of 0.0011 is used for TF and BPF solution.

TABLE 2.3

Absolute errors and performance indices for Example 2.3

t	Proposed method	Method in Ref. [9]	BPFs-based method
0.1	1.365055e-07	6.34E-5	2.154308e-06
0.3	2.274850e-07	1.80E-5	1.247001e-06
0.5	2.793437e-07	4.05E-6	8.998293e-07
0.7	2.079807e-07	1.88E-5	4.973795e-07
0.9	2.162055e-07	6.43E-5	2.449896e-07
		Performance Indices	
CT	9.186632	-	708.471841
L1	8.821899e-07	6.430000e-05	8.757334e-06
L2	5.973783e-06	9.406323e-05	2.691921e-05

The step size of 0.002 is used for TF and BPF solution.

The exact solution of Equation (2.61) is $x(t) = \sqrt{t(1-t)}$.

The results obtained in [9] and the ones achieved by TFs-based and BPFs-based numerical methods are presented in Table 2.3, from which one can see that the TF solution is the most accurate one.

Example 2.4: WS Volterra-Fredholm-Hammerstein IE of 2nd kind

The weakly singular Volterra-Fredholm-Hammerstein integral equation of second kind is

$$y(t) = t - \frac{32t^{5/2}}{15} - 2\sqrt{t} - \frac{12\sqrt{1-t}}{5} - \frac{8t\sqrt{1-t}}{15} - \frac{16t^2\sqrt{1-t}}{15} + \int_0^t \frac{y^2(s)}{\sqrt{|t-s|}}ds + \int_0^1 \frac{1+y^2(s)}{\sqrt{|t-s|}}ds. \tag{2.62}$$

It can simply be verified that $y(t) = t$ is the exact solution of Equation (2.62).

The results are given in Table 2.4. It is once more proved here that the proposed numerical method is more accurate and quicker than the BPFs-based numerical method.

Example 2.5: WS Volterra-Hammerstein IE of 2nd kind

The weakly singular Volterra-Hammerstein integral equation is

$$y(t) = t^{1/3} + \frac{4\Gamma\left(4/3\right)\Gamma\left(13/6\right)}{\sqrt{\pi}}t^{3/2} - \int_0^t \frac{s^{1/2}y^2(s)}{(t-s)^{2/3}}ds,\ t \in I,\ I = [a,b]. \tag{2.63}$$

TABLE 2.4

Performance of proposed and BPFs-based methods (Example 2.4)

	Proposed method			BPFs-based method		
m	L1	L2	CT	L1	L2	CT
50	0.00601590	0.01051159	0.882158	0.23068768	0.35366130	54.647130
100	0.00133433	0.00267476	1.267554	0.17995981	0.31947490	102.492798
150	0.00082322	0.00175374	1.860383	0.14731650	0.28441832	162.992119
200	0.00074459	0.00140626	4.328482	0.16455390	0.25339537	224.616251
250	0.00018869	0.00060195	8.112344	0.14920257	0.22508343	310.421609
300	2.7509e-07	5.0171e-07	5.741063	0.14732238	0.20484367	376.259100
350	9.6020e-08	1.2305e-07	8.429532	0.14324399	0.18840770	476.172420

TABLE 2.5

Maximum norm (L1) obtained by the proposed method, PEM, and PTM (Example 2.5)

	$t\in[0, 0.01]$			$t\in[0, 0.02]$			$t\in[0, 0.03]$		
		Ref. [10]			Ref. [10]			Ref. [10]	
m	PM	PEM	PTM	PM	PEM	PTM	PM	PEM	PTM
80	1.736e-07	6.60e-03	2.30e-05	2.548e-07	6.50e-03	2.30e-05	4.606e-07	6.30e-03	2.30e-05
160	6.295e-08	2.10e-03	8.75e-06	1.736e-07	2.10e-03	6.05e-06	2.053e-07	2.03e-03	8.75e-06
320	1.244e-08	1.50e-03	2.31e-06	6.295e-08	1.50e-03	1.58e-06	5.583e-08	1.50e-03	2.31e-06
640	7.632e-17	7.40e-04	6.03e-07	1.066e-08	7.20e-04	4.10e-07	2.921e-08	7.10e-04	6.03e-07

PM-proposed method

TABLE 2.6

Euclidean norm (L2) and CPU usage (CT) obtained by the proposed method for Example 2.5

	$t\in[0, 0.01]$		$t\in[0, 0.02]$		$t\in[0, 0.03]$	
m	L2	CT	L2	CT	L2	CT
80	1.736e-07	0.529962	2.548e-07	0.525487	4.606e-07	0.606266
160	6.940e-08	0.906945	1.736e-07	0.908630	2.053e-07	0.966618
320	7.373e-08	9.458081	6.940e-08	2.062113	5.583e-08	2.396263
640	1.411e-15	40.577329	8.016e-08	44.607859	4.176e-08	72.189409

TABLE 2.7

Performance of BPFs-based numerical method for Example 2.5

	b=0.01			b=0.02			b=0.03		
m	L1	L2	CT	L1	L2	CT	L1	L2	CT
80	0.0410046	0.0415221	84.122999	0.0465444	0.0493586	72.623420	0.0526442	0.0558277	87.405432
160	0.0273562	0.0292536	160.62293	0.0410046	0.0415221	156.667233	0.0472069	0.0481408	166.77392
320	0.0247354	0.0254096	286.46786	0.0273562	0.0292536	327.526297	0.0374269	0.0377372	347.50488
640	0.0181653	0.0189827	647.705641	0.0247354	0.0254096	770.199038	0.0305702	0.0306078	620.504602

TABLE 2.8

Numerical solution of Example 2.5 on $t\in[0, 1]$

	Proposed method			BPFs-based method		
m	L1	L2	CT	L1	L2	CT
80	1.46053e-04	1.55154e-04	0.688606	0.199321187	0.199468293	77.167354
160	5.35683e-05	5.67202e-05	2.002976	0.154640663	0.155369510	152.51406
320	1.88311e-05	2.00254e-05	2.714557	0.123629059	0.123818661	286.38760
640	6.69186e-06	7.15089e-06	17.20672	0.097318820	0.097604185	617.323506

The exact solution satisfying the above equation is $y(t) = t^{1/3}$.

The left end point of the interval I is fixed at 0. The right end point is varied in the interval [0.01, 0.03] with constant width of 0.01. The numerical solution of given problem obtained by product Euler's method (PEM) and the product trapezoidal method (PTM) in [10] are compared with the BPF and TF solution in Tables 2.5–2.7. The numerical method based on BPFs is not able to produce valid estimation even on the very small intervals. The PTM solution is close to the TF solution and is more accurate than the PEM solution. Based on the results in Table 2.8 (for larger interval [0, 1])), it can be stated that the proposed numerical method continues its triumph over the BPFs-based method in numerically solving weakly singular integral equations.

2.4.2 Numerical Stability Analysis

Example 2.6: WS linear Volterra-Fredholm IE of 2nd kind

The second kind weakly singular Volterra-Fredholm integral equation is

$$y(t) = f(t) + \int_0^t |t - s|^{-0.5} y(s)ds + \int_0^1 \left(t + y(t)\right)ds, \tag{2.64}$$

where $f(t) = -(4/3)t^{1.5} - 0.5$.

The exact solution is known as $y(t) = t$.

The given source function ($f(t)$) is purposefully corrupted with noise (Equation (2.65)) to demonstrate the capability of the proposed numerical method to simulate physical fractional order process models whose parameters are estimated from process data containing noise.

The corrupted source function is

$$\tilde{f}(t_j) = f(t_j) + \delta\,\theta_j,\ j \in [0, 1, 2, \ldots\ldots\ldots, m],\ m = 1000, \tag{2.65}$$

where θ_j is an arbitrary random number generated in the range [–1, 1], δ is the noise level.

The proposed numerical method is tested at different noise levels. The performance indices are calculated and tabulated in Table 2.9, and the TF solutions are displayed in Figures 2.1 and 2.2. The proposed method is able to yield the exact solution when the source function is free from noise. The TF solutions obtained for the noise level up to 10^{-3} are acceptable (Figure 2.1) whereas beyond this noise level, its inability is unveiled in Figure 2.2. However, more accurate TF solutions can be achieved with a smaller step size, which is possible only at the cost of significantly increased CPU time.

2.4.3 Application of Proposed Method to Physical Process Models

Application 2.1: Heat radiation in a semi-infinite solid

The second kind weakly singular Volterra-Hammerstein integral equation given below originates in heat transfer problems [11]:

TABLE 2.9

Robustness analysis of the proposed numerical method (Example 2.6)

	Proposed method		
δ	L1	L2	CT
0	0	0	337.649349
10^{-5}	2.693094679513e-07	4.688238418512e-07	132.297570
10^{-4}	1.199148287889e-06	1.204961347157e-06	121.312720
10^{-3}	0.001412316452155	0.006700778600268	1034.516118
10^{-2}	0.003639025461662	0.018657430336298	1106.537432
10^{-1}	0.052164047682997	0.264446472724325	1146.036168
10^{0}	0.26152701561128	0.801555549758445	3155.849109

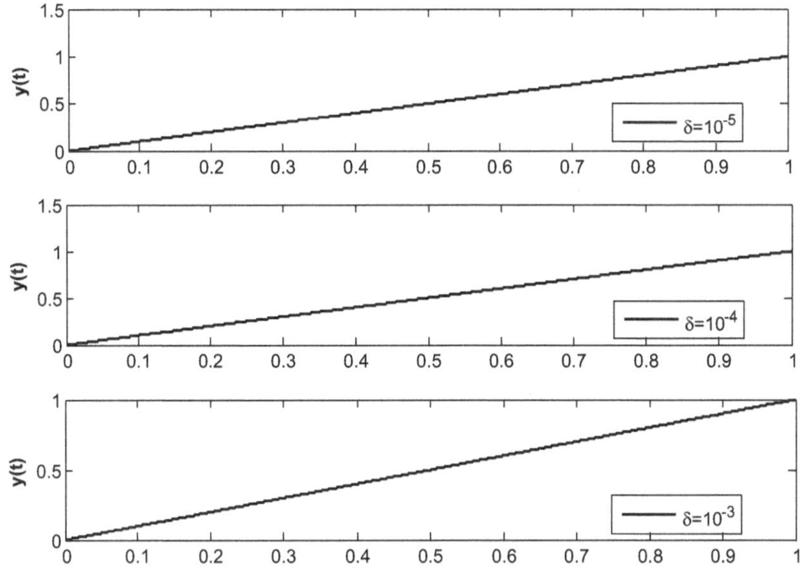

FIGURE 2.1
TF solution for $\delta=10^{-5}$, 10^{-4}, 10^{-3}

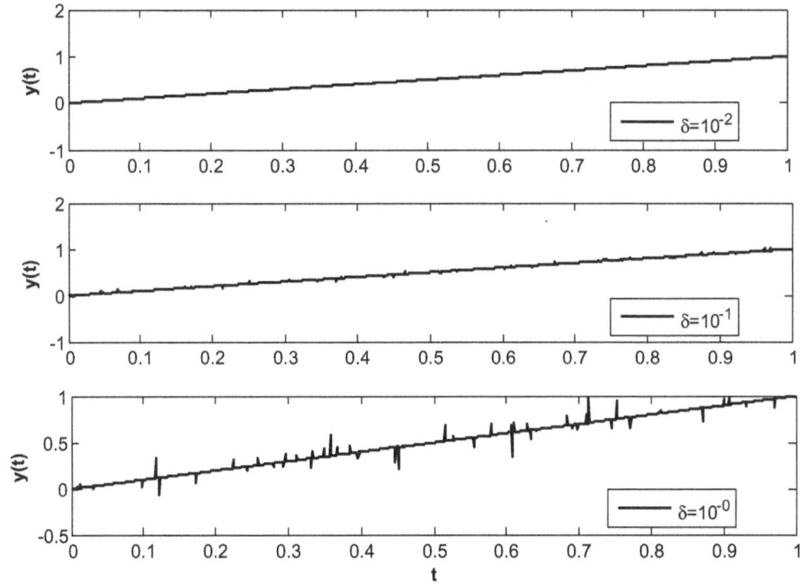

FIGURE 2.2
TF solution for $\delta=10^{-2}$, 10^{-1}, 10^{-0}

$$y(x) = \frac{1}{\sqrt{\pi}} \int_{0}^{t} \frac{f(t) - y^n(t)}{\sqrt{x-t}} dt. \tag{2.66}$$

In the above equation, $y(t)$ signifies the temperature of a semi-infinite heat conducting solid whose surface is heated by a source at a rate proportional to a given function $f(t)$. The surface of the solid radiates energy at a rate proportional to $y^n(t)$, where $n = 1$ represents Newton's law of cooling, and $n = 4$ corresponds to Stefan's radiation law. The physical problem is to find out the temperature of the semi-infinite solid for the constant heat source $f(t) = 1$. This problem was numerically examined by Hoog and Weis [12] by using a hybrid block-by-block method-Newton iterations. A semianalytical series solution for Equation (2.66), which is unacceptable in the interval considered here (as shown in Table 2.10), was computed by Wazwaz [13]. The TF solution as well as the BPF solution are obtained with the step size of 0.001, and are compared with those produced in [12, 13]. Only the TF solution is in accordance with the solution in [12].

Application 2.2: Hydrodynamics

A fluid with zero viscosity is called superfluid, which was discovered by Pyotr Leonidovich Kapitsa. The superfluidity property of fluid allows it to flow without friction past any surface, thus the fluid continues to circulate over hindrances and through openings in containers which hold it, subject only to its own inertia. This phenomenon is generally observed in helium-3 and helium-4 when they are liquefied by cooling to cryogenic temperatures. The helium-4 acts as a normal and colorless liquid (which is called Helium I)

TABLE 2.10

Approximate solutions to Equation (2.66)

t	Method in Ref. [12]	Proposed method	BPFs-based method	ADM in Ref. [13]
0.1	0.353818448	0.353818312	0.354671967	0.353818300
0.2	0.488801735	0.488801586	0.489340041	0.488771033
0.3	0.57879044	0.578790340	0.579164476	0.577964546
0.4	0.642539175	0.642539126	0.642809206	0.634574298
0.5	0.689214802	0.689214786	0.689415610	0.645316353
0.6	0.724383123	0.724383123	0.724536633	0.553516171
0.7	0.751600626	0.751600633	0.751721022	0.225608554
0.8	0.773187322	0.773187331	0.773283951	–0.596196471
0.9	0.790686089	0.790686099	0.790765239	–2.355520166
1	0.805145339	0.805145347	0.805211334	–5.758583719

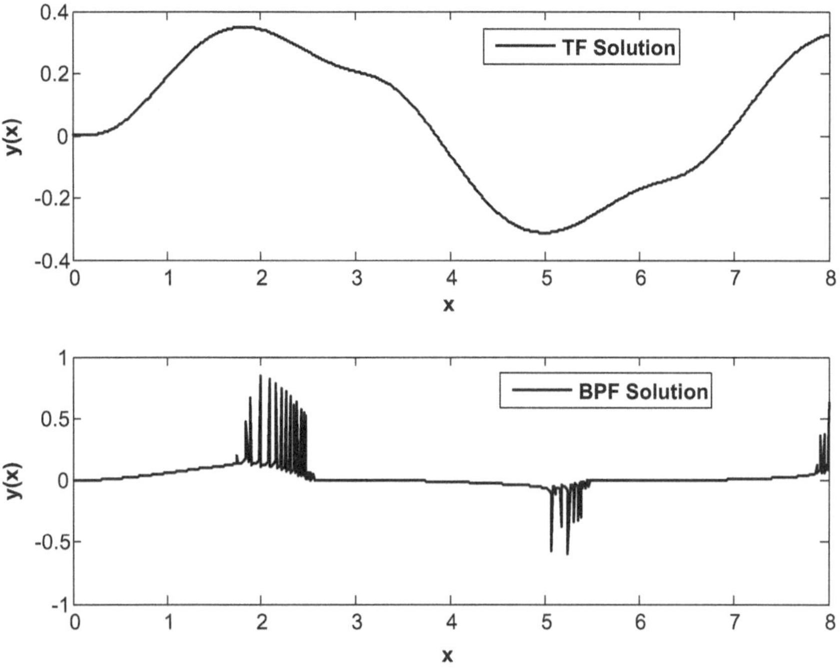

FIGURE 2.3
Numerical solutions to Equation (2.67)

below its boiling point (4.21 K) and above a temperature of 2.1768 K (the lambda point of helium). When the helium-4 is cooled below the lambda point, a part of it enters state called helium II which is a superfluid, and when the helium-4 is further cooled, increasing parts of it are converted to the superfluid state.

The weakly singular Volterra integral equation given in Equation (2.67) arises in the theory of superfluidity of helium II [14–16]:

$$y(x) = -\frac{1}{\sqrt{\pi}}\int_0^x (x-t)^{-0.5}\left(y(t) - \sin t\right)^3 dt,\ x \in [0, 8]. \tag{2.67}$$

The above process model holds the exact solution y(8)=0.323641294 and $y(x) = O\left(x^{3.5}\right)$ as $x \to 0$.

The physical process model in Equation (2.67) is solved by using the proposed method and the BPFs-based numerical method. The step size of 0.008 is used in simulations. The BPF solution produces oscillatory behavior (Figure 2.3) which in no way depicts the actual behavior of process model whereas the TF solution precisely mimics the function $y(x)$. At the end of the interval [0, 8], the absolute error achieved by the proposed method is 1.2899510e-06 and by the BPFs-based method is 0.3054689809.

Application 2.3: Lighthill singular integral equation

Lighthill [17] derived the nonlinear singular integro-differential equation (Equation (2.68)) describing the temperature distribution of the surface of a moving projectile through a laminar layer, and obtained the series solution given in Equation (2.69), which is valid for $0 \le z < R$, where $R \approx 0.106$.

The Lighthill singular integral equation is

$$\left(F(z)\right)^4 = \frac{1}{2\sqrt{z}} \int_0^z \frac{F'(s)}{\left(z^{3/2} - s^{3/2}\right)^{1/3}} ds, \; F(0) = 1, \; \lim_{t \to \infty} F(t) = 0. \tag{2.68}$$

The series solution, determined by Lighthill [17], for the above physical model is

$$\begin{aligned} F(z) &= 1 - 1.461z + 7.252z^2 - 46.460z^3 \\ &\quad + 332.9z^4 - 2538z^5 + 20120z^6 - \ldots\ldots\ldots \end{aligned} \tag{2.69}$$

After performing suitable variable transformations, the authors in [18] obtained the weakly singular Volterra integral equation of second kind:

$$y(x) = 1 - \frac{\sqrt{3}}{\pi} \int_0^x \frac{t^{1/3}\left(y(t)\right)^4}{(x-t)^{2/3}} dt, \; x \in [a, b], \tag{2.70}$$

where $y(x) = F\left(x^{2/3}\right)$.

TABLE 2.11

Approximate solution of Equation (2.70) at the initial point

m	NT in Ref. [20]	NS in Ref. [19]	Proposed method	BPFs-based method
10	0.9387219650	0.9528281752	0.9999999999	0.99956890622
20	0.993963696	0.9791345916	0.9999999999	0.99972825447
40	1.00189888	0.9912994100	0.9999999999	0.99982874221
80	0.999499586	0.9964753672	0.9999999999	0.99989208701

TABLE 2.12

Maximum norms via different methods (Lighthill singular integral equation)

	$x \in I_2$, $I_2 \in [0, 0.001]$				$x \in I_3$, $I_3 \in [0, 0.002]$			
m	PEM	PTM	PM	BBM	PEM	PTM	PM	BBM
40	8.44e-02	1.78e-03	1.87e-04	7.92e-04	6.60e-02	1.37e-03	2.97e-04	0.0013
80	2.66e-03	5.45e-04	1.18e-04	5.00e-04	2.35e-03	2.98e-04	1.87e-04	7.92e-04
160	1.42e-03	1.63e-04	7.46e-05	2.15e-04	1.22e-03	1.11e-04	1.18e-04	5.00e-04
320	6.64e-04	4.57e-05	4.70e-05	1.98e-04	5.70e-04	2.95e-05	7.46e-05	2.15e-04
640	2.16e-04	-	2.96e-05	1.25e-04	2.73e-04	-	4.70e-05	1.98e-04
	$x \in I_4$, $I_4 \in [0, 0.003]$				$x \in I_5$, $I_5 \in [0, 0.0006]$			
m	PEM	PTM	PM	BBM	PEM	PTM	PM	BBM
40	5.50e-02	1.11e-03	2.88e-04	0.0016	10	1.34e-02	2.35e-04	0.0014
80	2.14e-03	2.15e-04	2.45e-04	0.0010	80	2.60e-03	8.43e-05	2.55e-04
160	1.09e-03	8.50e-05	1.55e-04	6.54e-04	320	8.31e-04	2.35e-05	1.41e-04
320	5.11e-04	2.20e-05	9.78e-05	4.12e-04	640	2.08e-04	2.11e-05	8.90e-05
640	2.46e-04	-	6.16e-05	2.60e-04	1280	8.12e-05	1.33e-05	5.61e-05

PM: proposed method, BBM: BPFs-based method

Equation (2.70) bears the semianalytical solution within the interval of convergence, $[0, R_1]$ ($R_1 < R^{3/2}$), which can be obtained by replacing z with $x^{2/3}$ in Equation (2.69),

$$y(x) = F\left(x^{2/3}\right) = 1 - 1.461x^{2/3} + 7.252x^{4/3} - 46.460x^2 + 332.9x^{8/3} - 2538x^{10/3} + 20120x^4 - \ldots \tag{2.71}$$

Equation (2.70) is solved by means of TFs-based and BPFs-based numerical methods at the initial point $x = 0$, and the estimated values are compared (Table 2.11) with those obtained by Navot-Simpson's algorithm (NS) in [19], and by Navot-Trapezoidal algorithm using extrapolation (NT) in [20]. The actual value of temperature at $x = 0$ is $y(0) = 1$. It is seen from Table 2.11 that the TF estimate is the most accurate one.

The TF as well as BPF solution are computed on different small intervals near the initial point or origin, and are matched with the exact series solution in Equation (2.71). The infinity norms calculated are compared in Table 2.12 with those obtained by the PEM and the PTM in [10]. As usual, the BPFs-based numerical method cannot compete with the proposed method but shows better performance than the

TABLE 2.13

Approximate solutions to Lighthill singular integral equation ($x \in I_6$, $I_6 = [0, 1]$)

x	$m = 80$			$m = 160$			$m = 320$		
	PEM	**PM**	**BBM**	**PEM**	**PM**	**BBM**	**PEM**	**PM**	**BBM**
0	1	1	0.957823	1	1	0.971840	1	1	0.981541
0.1	0.835743	0.834516	0.829772	0.83471	0.834020	0.831705	0.834191	0.833825	0.832694
0.2	0.78719	0.786793	0.784198	0.786799	0.786563	0.785296	0.786600	0.786470	0.785851
0.3	0.757239	0.757067	0.755281	0.757030	0.756920	0.756048	0.756924	0.756861	0.756434
0.4	0.735494	0.735414	0.734056	0.735366	0.735308	0.734644	0.735299	0.735265	0.734940
0.5	0.718428	0.718394	0.717299	0.718342	0.718311	0.717776	0.718297	0.718277	0.718015
0.6	0.704395	0.704386	0.703471	0.704334	0.704318	0.703871	0.704302	0.704291	0.704072
0.7	0.692489	0.692495	0.691711	0.692445	0.692439	0.692055	0.692422	0.692416	0.692228
0.8	0.682160	0.682176	0.681489	0.682127	0.682127	0.681792	0.682110	0.682107	0.681943
0.9	0.673046	0.673067	0.672457	0.673021	0.673024	0.672726	0.673008	0.673007	0.672861
1	0.664895	0.664920	0.664372	0.664876	0.664882	0.664615	0.664866	0.664867	0.664736

x	$m = 640$			$m = 1280$		
	PEM	PM	BBM	PEM	PM	BBM
0	1	1	0.988060	1	1	0.992346
0.1	0.833937	0.833748	0.833194	0.833814	0.833717	0.833445
0.2	0.786502	0.786433	0.786130	0.786454	0.786419	0.786269
0.3	0.756871	0.756837	0.756628	0.756846	0.756827	0.756724
0.4	0.735267	0.735248	0.735088	0.735251	0.735241	0.735162
0.5	0.718275	0.718264	0.718135	0.718265	0.718258	0.718195
0.6	0.704287	0.704280	0.704172	0.704280	0.704275	0.704222
0.7	0.692411	0.692407	0.692314	0.692406	0.692403	0.692357
0.8	0.682102	0.682099	0.682018	0.682098	0.682096	0.682056
0.9	0.673001	0.673000	0.672928	0.672998	0.672997	0.672962
1	0.664861	0.664861	0.664796	0.664859	0.664858	0.664826

PEM. In this particular case (i.e., solving Equation (2.70)), the TFs-based numerical method is more accurate than PEM and is in good accordance with PTM.

Table 2.13 compares the approximate solutions via EM, TF, and BPF on I_6. Because the exact solution of Equation (2.70) in I_6 is not known, the TF and BPF solution are compared with the EM solution. As EM exhibited poor performance in smaller intervals (I_1 through I_5), it is believed that TF solution is, perhaps, the best one.

2.5 MATLAB® Codes for Numerical Experiments

This section offers source codes written in MATLAB for the examples and physical process models solved in the preceding sections. The initial guess x0 is required to be specified for each program.

Program 2.1

```
%%%%%%%%%%%%%%%% MATLAB code for Example 2.1 %%%%%%%%%%%%%
function [Norm_inf, Norm_Euclidean,EE1,EE2]=Example_2_1(x0)
t0=0;T=1;aa=50;m=10*aa;h=(T-t0)/m;t=[t0:h:T];
I=ones(1,length(t));
CC1=sqrt(t.*(I-t))+(16/15)*(t.^(5/2))+2*(t.^2).
*sqrt(I-t)+...
(4/3)*t.*((I-t).^(3/2))+(2/5)*((I-t).^(5/2))-(4/3)*
(t.^(3/2))-...
2*t.*sqrt(I-t)-(2/3)*((I-t).^(3/2));
C01=CC1(1:end-1);D01=CC1(2:end);
tic
[P1alph,P2alph]=TOF1(t0,T,m,1/2);[P3alph,P4alph]=TOF12
  (t0,T,m,1/2);
P1alphr=zeros(m,m);P2alphr=zeros(m,m);P3alphr=zeros
(m,m);P4alphr=zeros(m,m);
for j=1:m
P1alphr(j,:)=-wrev(P1alph(m-j+1,:));P2alphr(j,:)=-wrev
  (P2alph(m-j+1,:));
P3alphr(j,:)=-wrev(P3alph(m-j+1,:));P4alphr(j,:)=-wrev
  (P4alph(m-j+1,:));
end
[x,fval]=fsolve(@Problem_Fun,x0);
function E=Problem_Fun(x)
for i=1:1:length(t)
  C(i)=x(i);
end
C1=C(1:end-1);D1=C(2:end);
f01=gamma(1/2)*((C1.^2)*P1alphr+(D1.^2)*P3alphr);
f02=gamma(1/2)*((C1.^2)*P2alphr+(D1.^2)*P4alphr);
ff0=[f02(1) f01];
f1=C01+[gamma(1/2)*((C1.^2)*P1alph+(D1.^2)*P3alph)-ff0
  (1:end-1)]-C1;
f2=D01+[gamma(1/2)*((C1.^2)*P2alph+(D1.^2)*P4alph)-ff0
  (2:end)]-D1;
f=[f1 f2(end)];E=norm(f,inf);
end
toc
% Exact Solution
Exact=sqrt(t.*(I-t));
```

```
ee=abs(Exact-x);Norm_inf=norm(ee,inf);
Norm_Euclidean=norm(ee,2);
for i=1:11
bb=(i-1)*aa+1;EE1(i)=Exact(bb);
end
EE1=EE1';
for i=1:11
cc=(i-1)*aa+1;EE2(i)=x(cc);
end
EE2=EE2';
end
% Algorithm ends
```

Program 2.2

```
%%%%%%%%%%%% MATLAB code for Example 2.2 %%%%%%%%%%%%%%%
function    [Norm_inf,    Norm_Euclidean,EE1,EE2]=    Exam-
ple_2_2(x0)
t0=0;T=1;aa=90;m=10*aa;h=(T-t0)/m;t=[t0:h:T];
I=ones(1,length(t));
CC1=t-2*sqrt(I-t)-(4/3)*(t.^(3/2))+(4/3)*
  ((I-t).^(3/2));
C01=CC1(1:end-1);D01=CC1(2:end);
tic
[P1alph,P2alph]=TOF1(t0,T,m,1/2);[P3alph,P4alph]=TOF12
  (t0,T,m,1/2);
P1alphr=zeros(m,m);P2alphr=zeros(m,m);P3alphr=zeros
  (m,m);P4alphr=zeros(m,m);
for j=1:m
P1alphr(j,:)=-wrev(P1alph(m-j+1,:));P2alphr(j,:)=-wrev
  (P2alph(m-j+1,:));
P3alphr(j,:)=-wrev(P3alph(m-j+1,:));P4alphr(j,:)=-wrev
  (P4alph(m-j+1,:));
end
 [x,fval]=fsolve(@Problem_Fun,x0);
%dvenugopalarao%
function E=Problem_Fun(x)
for i=1:1:length(t)
   C(i)=x(i);
end
C1=C(1:end-1);D1=C(2:end);
f01=gamma(1/2)*((C1.^1)*P1alphr+(D1.^1)*P3alphr);
f02=gamma(1/2)*((C1.^1)*P2alphr+(D1.^1)*P4alphr);
ff0=[f02(1) f01];
f1=C01+[gamma(1/2)*((C1.^1)*P1alph+(D1.^1)*P3alph)-ff0
  (1:end-1)]-C1;
f2=D01+[gamma(1/2)*((C1.^1)*P2alph+(D1.^1)*P4alph)-ff0
  (2:end)]-D1;
f=[f1(1:end) f2(end)];E=norm(f,inf);
end
```

```
toc
% Exact Solution
Exact=t;ee=abs(Exact-x);Norm_inf=norm(ee,inf);
Norm_Euclidean=norm(ee,2);
for i=1:11
bb=(i-1)*aa+1;EE1(i)=Exact(bb);
end
EE1=EE1';
for i=1:11
cc=(i-1)*aa+1;EE2(i)=x(cc);
end
EE2=EE2';
end
% Algorithm ends
```

Program 2.3

```
%%%%%%%%%%%%%%%%%%%% MATLAB code for Example 2.3 %%%%%%%
%%%%%%%%%%%%%
function [Norm_inf, Norm_Euclidean]= Example_2_3(x0)
  t0=0;T=1;m=500;h=(T-t0)/m;t=[t0:h:T];
I=ones(1,length(t));
CC1=(9/40)*[((I-t).^(2/3))+2*t.*((I-t).^(2/3))+4*
(t.^(5/3))-...
3*(t.^2).*((I-t).^(2/3))-3*(t.^(8/3))];
C01=CC1(1:end-1);D01=CC1(2:end);
tic
[P1alph,P2alph]=TOF1(t0,T,m,2/3);[P3alph,P4alph]=TOF12
  (t0,T,m,2/3);
P1alphr=zeros(m,m);P2alphr=zeros(m,m);P3alphr=zeros
  (m,m);P4alphr=zeros(m,m);
for j=1:m
P1alphr(j,:)=-wrev(P1alph(m-j+1,:));P2alphr(j,:)=-wrev
  (P2alph(m-j+1,:));
P3alphr(j,:)=-wrev(P3alph(m-j+1,:));P4alphr(j,:)=-wrev
  (P4alph(m-j+1,:));
end
[x,fval]=fminunc(@Problem_Fun,x0);
function E=Problem_Fun(x)
for i=1:1:length(t)
    C(i)=x(i);
end
C1=C(1:end-1);D1=C(2:end);
f01=gamma(2/3)*((C1.^2)*P1alphr+(D1.^2)*P3alphr);
f02=gamma(2/3)*((C1.^2)*P2alphr+(D1.^2)*P4alphr);
ff0=[f02(1) f01];
f1=-C01+[gamma(2/3)*((C1.^2)*P1alph+(D1.^2)*P3alph)-ff0
  (1:end-1)];
f2=-D01+[gamma(2/3)*((C1.^2)*P2alph+(D1.^2)*P4alph)-ff0
  (2:end)];
```

```
f=[f1 f2];E=norm(f,inf);
end
toc
% Exact Solution
Exact=sqrt(t.*(I-t));
ee=abs(Exact-x);Norm_inf=norm(ee,inf);
Norm_Euclidean=norm(ee,2);
end
% Algorithm ends
```

Program 2.4

```
%%%%%%%%%%%%%%%%%%%% MATLAB code for Example 2.4 %%%%%%%
%%%%%%%%%%%%%
function [Norm_inf, Norm_Euclidean]= Example_2_4(x0)
t0=0;T=1;m=50;h=(T-t0)/m;t=[t0:h:T];
I=ones(1,length(t));
CC1=t-(32/15)*(t.^(5/2))-2*sqrt(t)-(12/5)*
  sqrt(I-t)-...
(8/15)*t.*sqrt(I-t)-(16/15)*(t.^2).*sqrt(I-t);
C01=CC1(1:end-1);D01=CC1(2:end);
tic
[P1alph,P2alph]=TOF1(t0,T,m,1/2);[P3alph,P4alph]=TOF12
  (t0,T,m,1/2);
P1alphr=zeros(m,m);P2alphr=zeros(m,m);P3alphr=zeros
  (m,m);P4alphr=zeros(m,m);
for j=1:m
P1alphr(j,:)=-wrev(P1alph(m-j+1,:));P2alphr(j,:)=-wrev
  (P2alph(m-j+1,:));
P3alphr(j,:)=-wrev(P3alph(m-j+1,:));P4alphr(j,:)=-wrev
  (P4alph(m-j+1,:));
end
[x,fval]=fminunc(@Problem_Fun,x0);
function f=Problem_Fun(x)
for i=1:1:length(t)
C(i)=x(i);
end
C1=C(1:end-1);D1=C(2:end);CC=I+C.*C;CC1=CC(1:end-1);
DD1=CC(2:end);
f01=gamma(1/2)*((CC1)*P1alphr+(DD1)*P3alphr);
f02=gamma(1/2)*((CC1)*P2alphr+(DD1)*P4alphr);
ff0=[f02(1) f01];
f1=C01+[gamma(1/2)*((CC1)*P1alph+(DD1)*P3alph)-ff0
  (1:end-1)]+ ...
[gamma(1/2)*((C1.^2)*P1alph+(D1.^2)*P3alph)]-C1;
f2=D01+[gamma(1/2)*((CC1)*P2alph+(DD1)*P4alph)-ff0
  (2:end)]+ ...
[gamma(1/2)*((C1.^2)*P2alph+(D1.^2)*P4alph)]-D1;
f=[f1 f2];E=norm(f,inf);
end
```

```
toc
% Exact Solution
Exact=t;
ee=abs(Exact-x);Norm_inf=norm(ee,inf);
Norm_Euclidean=norm(ee,2);
end
% Algorithm ends
```

Program 2.5

```
%%%%%%%%%%%%%%%%%%%% MATLAB code for Example 2.5 %%%%%%%
%%%%%%%%%%%%
function [Norm_inf, Norm_Euclidean]= Example_2_5(x0)
  t0=0;T=1;m=80;h=(T-t0)/m;t=[t0:h:T];
I=ones(1,length(t));
CC1=(t.^(1/3))+((4*gamma(4/3)*gamma(13/6))/(sqrt(pi)))*
  (t.^(3/2));
C01=CC1(1:end-1);D01=CC1(2:end);
tic
[P1alph,P2alph]=TOF1(t0,T,m,1/3);[P3alph,P4alph]=TOF12
  (t0,T,m,1/3);
[x,fval]=fminunc(@Problem_Fun,x0);
function E=Problem_Fun(x)
for i=1:1:length(t)
    C(i)=x(i);
end
C1=C(1:end-1);D1=C(2:end);CC=(t.^(1/2)).*C.*C;CC1=CC
  (1:end-1);
DD1=CC(2:end);
f1=C01-[gamma(1/3)*((CC1)*P1alph+(DD1)*P3alph)]-C1;
f2=D01-[gamma(1/3)*((CC1)*P2alph+(DD1)*P4alph)]-D1;
f=[f1 f2];E=norm(f,inf);
end
toc % CPU time is computed
% Exact Solution
Exact=t.^(1/3);
ee=abs(Exact-x);Norm_inf=norm(ee,inf);
Norm_Euclidean=norm(ee,2);
end
% Algorithm ends
```

Program 2.6

```
%%%%%%%%%%%%%%%%%%%% MATLAB code for Example 2.6 %%%%%%%
%%%%%%%%%%%%%
function [Norm_inf, Norm_Euclidean]= Example_2_6(x0)
t0=0;T=1;m=1000;h=(T-t0)/m;t=[t0:h:T];
I=ones(1,length(t));
tic
aa1=-1;bb=1;theta=zeros(1,length(t));
```

```
for ii=1:length(t)
   theta(ii)=(bb-aa1)*rand+aa1;
end
nn=1; % without noise-0, 1-with noise
if nn==0
  delta=0;
elseif nn==1
   delta=10^-1;
end
CC10=-(4/3)*(t.^(3/2))-0.5*I;CC10=CC10+delta*theta;
C01=CC10(1:end-1);
D01=CC10(2:end);C02=I(1:end-1);D02=I(2:end);
C03=t(1:end-1);D03=t(2:end);
[P1alph1,P2alph1]=TOF1(t0,T,m,1/2);[P3alph1,P4alph1]
 =TOF12(t0,T,m,1/2);
[P1alph2,P2alph2]=TOF1(t0,T,m,1);[P3alph2,P4alph2]
 =TOF12(t0,T,m,1);
[x,fval]=fminunc(@Problem_Fun,x0);
function E=Problem_Fun(x)
for i=1:1:length(t)
   C(i)=x(i);
end
C1=C(1:end-1);D1=C(2:end);f01=((C02)*P1alph2+(D02)
 *P3alph2);
f02=((C02)*P2alph2+(D02)*P4alph2);ff0=[f01 f02(end)];
 c1=ff0(end)-ff0(1);
f011=((C1)*P1alph2+(D1)*P3alph2);f021=((C1)*P2alph2
 +(D1)*P4alph2);
ff01=[f011 f021(end)];c2=ff01(end)-ff01(1);
f1=C01+[gamma(1/2)*((C1)*P1alph1+(D1)*P3alph1)]+c1*C03
 +c2*C02-C1;
f2=D01+[gamma(1/2)*((C1)*P2alph1+(D1)*P4alph1)]+c1*D03
 +c2*D02-D1;
f=[f1 f2];E=norm(f,inf);
end
toc
% Exact Solution
Exact=t;
ee=abs(Exact-x);Norm_inf=norm(ee,inf);
Norm_Euclidean=norm(ee,2);
end
% Algorithm ends
```

Program 2.7

```
%%%%%%%%%%%%%%%%%%%%%% MATLAB code for Application 2.1 %%%%
%%%%%%%%%%%%%%%%%%
function [Norm_inf, Norm_Euclidean]= Application_2_1(x0)
t0=0;T=1;m=1000;h=(T-t0)/m;t=[t0:h:T];
I=ones(1,length(t));
```

```
% Exact Solution
Exact=[0;2.53818448e-01;
4.88801735e-01;
5.78790440e-01;
6.42539175e-01;
6.89214802e-01;
7.24383123e-01;
7.51600626e-01;
7.73187322e-01;
7.90686089e-01;
8.05145339e-01];
tic
[P1alph,P2alph]=TOF1(t0,T,m,1/2);[P3alph,P4alph]=TOF12
  (t0,T,m,1/2);
[x,fval]=fsolve(@Problem_Fun,x0);
function f=Problem_Fun(x)
for i=1:1:length(t)
    CC1(i)=x(i);
end
C1=CC1(1:end-1);D1=CC1(2:end);CC0=(I-(CC1.^4));C01=CC0
  (1:end-1);
D01=CC0(2:end);
f1=(gamma(1/2)/sqrt(pi))*((C01)*P1alph+(D01)
  *P3alph)-C1;
f2=(gamma(1/2)/sqrt(pi))*((C01)*P2alph+(D01)*P4alph)-
D1;f=[f1 f2];
end
toc
ee=abs(Exact-x);Norm_inf=norm(ee,inf);Norm_Euclidean=-
norm(ee,2);
end
% Algorithm ends
```

Program 2.8

```
%%%%%%%%%%%%%%%%%%%%%% MATLAB code for Application 2.2 %%%%
%%%%%%%%%%%%%%%%%
function [error]= Application_2_2(x0)
t0=0;T=8;m=1000;h=(T-t0)/m;t=[t0:h:T];
I=ones(1,length(t));
tic
[P1alph,P2alph]=TOF1(t0,T,m,1/2);[P3alph,P4alph]=TOF12
  (t0,T,m,1/2);
[x,fval]=fsolve(@Problem_Fun,x0);
function f=Problem_Fun(x)
for i=1:1:length(t)
    C(i)=x(i);
end
C1=C(1:end-1);D1=C(2:end);I=ones(1,length(t));
CC1=(C-sin(t)).^3;C02=CC1(1:end-1);D02=CC1(2:end);
```

```
f1=-C1-(1/sqrt(pi))*gamma(0.5)*(C02*P1alph+D02*P3alph);
f2=-D1-(1/sqrt(pi))*gamma(0.5)*(C02*P2alph+D02*P4alph);
f=[f1 f2];
end
toc
error=abs(0.3236412904 -x(end));
end
```

Program 2.9

```
%%%%%%%%%%%%%%%%%%%%%% MATLAB code for Application 2.3 %%%%
%%%%%%%%%%%%%%%%%%
function [Norm_inf, Norm_Euclidean]= Application_2_3(x0)
t0=0;T=1;m=1000;h=(T-t0)/m;t=[t0:h:T];
I=ones(1,length(t));
% exact solution for t=[0, R], R<(0.106)^(2/3)
exact=1-1.460998*(t.^(2/3))+7.249416*(t.^(4/3))-
46.449738*(t.^2)+332.7552*(t.^(8/3))-2536.8206*
  (t.^(10/3))+20120.0609*(t.^4)-163991.7634*(t.^(14/3))
+1.36357*(10^6)*(t.^(16/3))-1.14446*(10^7)*(t.^6);
tic
[P1alph,P2alph]=TOF1(t0,T,m,1/3);[P3alph,P4alph]=TOF12
  (t0,T,m,1/3);
[x,fval]=fsolve(@Problem_Fun,x0);
function f=Problem_Fun(x)
for i=1:1:length(t)
    C(i)=x(i);
end
C1=C(1:end-1);D1=C(2:end);I=ones(1,length(t));CC1=I;
C01=CC1(1:end-1);
D01=CC1(2:end);CC1=(t.^(1/3)).*(C.^4);C02=CC1(1:end-1);
D02=CC1(2:end);
f11=-C1+C01-(sqrt(3)/pi)*gamma(1/3)*(C02*P1alph
  +D02*P3alph);
f22=-D1+D01-(sqrt(3)/pi)*gamma(1/3)*(C02*P2alph
  +D02*P4alph);f=[f11 f22];
end
toc
ee=abs(Exact-x);Norm_inf=norm(ee,inf);
Norm_Euclidean=norm(ee,2);
end
% Algorithm ends
```

2.6 Summary of Deliverables

The following is a brief summary of deliverables, conclusions, and directions for future work covered in this chapter.

- The proposed TFs-based numerical method is able to solve Fredholm, Volterra, and Volterra-Fredholm integral equations with weak singular kernels. It can also be used to get the accurate numerical solution to their counterpart, that is, classical integral equations consisting of no singular kernels.
- No matter whether the given mathematical model, which involves weakly singular integral equations, is linear or nonlinear (simple or intricate), it can better be simulated in the TF domain than in the BPF domain.
- It is concluded that the BPFs-based numerical method may not be able to provide an accurate approximate solution, in reasonable computational time, to complex integral equations (with or without weak singularity) representing real-world phenomena.
- The proposed numerical method contains attributes such as stability and convergence, indispensable for acceptance by the mathematics and applied mathematics fraternity.
- It is believed that the proposed method is capable enough to simulate weakly singular integral equations originating in different fields.
- Many initial value problems, boundary value problems, and initial-boundary value problems (parabolic boundary value problems) associated with fractional order differential or partial differential equations can be cast into equivalent weakly singular Volterra, Fredholm, and Volterra-Fredholm integral equations; thus, they can be solved by the proposed numerical method.
- The applications of the proposed method can be extended to two-dimensional weakly singular integral equations, systems of stiff and nonstiff weakly singular integral equations, and stiff weakly singular integral-algebraic equations.
- The triangular functions may also find applications in numerical solutions of fuzzy integral equations with or without weak singular kernels.

References

[1] I. Fredholm (1900). On a new method for solving the Dirichlet problem (Sur une nouvelle méthode pour la résolution du problème de Dirichlet) (in French). *Stockh. Ofv.*, vol. 57, pp. 39–46.

[2] E.I. Fredholm (1903). Sur une classe d'équations fonctionnelles. *Acta Math.*, vol. 27, pp. 365–390.

[3] A.D. Polyanin, A.V. Manzhirov (2008). *Handbook of Integral Equations*. Boca Raton, FL: Chapman & Hall/CRC, Taylor & Francis Group.

[4] R. Estrada, R. Kanwal (2000). *Singular Integral Equations*. Berlin: Birkhauser.

[5] H. Adibi, P. Assari (2010). On the numerical solution of weakly singular Fredholm integral equations of the second kind using Legendre wavelets. *J. Vibr. Contr.*, vol. 17(5), pp. 689–698.

[6] U. Lepik, E. Tamme (2007). Solution of nonlinear Fredholm integral equations via the Haar wavelet method. *Proc. Estonian Acad. Sci. Phys. Math.*, vol. 56, pp. 17–27.

[7] A. Pedas, G. Vainikko (1997). Superconvergence of piecewise polynomial collocations for nonlinear weakly singular integral equations. *J. Integral. Equ. Appl.*, vol. 9(4), pp. 379–406.

[8] H. Kaneko, Y. Xu (1991). Numerical solutions for weakly singular Fredholm integral equations of the second kind. *Appl. Numer. Math.*, vol. 7, pp. 167–177.

[9] K. Maleknejad, A. Ostadi (2017). Using Sinc-collocation method for solving weakly singular Fredholm integral equations of the first kind. *Appl. Anal.*, vol. 96(4), pp. 702–712.

[10] S.S. Allaei, T. Diogo, M. Rebelo (2017). Analytical and computational methods for a class of nonlinear singular integral equations. *Appl. Numer. Math.*, vol. 114, pp. 2–17.

[11] J.B. Keller, W.E. Olmstead (1971-72). Temperature of a nonlinearly radiating semi-infinite solid. *Quart. Appl. Math.*, vol. 29, pp. 559–566.

[12] F.D. Hoog, R. Weis (1974). High order methods for a class of Volterra integral equations with weakly singular kernels. *SIAM J. Numer. Anal.*, vol. 11(6), pp. 1166–1180.

[13] A.M. Wazwaz (1996). A reliable technique for solving the weakly singular second-kind Volterra type integral equations. *Appl. Math. Comput.*, vol. 80, pp. 287–299.

[14] C.C. Lin (1959). Hydrodynamics of liquid helium II. *Phys. Hev. Letters*, vol. 2, pp. 245–246.

[15] A.P. Orsi (1996). Product integration for Volterra integral equations of the second kind with weakly singular kernels. *Math. Comp.*, vol. 65, pp. 1201–1212.

[16] N. Levinson (1960). A nonlinear Volterra equation arising in the theory of superfluidity. *J. Math. Anal. Appl.*, vol. 1, pp. 1–11.

[17] J.M. Lighthill (1950). Contributions to the theory of the heat transfer through a laminary boundary layer. *Proc. Roy. Soc.*, vol. 202, pp. 359–377.

[18] T. Diogo, P. Lima, M. Rebelo (2006). Numerical solution of a nonlinear Abel type Volterra integral equation. *Commun. Pure Appl. Anal.*, vol. 5, pp. 277–288.

[19] M.A.F. Araghi, H.D. Kasmaei (2008). Numerical solution of the second kind singular Volterra integral equations by modified Navot-Simpson's quadrature. *Int. J. Open Problems Compt. Math.*, vol. 1(3), pp. 201–212.

[20] L. Tao, H. Yong (2006). Extrapolation method for solving weakly singular nonlinear Volterra integral equations of the second kind. *J. Anal and App. Math.*, vol. 324, pp. 225–237.

3

Numerical Method for Simulation of Physical Processes Modeled by Abel's Integral Equations

The standard (first-kind) Abel's integral equation,

$$f(x) = \int_0^x (x - t)^{-\alpha} y(t)dt,\ \alpha = 0.5,\ x \in [0, T], \tag{3.1}$$

is a singular integral equation (a particular case of the first-kind weakly singular Volterra integral equation) derived by Norwegian mathematician Niles Abel in 1823 to describe the motion of a particle sliding down along a smooth unknown curve, in a vertical plane, under the influence of the gravity. In Equation (3.1), $f(t)$ is the time taken by the particle to move from the highest point of vertical height x to the lowest point 0 on the smooth curve. Among the integral equations such as Fredholm integral equation, Volterra integral equation, Cauchy integral equation, and so on, Abel's integral equation is the most important one, the first integral equation ever treated, which directed mathematicians to the study of integral equations.

If α in Equation (3.1) varies in the range $(0, 1)$, Equation (3.1) is called the generalized first-kind Abel's integral equation. In addition to the standard form, Abel's integral equation often appears in the second form (Equation (3.2)) which is equally important as the standard form [1].

The second-kind Abel's integral equation is

$$y(x) = f(x) + \int_0^x (x - t)^{-\alpha} y(t)dt, \alpha \in (0, 1),\ 0 \le x \le T. \tag{3.2}$$

Abel's integral equations in Equations (3.1) and (3.2) arise in many branches of scientific fields such as seismology, radio astronomy, electron emission, atomic scattering, radar ranging, plasma diagnostics, X-ray radiography, fluid mechanics, semiconductors, metallurgy, satellite photometry of airglows, heat conduction, astrophysics, stellar winds, and so on [1, 2]. In the aforementioned applications, the first-kind Abel's integral equation relates the physical quantity ($f(x)$) accessible to measurement to the physically important but experimentally inaccessible quantity $y(x)$. To find the experimentally unmeasurable physical quantity $y(x)$, the inversion for Equation (3.1) was derived in [3] as

$$y(x) = \frac{1}{\Pi} \int_0^x (x-t)^{\alpha} f^{(1)}(t)dt. \quad (3.3)$$

If the input data function, $f(x)$, contains noise, the derivative operator in Equation (3.3) amplifies the noise in $f(x)$; consequently, the inversion formula provides unstable solution. Therefore, Equation (3.3) is not useful in practical situations where the input data function is associated with noise [4]. To avoid this problem, Deutsch and Beniaminy [5] derived a practically applicable derivative-free inversion formula. The inversion formula in Equation (3.3) and the one proposed in [5] may not be suitable to find the closed-form solutions of some of the first-kind Abel's integral equations such as highly nonlinear first-kind Abel's integral equations and systems of linear and nonlinear first-kind Abel's integral equations.

The first-kind Abel's integral equation is more problematic to solve either analytically or numerically than the second-kind Abel's integral equation. Hence, efficient and stable numerical methods are required to estimate the experimentally unmeasurable physical quantities in the aforesaid practical applications of first-kind Abel's integral equations as well as to numerically solve second-kind Abel's integral equations. In this regard, analytical, semi-analytical, and numerical techniques were developed. Malinowski and Smarzewski proposed a numerical method based on the Abel inversion formula derived in [6] and using polynomial interpolating spline functions for solving the Abel integral equations [7]. Lubich [8, 9] developed Runge-Kutta theory and fractional linear multistep methods for solving Abel-Volterra integral equations of the second kind. A modification of Mikusinski operational calculus was used by Gorenflo and Luchko [10] to explicitly express the solution of the generalized Abel integral equation with Riemann-Liouville fractional integral operator in terms of Mittag-Leffler function of several variables. Sizikov and Sidorov [11] proposed generalized quadrature methods and applied in infrared tomography problem. Semianalytical techniques like Homotopy perturbation method and Adomian decomposition method and their modified versions have been used by Pandey et al. [12] to solve singular integral equations of Abel type. Sunil Kumar et al. [13] formulated a new method called the homotopy perturbation transform method by merging the Laplace transform method and the homotopy perturbation method to solve analytically Abel's integral equations arising in astrophysics. A similar effort has been made by Mohamed et al. [14] to propose an optimal homotopy analysis transform method to solve a system of Abel's integral equations. The Laplace decomposition method (Adomian decomposition method in conjugation with Laplace transform method) and its variants are extensively employed in solving applied physical problems. But the method and its modified versions suffer from the problem of selecting the initial guess. To circumvent this hurdle, Khan and Gondal [15] developed a two-step Laplace decomposition method that explains how the initial approximation can be chosen without having noise terms. Apart from semianalytical techniques, there are other numerical methods based on orthogonal functions such as block pulse functions, hat functions, Chebyshev polynomials, shifted Chebyshev polynomials and Bernstein polynomials, wavelets like Legendre wavelets and B-spline wavelets, and so on are available in the literature [16–24].

In this chapter, we treat the general form of the system of Abel's integral equations

$$\lambda_1 y_i(x) = \lambda_2 f_i(x) + \lambda_3 \int_0^x \left(\frac{k_i(x,t)}{(x-t)^{\alpha}} F_i(t, y_1(t), y_2(t), \ldots, y_n(t)) \right) dt, \qquad i \in [1,n],\ 0 \le x,t \le 1,\ \alpha \in (0,1), \tag{3.4}$$

where $y_i(x)$ is the ith unknown function, $f_i(x)$ and $k_i(x,t)$ are known functions, and F_i can be linear or nonlinear. For the appropriate choices of λ_1, λ_2, λ_3, $k_i(x,t)$ and F_i, one can get the first- and second-kind Abel's integral equations from Equation (3.4). One of the aims of this chapter is to find the numerical solution to Equation (3.4) through the fractional calculus with the help of triangular orthogonal functions. The triangular functions (TFs) find numerous applications in the solution of classical integral equations [25–33]. Saeedi and Chuev [34] employed TFs to solve nonlinear fractional Volterra integral equations.

Equation (3.4) can be written as

$$\begin{aligned}\lambda_1 y_i(x) &= \lambda_2 f_i(x) + \lambda_3 \frac{\Gamma(1-\alpha)}{\Gamma(1-\alpha)} \int_0^x \left((x-t)^{1-\alpha-1} k_i(x,t) F_i(t, y_1(t), y_2(t), \ldots\ldots, y_n(t)) \right) dt, \\ &= \lambda_2 f_i(x) + \lambda_3 \Gamma(1-\alpha) J^{1-\alpha} k_i(x,t) F_i(x, y_1(x), y_2(x), \ldots\ldots, y_n(x)),\end{aligned} \tag{3.5}$$

where $J^{1-\alpha}$ is the Riemann-Liouville fractional integral of order $1-\alpha$.

The system of Abel's integral equations in Equation (3.5) can be transformed into a system of algebraic equations by using the generalized triangular function (TF) operational matrices derived in Equation (1.97) in place of the Riemann-Liouville fractional integral. The resulting system of algebraic equations may be solved by using any algebraic solver with less effort. Employing the generalized triangular function operational matrices in Equation (3.5) greatly reduces the complexity of the problem and saves CPU usage.

3.1 Existence and Uniqueness of Solution

For the sake of simplicity, we consider $\lambda_1 = 1$, $\lambda_2 = 1$, $\lambda_3 = 1$ and $n = 1$. So Equation (3.4) becomes

$$y(x) = f(x) + \int_0^x \left(\frac{k(x,t)}{(x-t)^{\alpha}} F(t, y(t))\right) dt, \tag{3.6}$$

where $x, t \in [0, 1]$, $y(x), f(x) \in C[0, 1]$, $k(x,t){:}[0,1]\times[0,1]\rightarrow[0,\infty)$, $F(t, y(t)) : [0, 1] \times [0, \infty) \rightarrow [0, \infty)$, $M - 1 \leq \alpha < M$, $M \in Q^+$ (Q^+ is the collection of positive integers).

Equation (3.6) can be written as the fixed point equation form $Ay(x) = y(x)$, where A is defined as

$$Ay(x) = f(x) + \int_0^x \left(\frac{k(x,t)}{(x-t)^{\alpha}} F(t, y(t))\right) dt. \tag{3.7}$$

Let us define Banach space as $C : [0, 1] \times [0, \infty)$ with $d : [0, \infty) \times [0, \infty) \rightarrow [0, \infty)$, $d(\tilde{y}(x), y(x)) = \|\tilde{y}(x) - y(x)\|_{\infty}$.

We also assume that the function $F(x, y(x))$ satisfies, with respect to the second argument, the following condition of Lipschitz continuity:

$$\|F(x, \tilde{y}(x)) - F(x, y(x))\|_{\infty} \leq L\|\tilde{y}(x) - y(x)\|_{\infty}, \textit{ where L is a Lipschitz constant, } L \in (0,1). \tag{3.8}$$

Let $A : C[0, 1] \rightarrow C[0, 1]$ such that

$$Ay(x) = f(x) + \int_0^x \left(\frac{k(x,t)}{(x-t)^{\alpha}} F(t, y(t))\right) dt. \tag{3.9}$$

Therefore, $x > 0$, we have

$$\begin{aligned} \|A\tilde{y}(x) - Ay(x)\|_{\infty} &= \left\| \int_0^x \left(\frac{k(x,t)}{(x-t)^{\alpha}} (F(t, \tilde{y}(t)) - F(t, y(t)))\right) dt \right\|_{\infty}, \\ &\leq \int_0^x \left(\frac{\|k(x,t)\|_{\infty}}{|x-t|^{\alpha}} \|F(t, \tilde{y}(t)) - F(t, y(t))\|_{\infty}\right) dt. \end{aligned} \tag{3.10}$$

We assume

$$M_1 = \max_{x,t \in [0,1]} |k(x,t)| = \|k(x,t)\|_\infty, \text{ where } M_1 \text{ is an arbitrary positive number.} \tag{3.11}$$

Employing Equation (3.11) in Equation (3.10),

$$\begin{aligned} \|A\tilde{y}(x) - Ay(x)\|_\infty &\le M_1 L \|\tilde{y}(x) - y(x)\|_\infty \left\| \int_0^x |x-t|^{-\alpha} dt \right\| \\ &= M_1 L \|\tilde{y}(x) - y(x)\|_\infty \frac{\left(\|x\|_\infty\right)^{1-\alpha}}{1-\alpha}. \end{aligned} \tag{3.12}$$

Since $\|x\|_\infty = 1$,

$$\|A\tilde{y}(x) - Ay(x)\|_\infty \le \Omega \|\tilde{y}(x) - y(x)\|_\infty,\ \Omega = \tfrac{M_1 L}{(1-\alpha)}. \tag{3.13}$$

If $\Omega < 1$, then, by contraction mapping theorem, Equation (3.6) has a unique solution in $C[0,1]$.

3.2 The Proposed Numerical Method

In this section, we propose a generalized solution procedure using triangular orthogonal functions to find the numerical solution of generalized Abel's integral equation of the first and second kind.

Without loss of generality, we consider the following general form of the system of generalized Abel's integral equations:

$$\lambda_1 y_i(x) = \lambda_2 f_i(x) + \lambda_3 \int_0^x \left(\frac{k_i(x,t)}{(x-t)^\alpha} F_i(t, y_1(t), y_2(t), \ldots, y_n(t)) \right) dt, \tag{3.14}$$
$$i \in [1,n],\ x \in [0,1),\ t \in [0,1).$$

Here $\alpha \in (0,1)$, $y_i(x)$ is the ith unknown function, $f_i(x)$ and $k(x,t)$ are known continuous functions, $F_i(x, y_1(x), y_2(x), \ldots, y_n(x))$ can be linear or nonlinear.

In the TF domain,

$$y_i(x) \approx C_i^T T1_m(x) + D_i^T T2_m(x), f_i(x) \approx C_0^T T1_m(x) + D_0^T T2_m(x), \ \forall i \in [1, n], \tag{3.15}$$

$$k_i(x,t) \approx T1_m^T(x)(F_1 T1_m(t) + F_2 T2_m(t)) + T2_m^T(x)(F_3 T1_m(t) + F_4 T2_m(t)), \tag{3.16}$$

$$F_i(x, y_1(x), y_2(x), \ldots, y_n(x)) \approx C_f^T T1_m(x) + D_f^T T2_m(x), \tag{3.17}$$

where $C_f^T = \begin{bmatrix} c_{f0}, c_{f1} & \cdots & c_{f(m-1)} \end{bmatrix}$, $D_f^T = \begin{bmatrix} d_{f0}, d_{f1} & \cdots & d_{f(m-1)} \end{bmatrix}$, $d_{fi} = c_{f(i+1)}$,

$$c_{fi} = F_i(ih, y_1(ih), y_2(ih), y_3(ih), \cdots\cdots, y_n(ih)).$$

Using Equations (3.15), (3.16) and (3.17) in Equation (3.14),

$$\lambda_1 \left(C_i^T T1_m(x) + D_i^T T2_m(x)\right) = \lambda_2 \left(C_0^T T1_m(x) + D_0^T T2_m(x)\right) + \lambda_3 \int_0^x \left(\frac{k_i(x,t)}{(x-t)^\alpha} F_i(t, y_1(t), y_2(t), \ldots, y_n(t)) \right) dt. \tag{3.18}$$

Multiplying Equation (3.16) by Equation (3.17),

$$\begin{aligned} & k_i(x,t) F_i(t, y_1(t), y_2(t), \ldots, y_n(t)) \\ & = \left(T1_m^T(x)(F_1 T1_m(t) + F_2 T2_m(t)) + T2_m^T(x)(F_3 T1_m(t) + F_4 T2_m(t))\right) \\ & \quad \times \left(T1_m^T(t) C_f + T2_m^T(t) D_f\right) = T1_m^T(x) F_1 T1_m(t) T1_m^T(t) C_f \\ & \quad + T1_m^T(x) F_2 T2_m(t) T1_m^T(t) C_f + T1_m^T(x) F_1 T1_m(t) T2_m^T(t) D_f \\ & \quad + T1_m^T(x) F_2 T2_m(t) T2_m^T(t) D_f + T2_m^T(x) F_3 T1_m(t) T1_m^T(t) C_f \\ & \quad + T2_m^T(x) F_4 T2_m(t) T1_m^T(t) C_f + T2_m^T(x) F_3 T1_m(t) T2_m^T(t) D_f \\ & \quad + T2_m^T(x) F_4 T2_m(t) T2_m^T(t) D_f. \end{aligned} \tag{3.19}$$

Employing Equations (1.69) to (1.71) in Equation (3.19),

$$\begin{aligned} & k_i(x,t) F_i(t, y_1(t), y_2(t), \ldots, y_n(t)) \\ & = T1_m^T(x) F_1 diag(T1_m(t)) C_f + T1_m^T(x) F_2 O C_f + T1_m^T(x) F_1 O D_f \\ & \quad + T1_m^T(x) F_2 diag(T2_m(t)) D_f + T2_m^T(x) F_3 diag(T1_m(t)) C_f \\ & \quad + T2_m^T(x) F_4 O C_f + \ T2_m^T(x) F_3 O D_f + T2_m^T(x) F_4 diag(T2_m(t)) D_f. \end{aligned} \tag{3.20}$$

Equation (3.20) can also be written as

$$\begin{aligned}&k_i(x,t)F_i(t,y_1(t),y_2(t),\ldots,y_n(t))\\&=T1_m^T(x)F_1diag(C_f)T1_m(t)+T1_m^T(x)F_2diag(D_f)T2_m(t)\\&\quad+T2_m^T(x)F_3diag(C_f)T1_m(t)+\ T2_m^T(x)F_4diag(D_f)T2_m(t).\end{aligned}\tag{3.21}$$

Taking the second term in the right side of Equation (3.14),

$$\begin{aligned}&\int_0^x\left(\frac{k_i(x,t)}{(x-t)^\alpha}F_i(t,y_1(t),y_2(t),\ldots,y_n(t))\right)dt=\frac{\Gamma(1-\alpha)}{\Gamma(1-\alpha)}\\&\int_0^x\left((x-t)^{1-\alpha-1}k_i(x,t)F_i(t,y_1(t),y_2(t),\ldots,y_n(t))\right)dt,\\&\quad=\Gamma(1-\alpha)_0J_t^{1-\alpha}\left(k_i(x,t)F_i(x,y_1(x),y_2(x),\ldots,y_n(x))\right).\end{aligned}\tag{3.22}$$

In Equation (3.22), the operator ${}_0J_t^{1-\alpha}$ is the Riemann-Liouville fractional integral of order $1-\alpha$. This allows us to use fractional calculus concepts in the computation of a numerical solution of Abel's integral equations.

From Equations (3.21) and (3.22),

$$\begin{aligned}&J^{1-\alpha}(k_i(x,t)F_i(x,y_1(x),y_2(x),\ldots,y_n(x)))\\&=(T1_m^T(x)F_1diag(C_f)+T2_m^T(x)F_3diag(C_f))J^{1-\alpha}(T1_m(t))\\&\quad+(\ T1_m^T(x)F_2diag(D_f)+\ T2_m^T(x)F_4diag(D_f))J^{1-\alpha}(T2_m(t)).\end{aligned}\tag{3.23}$$

Using Equation (3.14),

$$\begin{aligned}&J^{1-\alpha}(k_i(x,t)F_i(x,y_1(x),y_2(x),\ldots,y_n(x)))\\&=(T1_m^T(x)F_1diag(C_f)+T2_m^T(x)F_3diag(C_f))\,(P_1^\alpha T1_m(x)+P_2^\alpha T2_m(x))\\&\quad+(\ T1_m^T(x)F_2diag(D_f)+\ T2_m^T(x)F_4diag(D_f))\,(P_3^\alpha T1_m(x)+P_4^\alpha T2_m(x)),\\&=T1_m^T(x)F_1diag(C_f)P_1^\alpha T1_m(x)+T1_m^T(x)F_1diag(C_f)P_2^\alpha T2_m(x)\\&\quad+T2_m^T(x)F_3diag(C_f)P_1^\alpha T1_m(x)+T2_m^T(x)F_3diag(C_f)P_2^\alpha T2_m(x)\\&\quad+T1_m^T(x)F_2diag(D_f)P_3^\alpha T1_m(x)+T1_m^T(x)F_2diag(D_f)P_4^\alpha T2_m(x)\\&\quad+T2_m^T(x)F_4diag(D_f)P_3^\alpha T1_m(x)+T2^T(x)F_4diag(D_f)P_4^\alpha T2_m(x),\\&=T1_m^T(x)F_1diag(C_f)P_1^\alpha T1_m(x)+T2_m^T(x)F_3diag(C_f)P_2^\alpha T2_m(x)\\&\quad+T1_m^T(x)F_2diag(D_f)P_3^\alpha T1_m(x)+T2_m^T(x)F_4diag(D_f)P_4^\alpha T2_m(x).\end{aligned}\tag{3.24}$$

Performing matrix multiplication and then utilizing the properties of TFs given in Equation (1.69) to (1.78), the above equation becomes

$$J^{1-\alpha}\left(k_i(x,t)F_i\left(x,y_1(x),y_2(x),\ldots,y_n(x)\right)\right) = \left(\tilde{X}_1+\tilde{X}_2\right)T1_m(x)+\left(\tilde{X}_3+\tilde{X}_4\right)T2_m(x), \tag{3.25}$$

where $\tilde{X}_1 = diag\left(F_1 diag\left(C_f\right)P_1^\alpha\right)$, $\tilde{X}_2 = diag\left(F_2 diag\left(D_f\right)P_3^\alpha\right)$, $\tilde{X}_3 = diag\left(F_3 diag\left(C_f\right)P_2^\alpha\right)$,

$$\tilde{X}_4 = diag\left(F_4 diag\left(D_f\right)P_4^\alpha\right).$$

From Equations (3.18) and (3.25),

$$\lambda_1\left(C_i^T T1_m(x)+D_i^T T2_m(x)\right) = \lambda_2\left(C_0^T T1_m(x)+D_0^T T2_m(x)\right) + \lambda_3\Gamma(1-\alpha)\left(\left(\tilde{X}_1+\tilde{X}_2\right)T1_m(x)+\left(\tilde{X}_3+\tilde{X}_4\right)T2_m(x)\right). \tag{3.26}$$

Comparing the coefficients of the LHTF vector and the RHTF vector,

$$\lambda_1 C_i^T = \lambda_2 C_0^T + \lambda_3\Gamma(1-\alpha)\left(\tilde{X}_1+\tilde{X}_2\right), \tag{3.27}$$

$$\lambda_1 D_i^T = \lambda_2 D_0^T + \lambda_3\Gamma(1-\alpha)\left(\tilde{X}_3+\tilde{X}_4\right). \tag{3.28}$$

The system of nonlinear algebraic equations in Equations (3.27) and (3.28) may be solved by using any algebraic equations solver to get the coefficient vectors: C_i^T and D_i^T. From Equation (3.15), we can have the numerical solution of the system of generalized Abel's integral equations of the second kind.

If $\lambda_1 = 0$ and $\lambda_2 = -1$, Equation (3.14) becomes the system of generalized Abel's integral equations of the first kind, and its numerical solution can be obtained from the solution of Equations (3.27) and (3.28).

3.3 Convergence Analysis

In this section, we prove that the piecewise linear TF solution of the system of generalized Abel's integral equations in Equation (3.29) approaches the exact solution in the limit of step size (h) tends to zero.

$$\lambda_1 y_i(x) = \lambda_2 f_i(x) + \lambda_3 \int_0^x \left(\frac{k_i(x,t)}{(x-t)^\alpha}F_i(t,y_1(t),y_2(t),\ldots\ldots,y_n(t))\right)dt, \quad i\in[1,n],\ 0\le x,t\le 1,\ \alpha\in(0,1). \tag{3.29}$$

Let us represent the actual solution of Equation (3.29) as

$$Y(t) = [y_1(t), \quad y_2(t), \quad y_3(t), \quad \cdots \quad \cdots, \quad y_n(t)] \tag{3.30}$$

and the TF solution as

$$\tilde{Y}(t) = [\tilde{y}_1(t), \quad \tilde{y}_2(t), \quad \tilde{y}_3(t), \quad \cdots \quad \cdots, \quad \tilde{y}_n(t)]. \tag{3.31}$$

We suppose that the nonlinear function in Equation (3.29) satisfies the following condition of Lipschitz continuity.

$$\left\| F(t, Y(t)) - F\left(t, \tilde{Y}(t)\right) \right\| \le L \left\| Y(t) - \tilde{Y}(t) \right\|, \tag{3.32}$$

where L is a Lipschitz constant that lies in the range $(0, 1)$.

Firstly, we derive the error between the original source term $f_i(x)$ and its TF estimate $\tilde{f}_i(x)$ in the jth interval $[t_j, t_{j+1}) = [jh, (j+1)h)$.

Approximating $f_i(x)$ in the jth interval,

$$\begin{aligned} \tilde{f}_i(x) &= f_i(jh)\left(1 - \left(\frac{x - jh}{h}\right)\right) + f_i((j+1)h)\left(\frac{x - jh}{h}\right), \\ &= f_i(jh) + \left(\frac{f_i((j+1)h) - f_i(jh)}{h}\right)\left(\frac{x - jh}{h}\right), \\ &= f_i(jh) + \left.\frac{df_i(x)}{dx}\right|_{x=jh}\left(\frac{x - jh}{h}\right). \end{aligned} \tag{3.33}$$

The Taylor series expansion of $f_i(x)$ around the point $x_j = jh$,

$$\begin{aligned} f_i(x) \approx f_i(jh) &+ \frac{df_i(x)}{dx}\Big|_{x=jh}(x - jh) \\ &+ \frac{d^2 f_i(x)}{dx^2}\Big|_{x=jh}\frac{(x - jh)^2}{2} + \frac{d^3 f_i(x)}{dx^3}\Big|_{x=jh}\frac{(x - jh)^3}{3} + \cdots\cdots \end{aligned} \tag{3.34}$$

Truncating the above series to the first three terms,

$$f_i(x) \approx f_i(jh) + \frac{df_i(x)}{dx}\Big|_{x=jh}(x - jh) + \frac{d^2 f_i(x)}{dx^2}\Big|_{x=jh}\frac{(x - jh)^2}{2}. \tag{3.35}$$

Subtracting Equation (3.35) from Equation (3.33),

$$f_i(x) - \tilde{f}_i(x) = \left.\frac{d^2 f_i(x)}{dx^2}\right|_{x=jh}\frac{(x - jh)^2}{2}. \tag{3.36}$$

Because $x \in [jh, (j+1)h)$, the term $(x - jh)$ is always less than h. So we rewrite Equation (3.36) as follows:

$$f_i(x) - \tilde{f}_i(x) = \left.\frac{d^2 f_i(x)}{dx^2}\right|_{x=jh} \frac{h^2}{2}. \tag{3.37}$$

Taking the norm,

$$\left\|f_i(x) - \tilde{f}_i(x)\right\| \leq M_1 \frac{h^2}{2}, \textit{ where } M_1 = \left\|\left.\frac{d^2 f_i(x)}{dx^2}\right|_{x=jh}\right\|, \ M_1 \textit{ is a positive number.} \tag{3.38}$$

We now derive the expression for the error between the function $k_i(x,t)$ and its TF approximation in the jth interval.

By using Equation (1.78), we get the following.

$$\begin{aligned} k_i(x,t) &= k_i(jh,jh)T1_j(x)T1_j(t) + k_i(jh,(j+1)h)T1_j(x)T2_j(t) \\ &\quad + k_i((j+1)h,jh)T2_j(x)T1_j(t) + k_i((j+1)h,(j+1)h)\ T2_j(x)T2_j(t)\cdot \end{aligned} \tag{3.39}$$

From Equations (1.65) and (1.66),

$$\begin{aligned} \tilde{k}_i(x,t) &= k_i(jh,jh)\left(1-\left(\frac{x-jh}{h}\right)\right)\left(1-\left(\frac{t-jh}{h}\right)\right) \\ &\quad + k_i(jh,(j+1)h)\left(1-\left(\frac{x-jh}{h}\right)\right)\left(\frac{t-jh}{h}\right) \\ &\quad + k_i((j+1)h,jh)\left(\frac{x-jh}{h}\right)\left(1-\left(\frac{t-jh}{h}\right)\right) \\ &\quad + k_i((j+1)h,(j+1)h)\ \left(\frac{x-jh}{h}\right)\left(\frac{t-jh}{h}\right). \end{aligned} \tag{3.40}$$

Performing multiplication in the right-hand side of Equation (3.40) and rearranging the resulting equation gives

$$\begin{aligned} \tilde{k}_i(x,t) &= k_i(jh,jh) + \left(\frac{k_i((j+1)h,jh) - k_i(jh,jh)}{h}\right)(x-jh) \\ &\quad + \left(\frac{k_i(jh,(j+1)h) - k_i(jh,jh)}{h}\right)(t-jh) \\ &\quad + \left(\frac{k_i((j+1)h,(j+1)h) - k_i(jh,(j+1)h) - k_i((j+1)h,jh) + k_i(jh,jh)}{h^2}\right) \\ &\quad (x-jh)(t-jh)\cdot \end{aligned} \tag{3.41}$$

Use of the forward difference formula in Equation (3.41) yields

$$\tilde{k}_i(x,t) = k_i(jh,jh) + \left(\frac{dk_i(x,t)}{dx}\right)_{x=jh,t=jh}(x-jh) + \left(\frac{dk_i(x,t)}{dt}\right)_{x=jh,t=jh}(t-jh) + \left(\frac{d^2k_i(x,t)}{dxdt}\right)_{x=jh,t=jh}(x-jh)(t-jh). \tag{3.42}$$

Similar to Equation (3.35), we now expand the actual function $k_i(x,t)$ by a Taylor series around the point (jh,jh), and consider the first six terms of the resulting series as shown in the next equation:

$$k_i(x,t) \approx k_i(jh,jh) + \left(\frac{dk_i(x,t)}{dx}\right)_{x=jh,t=jh}(x-jh) + \left(\frac{dk_i(x,t)}{dt}\right)_{x=jh,t=jh}(t-jh) + \left(\frac{d^2k_i(x,t)}{dx^2}\right)_{x=jh,t=jh}\frac{(x-jh)^2}{2} + \left(\frac{d^2k_i(x,t)}{dt^2}\right)_{x=jh,t=jh}\frac{(t-jh)^2}{2} + \left(\frac{d^2k_i(x,t)}{dxdt}\right)_{x=jh,t=jh}(x-jh)(t-jh). \tag{3.43}$$

From Equations (3.42) and (3.43),

$$k_i(x,t) - \tilde{k}_i(x,t) = \left(\frac{d^2k_i(x,t)}{dx^2}\right)_{x=jh,t=jh}\frac{(x-jh)^2}{2} + \left(\frac{d^2k_i(x,t)}{dt^2}\right)_{x=jh,t=jh}\frac{(t-jh)^2}{2}. \tag{3.44}$$

Assuming the following,

$$M_2 = \left\| \left(\frac{d^2k_i(x,t)}{dx^2}\right)_{x=jh,t=jh} \right\|, \quad M_3 = \left\| \left(\frac{d^2k_i(x,t)}{dt^2}\right)_{x=jh,t=jh} \right\|, \ M_2 \textit{ and } M_3 \textit{ are positive numbers.} \tag{3.45}$$

Using Equation (3.45) in Equation (3.44),

$$\left\| k_i(x,t) - \tilde{k}_i(x,t) \right\| \leq M_2 \frac{(x-jh)^2}{2} + M_3 \frac{(t-jh)^2}{2}. \tag{3.46}$$

As $x, t \in [jh, (j+1)h)$, the terms; $(x - ih)$ and $(t - ih)$ are always less than h.

$$\left\| k_i(x,t) - \tilde{k}_i(x,t) \right\| \leq M_2 \frac{h^2}{2} + M_3 \frac{h^2}{2}. \tag{3.47}$$

In the TF domain, Equation (3.29) is rewritten as

$$\lambda_1 \tilde{y}_i(x) = \lambda_2 \tilde{f}_i(x) + \lambda_3 \int_0^x \left(\frac{\tilde{k}_i(x,t)}{(x-t)^\alpha} F_i\left(t, \tilde{y}_1(t), \tilde{y}_2(t), \ldots\ldots, \tilde{y}_n(t)\right) \right) dt. \tag{3.48}$$

We now define the error between $y_i(x)$ and $\tilde{y}_i(x)$ in the jth interval

$$\varepsilon_i = \left\| y_i(x) - \tilde{y}_i(x) \right\|. \quad 3.49$$

From Equations (3.38) and (3.49),

$$\begin{aligned}
\lambda_1 \left\| y_i(x) - \tilde{y}_i(x) \right\| &= \lambda_2 \left\| f_i(x) - \tilde{f}_i(x) \right\| \\
&\quad + \lambda_3 \int_{x_j}^{x_{j+1}} \frac{\left\| k_i(x,t) - \tilde{k}_i(x,t) \right\|}{(x_{j+1}-t)^\alpha} \left\| F(t, Y(t)) - F\left(t, \tilde{Y}(t)\right) \right\| dt, \\
&\leq \lambda_2 M_1 \frac{h^2}{2} + L\lambda_3 (M_2 + M_3) \frac{h^2}{2} \int_{x_j}^{x_{j+1}} \frac{1}{(x_{j+1}-t)^\alpha} \left\| Y(t) - \tilde{Y}(t) \right\| dt, \\
&\leq \lambda_2 M_1 \frac{h^2}{2} + L\lambda_3 (M_2 + M_3) \frac{h^2}{2} \int_{x_j}^{x_{j+1}} \frac{1}{(x_{j+1}-t)^\alpha} \left(\sum_{i=1}^n \varepsilon_i \right) dt, \\
&\leq \lambda_2 M_1 \frac{h^2}{2} + n\varepsilon_i L\lambda_3 (M_2 + M_3) \frac{h^2}{2} \int_{x_j}^{x_{j+1}} \frac{1}{(x_{j+1}-t)^\alpha} dt, \\
&\leq \lambda_2 M_1 \frac{h^2}{2} + n\varepsilon_i L\lambda_3 (M_2 + M_3) \frac{h^2}{2} \left(\frac{t^{1-\alpha}}{1-\alpha} \right)_{jh}^{(j+1)h}, \\
&\leq \lambda_2 M_1 \frac{h^2}{2} + n\varepsilon_i L\lambda_3 (M_2 + M_3) \frac{h^{3-\alpha}}{2(1-\alpha)} \left((j+1)^{1-\alpha} - j^{1-\alpha} \right).
\end{aligned} \tag{3.50}$$

In the limit of h tends to zero,

$$\lim_{h\to 0}(\lambda_1\varepsilon_i) = \lim_{h\to 0}\left(\lambda_2 M_1\frac{h^2}{2} + n\varepsilon_i L\lambda_3(M_2+M_3)\frac{h^{3-\alpha}}{2(1-\alpha)}\left((j+1)^{1-\alpha} - j^{1-\alpha}\right)\right) = 0. \tag{3.51}$$

3.4 Numerical Experiments

3.4.1 Investigation of Validity and Accuracy

In this subsection, the proposed numerical method is applied to some examples to demonstrate its applicability, and to compare its performance with some numerical methods reported in the literature. The error norms defined in Chapter 2 are employed here to quantify the performance of the numerical method. The step size or number of subintervals used and the CPU usage (in seconds) needed by TFs and BPFs for Example 3.1 through Example 3.3 are reported in Table 3.1.

Example 3.1

Abel's integral equation of the first kind is

$$\int_0^x \frac{f(t)}{\sqrt{x-t}}dt = \frac{2}{105}\sqrt{x}\left(105 - 56x^2 + 48x^3\right), x \in [0,1]. \tag{3.52}$$

The exact solution of the given problem is $f(x) = x^3 - x^2 + 1$.

The BPFs-based and proposed TFs-based numerical methods are applied to Equation (3.52) by using the step size of 0.001. The same problem was solved by Noeiaghdam et al. [35] with the homotopy analysis transform method, which was formulated by combining the homotopy analysis method and Laplace transform method, by Derili and Sohrabi [36] with Chebyshev wavelets, and by Suman et al. [37] with Bernstein polynomial multiwavelets. The absolute errors obtained by our method, BPFs-based method, Homotopy analysis transform method (HATM) [35], Chebyshev wavelets (CW) [36] and Bernstein polynomial multiwavelets (BPM) [37] and are compared in Table 3.2. The TF solution is the most accurate one among those provided in Table 3.2.

Example 3.2

The Abel's integral equation of the second kind

$$y(x) = x^2 + \frac{16}{15}x^{2.5} - \int_0^x \frac{y(t)}{\sqrt{x-t}}dt,\ x \in [0,1]. \tag{3.53}$$

TABLE 3.1

Computational time (seconds)

Example	m	BBM	PM
3.1	1000	691.470025	10.503069
3.2	1500	568.627016	68.065117
3.3	2000	717.000497	121.670334

BBM: BPFs-based method, PM: proposed method

TABLE 3.2

Absolute errors obtained via different methods (Example 3.1)

x	HATM in Ref. [35]	CW in Ref. [36]	BPM in Ref. [37]	BPFs-based method	Proposed method
0	0	0.003959999	1.045e-05	2.284190e-07	7.172509e-08
0.2	0.0000309186	0.000924999	2.587e-06	0.000140064	6.654351e-08
0.4	8:02707e-06	0.002295999	2.891e-06	0.000159985	3.180719e-08
0.6	4:31429e-06	0.020641000	7.011e-04	5.986042e-05	1.306284e-07
0.8	4:58399e-06	0.001298000	3.893e-04	0.000160251	2.293223e-07
1	4:76837e-07	NV	2.282e-03	0.000500334	9.521103e-08

NV indicates that no value is reported at the respective value of x in the referenced paper.

The exact solution of Equation (3.53) can be determined to be $y(x) = x^2$.

Table 3.3 presents the comparison between the absolute errors obtained by Bernstein polynomial wavelets [37], Chebyshev wavelets [36], Galerkin method with Hermite polynomials (GMHP) [38], hybrid of block pulse functions and Taylor expansion (BPF-TE) [39], the BPFs-based and the TFs-based numerical method. In this, too, the proposed method exhibits better performance.

Example 3.3 [36]

The linear Abel's integral equation of the first kind is

$$f(x) = 2\sqrt{x} - \int_0^x \frac{f(t)}{\sqrt{x-t}} dt. \tag{3.54}$$

The exact solution is $f(x) = 1 - e^{\pi x} erfc(\sqrt{\pi x})$, $x \in [0, 1]$.

The given problem is widely considered as a benchmark example for testing the performance of various numerical methods. In [39], the authors applied the hybrid of BPF and Taylor expansion to Equation (3.54). Chiquet et al. [40] first converted the given problem into an algebraic equation by using Laplace transforms, and the resulting algebraic equation was solved by employing the rational inverse formulae (RIF). A similar attempt was made by Yang [41] with the help of Laplace transforms but the inversion of the obtained algebraic equation was found in terms of power series, to which Pade approximants were applied to enhance the convergence rate and accuracy of the computed series. The numerical methods based on Chebyshev polynomials (CP) [42], shifted Legendre polynomials (SLP) [43], and Chebyshev wavelets (CW) [44] were devised for solving Equation (3.54). It is noticed from Table 3.4 that our results are better than those found via the BPFs-based method and methods reported in [39, 40, 42, 44]. It is also seen that agreement is achieved between our results and those presented in [41] and [43].

3.4.2 Numerical Stability Analysis

In the following example, it is shown that the numerical method maintains consistency in providing approximate solution with acceptable accuracy even when the source function $f(x)$ is corrupted with noise.

The corrupted source function is defined as

$$f(x_j) = f(x_j) + \delta\theta_j,\ j \in [0, m], \tag{3.55}$$

where δ is the noise, θ_j is a uniform random number in the range, $[-1, 1]$.

Example 3.4 [35]

The nonlinear first-kind Abel's integral equation is

TABLE 3.3

Absolute errors obtained via different methods (Example 3.2)

x	CW in Ref. [36]	BPM in Ref. [37]	GMHP in Ref. [38]	BPF-TE in Ref. [39]	BPFs-based method	Proposed method
0	1.8e-07	1.007e-08	NV	NV	1.46598e-07	2.65072e-11
0.1	1.2e-07	1.118e-11	3.57771e-03	1e-03	2.23500e-05	3.02134e-08
0.2	1.9e-07	1.147e-08	3.68249e-04	1e-03	3.45662e-05	3.72715e-08
0.3	3.1e-08	2.448e-05	3.69674e-04	2e-03	6.67845e-05	3.14601e-08
0.4	3.3e-07	3.909e-05	5.4428e-05	7e-03	8.90039e-05	3.43824e-08
0.5	1.8e-06	5.282e-05	1.38543e-04	1e-02	0.00011122	3.65922e-08
0.6	2.8e-06	6.386e-05	2.25302e-05	1e-02	0.00013344	3.83480e-08
0.7	2.9e-06	5.169e-05	6.77552e-05	3e-03	0.00015566	3.97915e-08
0.8	1.8e-06	1.068e-05	2.66823e-05	5e-03	0.00017788	5.10081e-08
0.9	3.8e-08	3.461e-05	5.72100e-05	1e-02	0.00020010	5.20545e-08
1	NV	2.854e-05	1.29771e-04	NV	0.00022232	5.35892e-08

TABLE 3.4

Absolute errors obtained via different methods (Example 3.3)

x	BPF-TE in Ref. [39]	RIF in Ref. [40]	Method in Ref. [41]	CP in Ref. [42]	SLP in Ref. [43]	CW in Ref. [44]	BPFs-based method	Proposed method
0	NV	NV	NV	NV	3.8e-03	NV	0.02895108	1.278543e-10
0.1	1.0e-02	5.31211e-06	3.33846e-09	0.000350000	1.2e-07	1.62983e-03	0.00032999	6.251860e-06
0.2	1.0e-03	7.40412e-06	3.87786e-08	8.999999e-05	2.7e-06	2.82352e-03	0.00017277	3.261511e-06
0.3	1.0e-03	9.26409e-06	1.82276e-07	0	1.4e-06	1.89633e-03	0.00011420	2.151253e-06
0.4	3.0e-03	1.05841e-05	3.42272e-07	1.000000e-05	1.8e-06	1.43922e-03	8.373281e-05	1.574970e-06
0.5	3.0e-03	1.17567e-05	8.53771e-07	1.000000e-05	6.0e-07	1.32002e-03	6.519544e-05	1.224937e-06
0.6	1.0e-03	1.27792e-05	1.43214e-06	0	1.0e-06	1.21446e-03	5.281577e-05	9.914881e-07
0.7	3.0e-04	1.39448e-05	2.18575e-06	1.000000e-05	6.0e-07	9.86938e-04	3.402130e-05	8.257262e-07
0.8	5.0e-04	1.49093e-05	3.11805e-06	2.999999e-05	6.3e-07	2.45968e-04	3.747779e-05	7.025891e-07
0.9	1.0e-03	1.056083e-05	3.22898e-06	9.9999999e-06	3.8e-07	2.45968e-04	3.244303e-05	6.079299e-07
1	NV	NV	NV	3.000000e-05	3.5e-07	NV	2.846566e-05	5.327439e-07

TABLE 3.5

Results for Example 3.4 ($m = 10$)

	Absolute error $\varepsilon(x) = \lvert y(x) - \tilde{y}(x)\rvert$	
x	$\delta = 0$	$\delta = 0.001$
0	0	2.2749170547e-08
0.1	0	1.2353035927e-08
0.2	0	7.0580798983e-09
0.3	3.4408920985e-16	1.1252156983e-08
0.4	3.4408920985e-16	1.9598784817e-08
0.5	1.3322676295e-15	2.3231112589e-09
0.6	1.3322676295e-15	9.6520667014e-09
0.7	8.8817841970e-16	1.7549293129e-08
0.8	3.5527136788e-15	6.2571938741e-09
0.9	5.5511151231e-15	6.8274230713e-09
1	1.5543122344e-15	2.2606549965e-09
	Performance Indices	
L1	7.1123628666e-15	3.1570131357e-08
L2	5.5511151231e-15	2.2749170547e-08
CT	0.228813	0.232829

$$f(x) = \int_0^x \left(\frac{1}{(x-t)^{1/5}} \ln(y(t))dt \right), \; x \in [0, 1]. \tag{3.56}$$

The source function is given as $f(x) = \frac{5}{36} x^{4/5}(9 - 5x)$.

The exact solution is $y(x) = e^{1-x}$.

The numerical results for the step size of 0.1 are given in Table 3.5.

3.4.3 Application to Physical Process Models Involving Abel's Integral Equations

Here it is proved that the proposed numerical method is suitable to simulate physical process models represented by Abel's integral equations. The comparison between the TF solution and the BPF solution is not made for the reason that the performance of the BPFs-based numerical method is inferior to that of the proposed method.

Application 3.1: Cyclic voltammetry for the reversible deposition of metals on a solid planar macroelectrode [45]

Cyclic voltammetry function for the reversible deposition $A + e^- \rightleftarrows B_{\text{dep}}$ of metals on a solid planar macroelectrode, assuming a constant activity of B_{dep} (dep refers to deposition). The current function ($\chi(t)$) can be shown to obey the following Abel's integral equation of the first kind:

$$\int_0^t \left[(t-\tau)^{-0.5}\chi(\tau)\right]d\tau = 1 - \exp[-S(t,t_s)], \tag{3.57}$$

where t_s is potential sweep switching moments,

S is the Sawtooth function defined as $S(t,t_s) = \begin{cases} t & \text{for} t \le t_s \\ 2t_s - t & \text{for} t > t_s \end{cases}$.

The closed form expression for the current function, which satisfies the given Abel's integral equation, is provided here:

$$\chi(t) = \begin{cases} 2\pi^{-1} daw\left(t^{0.5}\right) & \text{for} t \le t_s \\ 2\pi^{-1} daw\left(t^{0.5}\right) - \pi^{-0.5}\exp(-t_s)\left(2\pi^{-0.5} daw\left((t-t_s)^{0.5}\right) + A\right) & \text{for} t > t_s \end{cases} \tag{3.58}$$

where $A = -erex\left((t-t_s)^{0.5}\right) + \exp(t-t_s)$, $erex(x) = \exp(x^2)erfc(x)$.

The absolute errors between the exact solution and the TF solution obtained by using the step size of (1/1500) are provided in Table 3.6. At the initial time point 0, the approximation is not highly accurate but at the remaining time points, the TF solution accurately predicts the original current function.

Application 3.2: Cyclic voltammetry for reversible charge transfer at a planar macroelectrode [45]

Cyclic voltammetry current function for the catalytic reaction mechanism $A + e^- \rightleftarrows B$, $B \rightarrow A$ involves a reversible charge transfer at a planar macroelectrode and a (pseudo) first-order irreversible homogeneous reaction. At the assumption of equal diffusion coefficients of A and B, the original Abel's integral equation of the first kind is

$$\int_0^t \frac{\exp(-\lambda(t-\tau))}{(t-\tau)^{0.5}}\chi(\tau)d\tau = \frac{1}{1+\exp(\mu - S(t,t_s))} - \frac{1}{1+\exp(\mu)}, \tag{3.59}$$

where λ has a meaning of a dimensionless rate constant of the homogeneous reaction, μ is a dimensionless parameter related to the difference between the starting potential and the half-wave potential of the charge-transfer reaction.

In the limit of $\lambda \rightarrow \infty$ (extremely fast homogeneous reaction), a steady state is approached for which an analytical solution exists.

TABLE 3.6

Results for Application 3.1

t	$\varepsilon(t) = \|\chi(t) - \tilde{\chi}(t)\|$
0	0.00122742880169
0.1	3.5891718411e-07
0.2	1.3337849752e-07
0.3	7.4232208491e-08
0.4	3.8290525744e-08
0.5	3.4053200648e-08
0.6	2.5197917252e-08
0.7	1.9239756177e-08
0.8	1.5011961118e-08
0.9	1.1895585405e-08
1	9.5361560870e-09
	Performance Indices
L1	0.0028352121629
L2	0.0032730091628
CT	35.671022

The steady state solution is

$$\chi(t) = \left(\frac{\lambda}{\pi}\right)^{0.5}\left(\frac{1}{1+\exp(\mu - S(t,t_s))} - \frac{1}{1+\exp(\mu)}\right). \tag{3.60}$$

The values of model parameters are considered as $\mu = 20$, $t_s = 30$ and $\lambda = 100$. The step size of (1/300) is used for simulations. As per Table 3.7, it can be asserted that the TF approximation solution is in good accordant with the original solution.

Application 3.3: Potential step chronoamperometry for an irreversible charge transfer at a spherical electrode [46].

The second-kind Abel's integral equation in Equation (3.61) relates the concentration $c(r_0, t)$ of a reactant at the surface of a spherical electrode with radius r_0 and the flux $f(r_0, t) = (\partial c(r,t)/\partial r)|_{r=r_0} = j(t)/(nF)$ to the electric current density $j(t)$.

$$c(r_0,t) = c^0 - D^{-1/2}\int_0^t \left\{[\pi(t-\tau)]^{-1/2} - \rho\left(erex\left[\rho(t-\tau)^{1/2}\right]\right)\right\} f(r_0,\tau)d\tau, \quad t \in [0,1], \tag{3.61}$$

where c^0 is the initial/bulk concentration, $\rho = (D^{0.5})/r$, D is the diffusion coefficient of the reactant, $erex(z) = \exp(z^2)(1 - erf(z))$.

Using $j(t)/(nF) = kc(r_0, t)$ (k is a (potential-dependent) rate constant), $t = t/t_{\text{step}}$ and $\rho = \rho t_{\text{step}}^{1/2}$ in Equation (3.61), we get

$$k_1 \left\{ 1 - \int_0^t [\pi(t-\tau)]^{-1/2} \psi(\tau) d\tau + \rho \int_0^t erex\left[\rho(t-\tau)^{1/2}\right] \psi(\tau) d\tau \right\} - \psi(t) = 0, \tag{3.62}$$

where $k_1 = k\left(D/t_{\text{step}}\right)^{-1/2}$, $\psi(t) = j(t)/\left[nFc^0\left(D/t_{\text{step}}\right)^{1/2}\right]$ is the unknown function to be determined.

Equation (3.62) has exact solution,

$$\psi(t) = \frac{k_1}{1+\frac{k_1}{\rho}} \left\{ 1 + \frac{k_1}{\rho} erex\left[(k_1 + \rho)t^{1/2}\right] \right\}. \tag{3.63}$$

The approximate TF solutions are compared with the exact solutions in Table 3.8.

Application 3.4: Cyclic voltammetry for an irreversible charge transfer at a spherical electrode [46]

The following linear Abel's integral equation of the second kind originates in the theory of linear potential sweep voltammetry for an irreversible charge transfer at spherical electrodes. Equation (3.64) can be derived based on Equation (3.61).

$$\begin{aligned} \int_0^t (t-\tau)^{-1/2} \chi(\tau) d\tau - \pi^{1/2} \rho \int_0^t erex\left[\rho(t-\tau)^{1/2}\right] \chi(\tau) d\tau \\ + \exp[\mu - S(t, ts)]\chi(t) - 1 = 0, \end{aligned} \tag{3.64}$$

where $\chi(t)$ is the dimensionless current function, t is the time normalized by $RT(\alpha_c nFv)^{-1}$, α_c is the cathodic charge transfer coefficient, $\rho = \left(D^{0.5}\left(\alpha_c nFv(RT)^{-1}\right)^{-0.5} \right)/r_0$, r_0 is the electrode radius and D is the diffusion coefficient of the reactant. The parameter μ is defined as $\mu = \alpha_c nF(RT)^{-1}(E_{\text{start}} - E^0) - \ln(k^0)$, where E_{start} is the starting potential, E^0 is the formal potential and k^0 is the conditional rate constant of the charge transfer reaction. The parameters t_s and E_{switch} (switching potential) are related as $t_s = \alpha_c nF(RT)^{-1}(E_{\text{start}} - E_{\text{switch}})$.

The singular integral equation in Equation (3.64) has the exact solution $\chi(t) = \rho\pi^{-1/2}\left[1 + \rho\pi^{-1/2}\exp[\mu - S(t, t_s)]\right]^{-1}$.

The parameters μ and t_s are fixed at constant values ($\mu = 15$, $t_s = 20$) and ρ is varied in the range [0.1, 10]. We can see from Table 3.9 that the original cyclic voltammetry current function, $\chi(t)$, and its approximate solution are almost equal.

Application 3.5: Cyclic voltammetry for the catalytic mechanism at a planar electrode [46]

Abel's integral equations model describing cyclic voltammetry for the catalytic mechanism involving irreversible charge transfer and irreversible homogeneous reaction, under the assumption of different diffusion coefficient, and for planar electrodes is

$$\int_0^t (t-\tau)^{-1/2}\chi(\tau)d\tau + \left(\frac{\pi\lambda\delta}{1-\delta}\right)^{1/2}\left\{\int_0^t \exp[-\lambda(t-\tau)]erex\left[\left(\frac{\lambda}{1-\delta}\right)^{1/2}(t-\tau)^{1/2}\right]\chi(\tau)d\tau\right\}$$
$$-\left(\frac{\pi\lambda\delta}{1-\delta}\right)^{1/2}\left\{\int_0^t erex\left[\left(\frac{\lambda\delta}{1-\delta}\right)^{1/2}(t-\tau)^{1/2}\right]\chi(\tau)d\tau\right\} + \exp[\mu - S(t,t_s)]\chi(t) - 1 = 0, \text{for}\delta < 1 \quad (3.65)$$

$$\int_0^t (t-\tau)^{-1/2}\chi(\tau)d\tau + 2\left(\frac{\lambda\delta}{\delta-1}\right)^{1/2}$$
$$\left\{\int_0^t \exp[-\lambda(t-\tau)]daw\left[\left(\frac{\lambda}{\delta-1}\right)^{1/2}(t-\tau)^{1/2}\right]\chi(\tau)d\tau\right\} \quad (3.66)$$
$$-2\left(\frac{\lambda\delta}{\delta-1}\right)^{1/2}\left\{\int_0^t daw\left[\left(\frac{\lambda\delta}{\delta-1}\right)^{1/2}(t-\tau)^{1/2}\right]\chi(\tau)d\tau\right\}$$
$$+\exp[\mu - S(t,t_s)]\chi(t) - 1 = 0, \text{for}\delta > 1.$$

If δ tends to 1, Equations (3.65) and (3.66) will converge to the following equation.

$$\int_0^t \exp[-\lambda(t-\tau)](t-\tau)^{1/2}\chi(\tau)d\tau + \exp[\mu - S(t,t_s)]\chi(t) - 1 = 0, \textit{for}\ \delta \to 1, \quad (3.67)$$

where λ is the dimensionless rate constant of the homogeneous reaction, δ is the ratio of diffusion coefficients. Parameters μ and t_s retain the same meaning as in Application 3.3.

The exact solution is $\chi(t) = (\lambda\pi^{-1}\delta)^{-1/2}\left[1 + (\lambda\pi^{-1}\delta)^{-1/2}\exp[\mu - S(t,t_s)]\right]^{-1}$.

Table 3.10 presents the absolute error between the original cyclic voltammetry current function and its approximation obtained by our numerical method.

TABLE 3.7

Results for Application 3.2

t	$\varepsilon(t) = \lvert\chi(t) - \tilde{\chi}(t)\rvert$
0	2.5264983208e-11
0.1	5.2048512012e-11
0.2	3.7904021600e-11
0.3	3.3323611154e-11
0.4	3.8261474734e-11
0.5	3.2666948778e-11
0.6	2.6484041388e-11
0.7	1.9650871949e-11
0.8	1.2099051803e-11
0.9	3.752999795e-12
1	5.4708295523e-12
Performance Indices	
L1	5.4577243121e-11
L2	6.0160549688e-10
CT	2.272341

TABLE 3.8

Results for Application 3.3

t	**Absolute error** $\varepsilon(t) = \lvert\psi(t) - \tilde{\psi}(t)\rvert$ **using** $h = 1/500$			
	$(k_1, \rho) = (1, 0.1)$	$(k_1, \rho) = (1, 1)$	$(k_1, \rho) = (1, 2)$	$(k_1, \rho) = (1, 5)$
0	7.5246309183e-10	7.9746398373e-10	8.3266726846e-15	8.8817841970e-15
0.1	2.1640346194e-05	2.4941821514e-06	3.3074482255e-05	0.0003679380977
0.2	1.2346008830e-05	3.3142669285e-06	3.4747578551e-05	0.0003616603542
0.3	8.6307463452e-06	5.1380266393e-06	3.4549582145e-05	0.0003580137041
0.4	6.5959238606e-06	5.9446508610e-06	3.4121063092e-05	0.0003556022439
0.5	5.3062414668e-06	6.3595456389e-06	3.3690403846e-05	0.0003538616026
0.6	3.4152816616e-06	6.5903073883e-06	3.3300034295e-05	0.0003525304074
0.7	3.7634201787e-06	6.7235238906e-06	3.2954390397e-05	0.0003514701996
0.8	3.2664443397e-06	6.8007899476e-06	3.2649222598e-05	0.0003506001390
0.9	2.8755841418e-06	6.8441312717e-06	3.2378795698e-05	0.0003498695085
1	2.5608634984e-06	6.8656635576e-06	3.2137761387e-05	0.0003492446862
Performance Indices				
CT	229.807499	230.088138	293.899394	306.799846
L1	2.9047327378e-04	2.6113691996e-04	2.0045749433e-04	3.7260775928e-04
L2	5.1718190721e-04	3.9096452096e-04	9.7910269050e-04	0.0079489173295

TABLE 3.9

Results for Application 3.4

	Absolute error $\varepsilon(t) = \lvert\chi(t) - \tilde{\chi}(t)\rvert$ **using** $h = 1/200$				
t	$\rho = 0.1$	$\rho = 1$	$\rho = 2$	$\rho = 5$	$\rho = 10$
0	1.6585865e-12	1.65859467e-13	8.29297562e-14	3.3171907e-14	1.6585955e-14
0.1	1.9577372e-12	1.48183487e-13	5.72671139e-14	1.3388876e-14	3.9832724e-15
0.2	2.3616449e-12	1.63915969e-13	5.97316283e-14	1.3073074e-14	5.0719080e-15
0.3	2.8599959e-12	1.87982702e-13	6.64111258e-14	1.4173362e-14	5.6788274e-15
0.4	3.4707313e-12	2.19585254e-13	7.61424038e-14	1.6068642e-14	6.5908724e-15
0.5	3.2175796e-12	2.59429631e-13	8.88894475e-14	1.8661218e-14	7.7874774e-15
0.6	5.1300640e-12	3.08868308e-13	1.04989727e-13	2.1990249e-14	9.2962214e-15
0.7	6.2444666e-12	3.69753289e-13	1.25004583e-13	2.6160043e-14	1.1168688e-14
0.8	7.6052105e-12	3.44444379e-13	1.49690046e-13	3.1322904e-14	1.3475537e-14
0.9	9.2666113e-12	5.35877378e-13	1.80007040e-13	3.7677423e-14	1.6306844e-14
1	1.1295041e-11	6.47669716e-13	2.17150956e-13	3.5473132e-14	1.9774663e-14
			Performance Indices		
CT	6.099794	6.455320	6.987790	7.445437	7.971979
L1	1.1295041e-11	6.47669716e-13	2.17150956e-13	3.5473132e-14	1.9774663e-14
L2	8.0090666e-11	3.76506535e-12	1.62501437e-12	3.4548491e-13	1.4671931e-13

TABLE 3.10

Results for Application 3.5 (for fixed $\mu = 15$ and $t_s = 20$)

	Absolute error $\varepsilon(t) = \lvert\chi(t) - \tilde{\chi}(t)\rvert$ **using** $h = 1/200$				
t	$(\lambda, \delta) = (0.1, 1)$	$(\lambda, \delta) = (1, 0.5)$	$(\lambda, \delta) = (0.1, 2)$	$(\lambda, \delta) = (1, 2)$	$(\lambda, \delta) = (3, 2)$
0	5.2449304e-13	2.34560656e-13	3.70872775e-13	1.1728036e-13	9.5759025e-14
0.1	5.7089620e-13	2.18368132e-13	3.83301905e-13	7.5867834e-14	3.9237495e-14
0.2	6.6617221e-13	2.38701782e-13	3.37122530e-13	6.6085414e-14	3.2889093e-14
0.3	7.8756860e-13	2.69687316e-13	5.07918410e-13	6.0778875e-14	1.9320232e-14
0.4	9.3787952e-13	3.10605736e-13	5.96459294e-13	5.7808035e-14	6.0338304e-15
0.5	1.1223428e-12	3.62471403e-13	7.05512191e-13	5.6483169e-14	8.1104037e-15
0.6	1.3479012e-12	3.27099822e-13	8.39002757e-13	5.6552770e-14	2.3921224e-14
0.7	1.6232595e-12	5.06952882e-13	1.00195173e-12	5.7962370e-14	3.2138016e-14
0.8	1.9591597e-12	6.05161151e-13	1.20060183e-12	6.0774660e-14	6.3537816e-14
0.9	2.3687804e-12	7.25618225e-13	1.44263717e-12	6.5140357e-14	8.8998192e-14
1	2.8682509e-12	8.73122459e-13	1.73747649e-12	7.1289725e-14	1.1954722e-13
			Performance Indices		
CT	0.604694	12.849861	7.040101	3.742449	3.661705
L1	2.8682509e-12	8.73122459e-13	1.73747649e-12	1.1728036e-13	1.1954722e-13
L2	2.0798184e-11	6.54083292e-12	1.28365879e-11	9.2121753e-13	7.4985875e-13

3.5 MATLAB® Codes for Numerical Experiments

The MATLAB codes for the examples and physical process models simulated in the preceding section are provided here. The initial guess (x0) should be supplied for all programs.

Program 3.1

```
%%%%%%%%%%%%%%%%%%%%%%%% MATLAB code for Example 3.1 %%%%%
  %%%%%%%%%%
function [EE1]= Example_3_1(x0)
t0=0;T=1;aa=100;m=10*aa;h=(T-t0)/m;t=[t0:h:T];I=ones
  (1,m+1);
tic
CC=(2/105)*sqrt(t).*(105*I-56*t.*t+48*(t.^3));C01=CC(1:
  end-1);D01=CC(2:end);
[P1alph,P2alph]=TOF1(t0,T,m,0.5);[P3alph,P4alph]=TOF12
  (t0,T,m,0.5);
x=fsolve(@Problem_Fun,x0);
function f=Problem_Fun(x)
for i=1:1:length(t)
  C(i)=x(i);
end
C1=C(1:end-1);D1=C(2:end);f1=C01-gamma(0.5)*(C1*P1alph
  +D1*P3alph);
f2=D01-gamma(0.5)*(C1*P2alph+D1*P4alph);f=[f1 f2];
end
toc
% Exact Solution
exact=(t.^3)-t.*t+I;ee=abs(exact-x);
for i=1:11
  bb=(i-1)*aa+1;EE1(i)=ee(bb);
end
EE1=EE1';
end
% Algorithm ends
```

Program 3.2

```
%%%%%%%%%%%%%%%%%%%%%%%% MATLAB code for Example 3.2 %%%%%
  %%%%%%%%%%
function [EE1]= Example_3_2(x0)
t0=0;T=1;aa=150;m=10*aa;h=(T-t0)/m;t=[t0:h:T];
tic
CC=t.*t+(16/15)*(t.^(5/2));C01=CC(1:end-1);D01=CC
  (2:end);
[P1alph,P2alph]=TOF1
  (t0,T,m,0.5);[P3alph,P4alph]=TOF12(t0,T,m,0.5);
x=fsolve(@Problem_Fun,x0);
```

```
function f=Problem_Fun(x)
for i=1:1:length(t)
  C(i)=x(i);
end
C1=C(1:end-1);D1=C(2:end);f1=C01-gamma(0.5)*(C1*P1alph
+D1*P3alph)-C1;
f2=D01-gamma(0.5)*(C1*P2alph+D1*P4alph)-D1;f=[f1 f2];
end
toc
% Exact Solution
exact=t.*t;ee=abs(exact-x);emax1=max(ee);
for i=1:11
  bb=(i-1)*aa+1;EE1(i)=ee(bb);
end
EE1=EE1';
end
% Algorithm ends
```

Program 3.3

```
%%%%%%%%%%%%%%%%%%%%%% MATLAB code for Example 3.3 %%%%%
 %%%%%%%%%%
function [EE1]= Example_3_3(x0)
t0=0;T=1;aa=200;m=10*aa;h=(T-t0)/m;t=[t0:h:T];
tic
CC=2*sqrt(t);C01=CC(1:end-1);D01=CC(2:end);
[P1alph,P2alph]=TOF1(t0,T,m,0.5);[P3alph,P4alph]=TOF12
(t0,T,m,0.5);
x=fsolve(@Problem_Fun,x0);
function f=TestProblem_Fun(x)
for i=1:1:length(t)
  C(i)=x(i);
end
C1=C(1:end-1);D1=C(2:end);f1=C01-gamma(0.5)*(C1*P1alph
 +D1*P3alph)-C1;
f2=D01-gamma(0.5)*(C1*P2alph+D1*P4alph)-D1;f=[f1 f2];
end
toc
% Exact Solution
exact=1-exp(pi*t).*erfc(sqrt(pi*t));ee=abs(exact-x);
emax1=max(ee);
for i=1:11
  bb=(i-1)*aa+1;EE1(i)=ee(bb);
end
EE1=EE1';
end
% Algorithm ends
```

Program 3.4

```
%%%%%%%%%%%%%%%%%%%%%% MATLAB code for Example 3.4 %%%%%
  %%%%%%%%%
function [x,norminf,norm2]= Example_3_4(x0)
t0=0;T=1;m=10;h=(T-t0)/m;t=[t0:h:T];
tic
[P1alph,P2alph]=TOF1(t0,T,m,4/5);[P3alph,P4alph]=TOF12
  (t0,T,m,4/5);
x=fsolve(@Problem_Fun,x0);
function f=Problem_Fun(x)
for i=1:1:length(t)
  C(i)=x(i);
end
C1=C(1:end-1);D1=C(2:end);I=ones(1,length(t));
CC1=(5/36)*(t.^(4/5)).*(9*I-5*t);C01=CC1(1:end-1);
D01=CC1(2:end);CC1=log(C);C02=CC1(1:end-1);D02=CC1
(2:end);
f1=C01-gamma(4/5)*(C02*P1alph+D02*P3alph);
f2=D01-gamma(4/5)*(C02*P2alph+D02*P4alph);f=[f1 f2];
end
toc
% Exact Solution
exact1=exp(1-t);error=abs(-x+exact1);        norminf=norm
(error,inf);norm2=norm(error,2);
end
```

Program 3.5

```
%%%%%%%%%%%%%%%%%%%%%% MATLAB code for Application 3.1 %%
  %%%%%%%%%%%%
function [EE,norminf,norm2]= Application_3_1(x0)
t0=0;T=1;aa=150;m=10*aa;h=(T-t0)/m;t=[t0:h:T];
I=ones(1,length(t));ts=2;
tic
if T<=ts
  Stts=t;
elseif T>ts
  Stts=[t1 2*ts-t2];
end
CC=exp((-Stts));[P1alph,P2alph]=TOF1(t0,T,m,0.5);
[P3alph,P4alph]=TOF12(t0,T,m,0.5);x=fsolve(@Proble-
m_Fun,x0);
function f=Problem_Fun(x)
for i=1:1:length(t)
  CC1(i)=x(i);
end
C1=CC1(1:end-1);D1=CC1(2:end);CC0=CC-I;C01=CC0(1:end-
  1);D01=CC0(2:end);
f1=(gamma(1/2))*((C1)*P1alph+(D1)*P3alph)+C01;
f2=(gamma(1/2))*((C1)*P2alph+(D1)*P4alph)+D01;f=[f1 f2];
```

```
end
toc
% Exact Solution
if T<=ts
  exact1=2*(1/pi)*dawson(t.^0.5);exact=exact1;
elseif T>ts
  exact1=2*(1/pi)*dawson(t1.^0.5);
   exact2=2*(1/pi)*dawson(sqrt(t2))-(pi^-0.5)*exp(-ts).*
[2*(pi^-0.5)*dawson(sqrt(t2-ts))-exp(t2-ts).*erfc((t2-
  ts).^0.5)+exp(t2-ts)];
  exact=[exact1 exact2];
end
E1=abs(-x+exact);
for i=1:11
  bb=(i-1)*aa+1;EE(i)=E1(bb);
end
EE=EE';norminf=norm(E1,inf);norm2=norm(E1,2);
end
```

Program 3.6

```
%%%%%%%%%%%%%%%%%%%%%%%%%% MATLAB code for Application 3.2 %%
  %%%%%%%%%%%%%%
function [EE,norminf,norm2]= Application_3_2(x0)
t0=0;T=1;aa=30;m=10*aa;h=(T-t0)/m;
t=[t0:h:T];I=ones(1,length(t));ts=30;mu=20;lambda=100;
Stts=t;
tic
CC=[(I./(I+exp(mu-Stts)))-(I./(I+exp(mu)*I))];
C01=CC(1:end-1);D01=CC(2:end);[P1alph,P2alph]=TOF1(t0,
  T,m,0.5);
[P3alph,P4alph]=TOF12(t0,T,m,0.5);x=fsolve(@Proble-
m_Fun,x0);
function E=Problem_Fun(x)
for i=1:1:length(t)
  C(i)=x(i);
end
C1=C(1:end-1);D1=C(2:end);
CC1=exp(lambda*t).*C;C02=CC1(1:end-1);D02=CC1(2:end);
C03=gamma(0.5)*(C02*P1alph+D02*P3alph);
D03=gamma(0.5)*(C02*P2alph+D02*P4alph);
CC2=exp(-lambda*t);C04=CC2(1:end-1);D04=CC2(2:end);
f1=C03.*C03-C01;f2=D03.*D03-D01;f=[f1 f2];
end
toc
% Exact Solution
exact=sqrt(lambda/pi)*[(I./(I+exp(mu-Stts)))-(I./(I+exp
(mu)*I))];
E1=abs(-x+exact);
for i=1:11
```

```
  bb=(i-1)*aa+1;EE(i)=E1(bb);
end
EE=EE';norminf=norm(E1,inf);norm2=norm(E1,2);
end
```

Program 3.7

```
%%%%%%%%%%%%%%%%%%%%%%%% MATLAB code for Application 3.3 %%
  %%%%%%%%%%%%%%
function [norminf,norm2]= Application_3_3(rhow,K,x0)
t0=0;T=1;aa=50;m=10*aa;h=(T-t0)/m;t=[t0:h:T];
tic
[P1alph1,P2alph1]=TOF1(t0,T,m,0.5);[P3alph1,P4alph1]
  =TOF12(t0,T,m,0.5);
[P1alph2,P2alph2]=TOF1(t0,T,m,1);[P3alph2,P4alph2]
  =TOF12(t0,T,m,1);
x=fsolve(@Problem_Fun,x0);
function f=Problem_Fun(x)
for i=1:1:length(t)
  C(i)=x(i);
end
C1=C(1:end-1);D1=C(2:end);I=ones(1,length(t));CC1=K*I;
C01=CC1(1:end-1);D01=CC1(2:end);m=length(t);XX=zeros
  (m,m);
for i=1:m
    XX(i,:)=exp(rhow*rhow*(abs(t-t(i)))).*erfc(rhow*sqrt
(abs(t-t(i))));
end
F1=XX(1:end-1,1:end-1);F2=XX(1:end-1,2:end);F3=XX(2:
end,1:end-1);
F4=XX(2:end,2:end);X1=F1*diag(C1.^1)*P1alph2;
X1=diag(X1)';
X2=F2*diag(D1.^1)*P3alph2;X2=diag(X2)';X11=X1+X2;
X3=F3*diag(C1.^1)*P2alph2;X3=diag(X3)';X4=F4*diag(D1.
  ^1)*P4alph2;
X4=diag(X4)';X22=X3+X4;
f1=C01-(K/sqrt(pi))*gamma(0.5)*((C1)*P1alph1+(D1)
*P3alph1)+K*rhow*(X11)-C1;
f2=D01-(K/sqrt(pi))*gamma(0.5)*((C1)*P2alph1+(D1)
*P4alph1)+K*rhow*(X22)-D1;
f=[f1 f2];
end
toc
%exact soln.
tt=(K+rhow)*sqrt(t);
exact=K*(1/(1+(K/rhow)))*[1+(K/rhow)*exp(tt.^2).
*erfc(tt)];
error=abs(-x+exact);norminf=norm(error,inf);norm2=norm
(error,2);
end
```

Program 3.8

```
%%%%%%%%%%%%%%%%%%%%%%%% MATLAB code for Application 3.4 %%
  %%%%%%%%%%%%%%
function [EE,norminf,norm2]= Application_3_4(x0,rhow)
t0=0;m=200;T=1;h=(T-t0)/m;t=[t0:h:T];mu=15;ts=20;
tic
[P1alph1,P2alph1]=TOF1(t0,T,m,0.5);[P3alph1,P4alph1]
=TOF12(t0,T,m,0.5);
[P1alph2,P2alph2]=TOF1(t0,T,m,1);[P3alph2,P4alph2]
=TOF12(t0,T,m,1);
x=fsolve(@Problem_Fun,x0);
function f=Problem_Fun(x)
for i=1:1:length(t)
  C(i)=x(i);
end
C1=C(1:end-1);D1=C(2:end);I=ones(1,length(t));CC1=exp
(mu-t).*C-I;
C01=CC1(1:end-1);D01=CC1(2:end);m=length(t);XX=zeros
(m,m);
for i=1:m
    XX(i,:)=exp(rhow*rhow*(abs(t-t(i)))).*erfc(rhow*sqrt
(abs(t-t(i))));
end
F1=XX(1:end-1,1:end-1);F2=XX(1:end-1,2:end);F3=XX(2:
end,1:end-1);
F4=XX(2:end,2:end);X1=F1*diag(C1.^1)*P1alph2;
X1=diag(X1)';
X2=F2*diag(D1.^1)*P3alph2;X2=diag(X2)';X11=X1+X2;
X3=F3*diag(C1.^1)*P2alph2;X3=diag(X3)';X4=F4*diag(D1.
^1)*P4alph2;
X4=diag(X4)';X22=X3+X4;
f1=gamma(0.5)*((C1)*P1alph1+(D1)*P3alph1)-sqrt(pi)
*rhow*(X11)+C01;
f2=gamma(0.5)*((C1)*P2alph1+(D1)*P4alph1)-sqrt(pi)
*rhow*(X22)+D01;
f=[f1 f2];
end
toc
% exact soln.
tt=t-mu;
exact=((rhow/sqrt(pi))*ones(1,m+1))./[1+(rhow/sqrt(pi))
*exp(-tt)];
E1=abs(-x+exact);
for i=1:11
  bb=(i-1)*aa+1;EE(i)=E1(bb);
end
EE=EE';norminf=norm(E1,inf);norm2=norm(E1,2);
end
```

Program 3.9

```
%%%%%%%%%%%%%%%%%%%%%%% MATLAB code for Application 3.5 %%
  %%%%%%%%%%%%%
function [EE,norminf,norm2]= Application_3_5(x0,lambda,
  delta)
t0=0;T=1;aa=20;m=10*aa;h=(T-t0)/m;t=[t0:h:T];mu=15;
  ts=20;
tic
[P1alph1,P2alph1]=TOF1(t0,T,m,0.5);[P3alph1,P4alph1]
  =TOF12(t0,T,m,0.5);
[P1alph2,P2alph2]=TOF1(t0,T,m,1);[P3alph2,P4alph2]
  =TOF12(t0,T,m,1);
x=fsolve(@Problem_Fun,x0);
function f=Problem_Fun(x)
for i=1:1:length(t)
  C(i)=x(i);
end
C1=C(1:end-1);D1=C(2:end);I=ones(1,length(t));
if delta<1
CC1=exp(mu-t).*C-I;C01=CC1(1:end-1);D01=CC1(2:end);
labdel=sqrt(lambda/(1-delta));m=length(t);XX=zeros
  (m,m);
for i=1:m
XX(i,:)=[exp(-lambda*(t(i)-t))].*[exp(labdel*labdel*(t
(i)-t))].*erfc(labdel*sqrt(abs(t(i)-t)));
end
F1=XX(1:end-1,1:end-1);F2=XX(1:end-1,2:end);F3=XX(2:
end,1:end-1);
F4=XX(2:end,2:end);X1=F1*diag(C1.^1)*P1alph2;
X1=diag(X1)';
X2=F2*diag(D1.^1)*P3alph2;X2=diag(X2)';X11=X1+X2;
X3=F3*diag(C1.^1)*P2alph2;X3=diag(X3)';X4=F4*diag(D1.
^1)*P4alph2;
X4=diag(X4)';X22=X3+X4;labdel1=sqrt((lambda*delta)/(1-
delta));
for i=1:m
XX(i,:)=[exp(labdel1*labdel1*(t(i)-t))].*erfc(lab-
del1*sqrt(abs(t(i)-t)));
end
F1=XX(1:end-1,1:end-1);F2=XX(1:end-1,2:end);F3=XX(2:
end,1:end-1);
F4=XX(2:end,2:end);X1=F1*diag(C1.^1)*P1alph2;
X1=diag(X1)';
X2=F2*diag(D1.^1)*P3alph2;X2=diag(X2)';X33=X1+X2;
X3=F3*diag(C1.^1)*P2alph2;
X3=diag(X3)';X4=F4*diag(D1.^1)*P4alph2;X4=diag(X4)';
  X44=X3+X4;
aa=sqrt((pi*lambda*delta)/(1-delta));
```

```
f1=gamma(0.5)*((C1)*P1alph1+(D1)*P3alph1)+aa*(X11-
  X33)+C01;
f2=gamma(0.5)*((C1)*P2alph1+(D1)*P4alph1)+aa*(X22-X44)
  +D01;f=[f1;f2];
elseif delta==1
CC1=exp(mu-t).*C-I;C01=CC1(1:end-1);D01=CC1(2:end);
CC1=exp(-lambda*t);C02=CC1(1:end-1);D02=CC1(2:end);
CC1=exp(lambda*t).*C;C03=CC1(1:end-1);D03=CC1(2:end);
f1=gamma(0.5)*C02.*((C03)*P1alph1+(D03)*P3alph1)+C01;
f2=gamma(0.5)*D02.*((C03)*P2alph1+(D03)*P4alph1)+D01;
  f=[f1;f2];
elseif delta>1
CC1=exp(mu-t).*C-I;C01=CC1(1:end-1);D01=CC1(2:end);
XX=zeros(m,m);
for i=1:m
     XX(i,:)=[exp(-lambda*(t(i)-t)).*dawson(sqrt(lambda/
(delta-1))*sqrt(abs(t(i)-t)))];
end
F1=XX(1:end-1,1:end-1);F2=XX(1:end-1,2:end);F3=XX(2:
  end,1:end-1);
F4=XX(2:end,2:end);X1=F1*diag(C1.^1)*P1alph2;
  X1=diag(X1)';
X2=F2*diag(D1.^1)*P3alph2;X2=diag(X2)';X11=X1+X2;
  X3=F3*diag(C1.^1)*P2alph2;
X3=diag(X3)';X4=F4*diag(D1.^1)*P4alph2;X4=diag(X4)';
X22=X3+X4;XX=zeros(m,m);
  for i=1:m
XX(i,:)=[dawson(sqrt((lambda*delta)/(delta-1))*sqrt(abs
  (t(i)-t)))];
end
F1=XX(1:end-1,1:end-1);F2=XX(1:end-1,2:end);F3=XX(2:
  end,1:end-1);
F4=XX(2:end,2:end);X1=F1*diag(C1.^1)*P1alph2;
X1=diag(X1)';
X2=F2*diag(D1.^1)*P3alph2;X2=diag(X2)';X33=X1+X2;
X3=F3*diag(C1.^1)*P2alph2;X3=diag(X3)';X4=F4*diag(D1.
  ^1)*P4alph2;
X4=diag(X4)';X44=X3+X4;aa=sqrt((lambda*delta)/
  (delta-1));
f1=gamma(0.5)*((C1)*P1alph1+(D1)*P3alph1)+2*aa*(X11-
X33)+C01;
f2=gamma(0.5)*((C1)*P2alph1+(D1)*P4alph1)+2*aa*(X22-
X44)+D01;f=[f1;f2];
end
toc
% exact soln.
bb=sqrt((lambda*delta)/pi);tt=t-mu;
exact=bb*ones(1,length((t)))./(1+bb*exp(-tt));
E1=abs(-x+exact);
```

```
for i=1:11
  bb=(i-1)*aa+1;EE(i)=E1(bb);
end
EE=EE';norminf=norm(E1,inf);norm2=norm(E1,2);
end
```

3.6 Concluding Remarks

The following conclusions are drawn from this study:

- The proposed numerical method has been successfully implemented on first-kind and second-kind Abel's integral equations.
- The stability of the TFs-based numerical method is verified by solving Abel's integral equations having corrupted source (input) functions. The noise presented in the source function cannot not hamper the effectiveness of the proposed method. Hence, the proposed numerical method is capable of filtering the noise associated with the source function.
- It is shown that the proposed numerical method can provide accurate numerical solutions to applications of Abel's integral equations.
- The developed numerical method is of immense worth for the solution of scientific and engineering problems involving Abel's integral equations at the present as well as in the future.

References

[1] R. Gorenflo, S. Vessella (1991). *Abel Integral Equations: Analysis and Applications.* Berlin: Springer.

[2] H. Brunner (1997). 1896-1996: One hundred years of Volterra integral equation of the first kind. *Appl. Numer. Math.*, vol. 24, pp. 83–93.

[3] F.G. Tricomi (1957). *Integral Equations*. New York: Interscience.

[4] D.A. Murio, D.G. Hinestroza, C.W. Mejia (1992). New stable numerical inversion of Abel's integral equation. *Comput. Math. Appl.*, vol. 11, pp. 3–11.

[5] M. Deutsch, I. Beniaminy (1982). Derivative-free inversion of Abel's integral equation. *Appl. Phys. Lett.*, vol. 41, pp. 27–28.

[6] M. Bocher (1914). *An introduction to the study of integral equations, 2nd edn.* London: Cambridge University Press.

[7] H. Malinowski, R. Smarzewski (1978). A numerical method for solving the Abel integral equation. *Appl. Math.*, vol. 16(2), pp. 275–281.

[8] C. Lubich (1983). Runge-Kutta theory for Volterra and Abel integral equations of the second kind. *Math. Comput.*, vol. 41(163), pp. 87–102.

[9] C. Lubich (1985). Fractional linear multistep methods for Abel-Volterra integral equations of the second kind. *Math. Comput.*, vol. 45(172), pp. 463–469.

[10] R. Gorenflo, Y. Luchko (1997). Operational method for solving generalized Abel integral equation of the second kind. *Integr. Transf. Spec. F.*, vol. 5, pp. 47–58.

[11] V. Sizikov, D. Sidorov (2016). Generalized quadrature for solving singular integral equations of Abel type in application to infrared tomography. *Appl. Numer. Math.*, vol. 106, pp. 69–78.

[12] R.K. Pandey, O.P. Singh, V.K. Singh (2009). Efficient algorithms to solve singular integral equations of Abel type. *Comput. Math. Appl.*, vol. 57, pp. 664–676.

[13] S. Kumar, A. Kumar, D. Kumar, J. Singh, A. Singh (2015). Analytical solution of Abel integral equation arising in astrophysics via Laplace transform. *J. Egyptian Math. Soc.*, vol. 23, pp. 102–107.

[14] M.S. Mohamed, K.A. Gepreel, F.A. Al-Malki, M. Al-Humyani (2015). Approximate solutions of the generalized Abel's integral equations using the extension Khan's Homotopy analysis transformation method. *J. Appl. Math.*, vol. 2015, Article ID 357861, 9 pages.

[15] M. Khan, M.A. Gondal (2012). A reliable treatment of Abel's second kind singular integral equations. *Appl. Math. Lett.*, vol. 25, pp. 1666–1670.

[16] M. Gulsu, Y. Ozturk, M. Sezer (2011). On the solution of the Abel equation of the second kind by the shifted Chebyshev polynomials. *Appl. Math. Comput.*, vol. 217, pp. 4827–4833.

[17] R.K. Pandey, S. Suman, K.K. Singh, O.P. Singh (2014). An approximate method for Abel inversion using Chebyshev polynomials. *Appl. Math. Comput.*, vol. 237, pp. 120–132.

[18] S. Dixit, O.P. Singh, S. Kumar (2012). A stable numerical inversion of generalized Abel's integral equation. *Appl. Numer. Math.*, vol. 62, pp. 567–579.

[19] M.P. Tripathi, R.K. Pandey, V.K. Baranwal, O.P. Singh (2013). Generalized Abel inversion using extended hat functions operational matrix. *Int. J. Anal.*, vol. 2013, Article ID 652541, 12 pages.

[20] M.N. Sahlan, H.R. Marasi, F. Ghahramani (2015). Block-pulse functions approach to numerical solution of Abel's integral equation. *Cogent Math.*, vol. 2 (1). DOI: 10.1080/23311835.2015.1047111.

[21] S.A. Yousefi (2006). Numerical solution of Abel's integral equation by using Legendre wavelets. *Appl. Math. Comput.*, vol. 175, pp. 574–580.

[22] K. Maleknejad, M. Nosrati, E. Najafi (2012). Wavelet Galerkin method for solving singular integral equations. *Comput. Math. Appl.*, vol. 31(2), pp. 373–390.

[23] X. Li, T. Tang (2012). Convergence analysis of Jacobi spectral collocation methods for Abel-Volterra integral equations of second kind. *Front. Math. China.*, vol. 7(1), pp. 69–83.

[24] M.A. Fariborzi-Araghi, G. Kazemi-Gelian (2014). The combined Sinc-Taylor expansion method to solve Abel's integral equation. *Theory. Approx. Appl.*, vol. 10(1), pp. 27–39.

[25] E. Babolian, H.R. Marzban, M. Salmani (2008). Using triangular orthogonal functions for solving Fredholm integral equations of the second kind. *Appl. Math. Comput.*, vol. 201, pp. 452–463.

[26] K. Maleknejad, Z. JafariBehbahani (2012). Applications of two-dimensional triangular functions for solving nonlinear class of mixed Volterra–Fredholm integral equations. *Math. Comput. Model.*, vol. 55, pp. 1833–1843.

[27] K. Maleknejad, H. Almasieh, M. Roodaki (2010). Triangular functions (TF) method for the solution of nonlinear Volterra–Fredholm integral equations. *Commun. Nonlinear. Sci.*, vol. 15, pp. 3293–3298.

[28] E. Babolian, K. Maleknejad, M. Roodaki, H. Almasieh (2010). Two-dimensional triangular functions and their applications to nonlinear 2D Volterra–Fredholm integral equations. *Comput. Math. Appl.*, vol. 60, pp. 1711–1722.

[29] F. Mirzaee, S. Piroozfar (2010). Numerical solution of the linear two-dimensional Fredholm integral equations of the second kind via two-dimensional triangular orthogonal functions. *J. King. Saud. Univ.*, vol. 22, pp. 185–193.

[30] F. Mirzaee, E. Hadadiyan (2016). Three-dimensional triangular functions and their applications for solving nonlinear mixed Volterra–Fredholm integral equations. *Alexandria Eng. J.*, vol. 53(3), pp. 2943–2952.

[31] M. Khodabin, K. Maleknejad, F.H. Shekarabi (2013). Application of triangular functions to numerical solution of stochastic Volterra integral equations. *IAENG Int. J. Appl. Math.*, vol. 43(1), pp. 1–9.

[32] Z. Sadati (2013). Numerical implementation of triangular functions for solving a stochastic nonlinear Volterra-Fredholm integral equation. *IAENG Int. J. Appl. Math.*, vol. 45(2), pp. 102–107.

[33] F. Mirzaee, M.K. Yari, E. Hadadiyan (2015). Numerical solution of two-dimensional fuzzy Fredholm integral equations of the second kind using triangular functions. *Beni-Suef Univ. J. Basic Appl. Sci.*, vol. 4, pp. 109–118.

[34] H. Saeedi, G.N. Chuev (2015). Triangular functions for operational matrix of nonlinear fractional Volterra integral equations. *J. Appl. Math. Comput.*, vol. 49, pp. 213–232.

[35] S. Noeiaghdam, E. Zarei, H.B. Kelishami (2016). Homotopy analysis transform method for solving Abel's integral equations of the first kind. *Aim Shams Eng. J.*, vol. 7, pp. 483–495.

[36] H. Derili, S. Sohrabi (2008). Numerical solution of singular integral equations using orthogonal functions. *Math. Sci.*, vol. 2(3), pp. 261–272.

[37] S. Suman, K.K. Singh, R.K. Pandey (2014). Approximate solution of integral equation using Bernstein polynomial multiwavelets. In: M Pant, K Deep, A Nagar, J Bansal (eds) *Proceedings of the Third International Conference on Soft Computing for Problem Solving. Advances in Intelligent Systems and Computing*, vol. 259, New Delhi, India: Springer, pp. 486–496.

[38] M.M. Rahman (2013). Numerical solutions of Volterra integral equations using Galerkin method with Hermite Polynomials. In *Proceedings of the 2013 International Conference on Applied Mathematics and Computational Methods in Engineering*, pp. 276–281.

[39] A. Shahsavaran (2011). Numerical approach to solve second kind Volterra integral equation of Abel type using Block-Pulse functions and Taylor expansion by collocation method. *Appl. Math. Sci.*, vol. 5(14), pp. 685–696.

[40] R.A. Chiquet, P. Jara, K.W. Zito (2015). Approximate solutions of Abel's equation using rational inversion of the Laplace transform. *Neur. Parallel Sci. Comput.*, vol. 23, pp. 169–178.

[41] C. Yang (2014). An efficient numerical method for solving Abel integral equation. *Appl. Math. Comput.*, vol. 227, pp. 656–661.

[42] Z. Avazzadeh, B. Shafiee, G.B. Loghmani (2011). Fractional calculus for solving Abel's integral equations using Chebyshev polynomials. *Appl. Math. Sci.*, vol. 5(45), pp. 2207–2216.
[43] A. Saadatmandi, M. Dehghan (2008). A collocation method for solving Abel's integral equations of first and second kinds. *Z. Naturforsch.*, vol. 63(12), pp. 752–756.
[44] S. Sohrabi (2011). Comparison Chebyshev wavelets method with BPFs method for solving Abel's integral equations. *Ain Shams Eng. J.*, vol. 2, pp. 249–253.
[45] L.K. Bieniasz (2008). Cyclic Voltammetric Current Functions Determined with a Prescribed Accuracy by the Adaptive Huber Method for Abel Integral Equations. *Anal Chem*, vol. 80, pp. 9659–9665.
[46] L.K. Bieniasz (2012). Automatic simulation of electrochemical transients by the adaptive Huber method for Volterra integral equations involving Kernel terms exp[−α(t − τ)] erex {[β(t − τ)] 1/2} and exp [−α(t − τ)] daw {[β(t − τ)] 1/2}. *J. Math. Chem.*, vol. 50, pp. 765–781.

4

Numerical Method for Simulation of Physical Processes Described by Fractional-Order Integro-Differential Equations

Fractional-order integro-differential equation (FIDE) is an equation that involves both fractional derivative and fractional integral, or fractional derivative and integer order integral, or integer order derivative and fractional integral. Fractional-order integro-differential equations have substantial applications in several fields like magnetic field induction in dielectric media, anomalous diffusion, velocity fluctuation of a hard-core Brownian particle, viscoelasticity, optimal control, heat transfer, thermodynamics, electrical conduction of polymers, nanotransistors, and so on. Analytical solutions of fractional-order integro-differential equations are hard to obtain; hence, devising effective numerical methods have been the focus of pure and applied mathematicians.

In this chapter, we deal with formulating a new numerical method for the solution of the system of Fredholm-Volterra-Hammerstein integro-differential equations of fractional order:

$$\lambda_1\left({}_0D_x^{\alpha_i}\, y_i(x)\right) = \lambda_2 g_i(x) + \lambda_3\left(\sum_{k=1}^{n}\int_0^1 k_{ik}(x,t)F_{ik}(t,y_1(t),y_2(t),\ldots,y_n(t))dt\right)$$
$$+\,\lambda_4\left(\sum_{j=1}^{n}\int_0^x \frac{h_{ij}(x,t)}{(x-t)^{\beta}}G_{ij}(t,y_1(t),y_2(t),\ldots,y_n(t))dt\right), \tag{4.1}$$

where $x,t\in[0,1]$, $m_i-1\le\alpha_i<m_i$, $\forall i\in[1,n]$, $m_i,n\in Q^+$ (Q^+ is the collection of positive integers), $\lambda_1,\lambda_2,\lambda_3,\lambda_4$ are arbitrary integers, β is an arbitrary number, $y_i(x),g_i(x)\in C[0,1]$, $k_{ik}(x,t)$ and $h_{ij}(x,t)$ are given sufficiently smooth kernel functions, F_{ik} and G_{ij} are continuous functions in $[0,1]$.

The triangular functions (TFs) discussed in Section 1.6 (Chapter 1) are the foundation of the proposed method. The important attribute of TFs aids in construction of the proposed method via transforming the fractional-order Fredholm-Volterra-Hammerstein integro-differential equations in Equation (4.1) into a system of algebraic equations. The obtained system of algebraic equations is easier to solve in comparison to Equation (4.1).

4.1 Existence and Uniqueness of Solution

For the sake of simplicity, we substitute $n = 1$, $\lambda_1 = 1$, $\lambda_2 = 1$, $\lambda_3 = 1$ and $\lambda_4 = 1$ in Equation (4.1). Then we have

$$ {}_0D_x^{\alpha} y(x) = g(x) + \int_0^1 k(x,t)F(t,y(t))dt + \int_0^x \frac{h(x,t)}{(x-t)^{\beta}} G(t,y(t))dt, \tag{4.2} $$

where $x, t \in [0,1]$, $y(x) : [0,1] \to [0,\infty)$, $g(x) : [0,1] \to [0,\infty)$, $k(x,t) : ([0,1],[0,1]) \to [0,\infty)$, $h(x,t) : ([0,1],[0,1]) \to [0,\infty)$, $F(x,y(x)) : ([0,1], [0,\infty)) \to [0,\infty)$, $G(x,y(x)) : ([0,1],[0,\infty)) \to [0,\infty)$, α is a noninteger number satisfying $M-1 \leq \alpha < M$, M is an integer greater than α, β is an arbitrary number.

Performing fractional integration on both sides of Equation (4.2) results in

$$ y(x) = Y0 + J^{\alpha}(g(x)) + J^{\alpha}\left(\int_0^1 k(x,t)F(t,y(t))dt\right) + J^{\alpha}\left(\int_0^x \frac{h(x,t)}{(x-t)^{\beta}} G(t,y(t))dt\right), \tag{4.3} $$

where $Y0 = \sum_{p=0}^{M-1} \frac{x^p}{p!} y^{(p)}(x)|_{x=0}$.

Equation (4.3) can be written in fixed point equation form $\Phi y(x) = y(x)$, where Φ is defined as

$$ \Phi y(x) = Y0 + J^{\alpha}(g(x)) + J^{\alpha}\left(\int_0^1 k(x,t)F(t,y(t))dt\right) + J^{\alpha}\left(\int_0^x \frac{h(x,t)}{(x-t)^{\beta}} G(t,y(t))dt\right). \tag{4.4} $$

Let us define Banach space as $C : [0,1] \times [0,\infty)$ with $d : [0,\infty) \times [0,\infty) \to [0,\infty)$, $d(\tilde{y}(x), y(x)) = \|\tilde{y}(x) - y(x)\|_{\infty}$.

Let us also suppose that the functions $F(x,y(x))$ and $G(x,y(x))$ satisfy the Lipschitz condition, with respect to the second argument, on $[0,1]$ as

$$ \begin{gathered} \|F(x,\tilde{y}(x)) - F(x,y(x))\|_{\infty} \leq L_1 \|\tilde{y}(x) - y(x)\|_{\infty}, \\ L_1 \textit{ is a Lipschitz constant}, L_1 \in (0,1), \end{gathered} \tag{4.5} $$

$$ \begin{gathered} \|G(x,\tilde{y}(x)) - G(x,y(x))\|_{\infty} \leq L_2 \|\tilde{y}(x) - y(x)\|_{\infty}, \\ L_2 \textit{ is a Lipschitz constant}, L_2 \in (0,1). \end{gathered} \tag{4.6} $$

Let $\Phi : C[0,1] \to C[0,1]$ such that

$$\Phi y(x) = Y0 + J^{\alpha}(g(x)) + J^{\alpha}\left(\int_0^1 k(x,t)F(t,y(t))dt\right) + J^{\alpha}\left(\int_0^x \frac{h(x,t)}{(x-t)^{\beta}}G(t,y(t))dt\right). \tag{4.7}$$

Let $\tilde{y}(x), y(x) \in C[0,1]$ and

$$\begin{aligned}\Phi\tilde{y}(x) - \Phi y(x) &= J^{\alpha}\left(\int_0^1 k(x,t)(F(t,\tilde{y}(t)) - F(t,y(t)))dt\right)\\ &+ J^{\alpha}\left(\int_0^x \frac{h(x,t)}{(x-t)^{\beta}}(G(t,\tilde{y}(t)) - G(t,y(t)))dt\right).\end{aligned} \tag{4.8}$$

Then for $x > 0$, we have

$$\begin{aligned}\left\|\Phi\tilde{y}(x) - \Phi y(x)\right\|_{\infty} &\le J^{\alpha}\left|\int_0^1 \|k(x,t)\|_{\infty}\left\|F(t,\tilde{y}(t)) - F(t,y(t))\right\|_{\infty}dt\right|\\ &+ J^{\alpha}\left|\int_0^x \frac{\|h(x,t)\|_{\infty}}{|(x-t)|^{\beta}}\left\|G(t,\tilde{y}(t)) - G(t,y(t))\right\|_{\infty}dt\right|.\end{aligned} \tag{4.9}$$

Let us consider $M_1 = \|k(x,t)\|_{\infty} = \max_{0 \le x,t \le 1}|k(x,t)|$, $M_2 = \|h(x,t)\|_{\infty} = \max_{0 \le x,t \le 1}|h(x,t)|$.

Equation (4.9) becomes

$$\begin{aligned}\left\|\Phi\tilde{y}(x) - \Phi y(x)\right\|_{\infty} &\le J^{\alpha}\left|\int_0^1 M_1L_1\left\|\tilde{y}(x) - y(x)\right\|_{\infty}dt\right| + J^{\alpha}\left|\int_0^x \frac{M_2}{|(x-t)|^{\beta}}L_2\left\|\tilde{y}(t) - y(t)\right\|_{\infty}dt\right|,\\ &\le M_1L_1\left\|\tilde{y}(x) - y(x)\right\|_{\infty}J^{\alpha}\left(|x|^0\right) + M_2L_2\left\|\tilde{y}(x) - y(x)\right\|_{\infty}\left|\frac{1}{1-\beta}\right|J^{\alpha}\left(|x|^{1-\beta}\right),\\ &\le M_1L_1\left\|\tilde{y}(x) - y(x)\right\|_{\infty}\frac{|x|^{\alpha}}{\alpha!} + M_2L_2\left\|\tilde{y}(x) - y(x)\right\|_{\infty}\left|\frac{1}{1-\beta}\right|\frac{|x|^{1-\beta+\alpha}}{(1-\beta+\alpha)!},\\ &\le \Omega\left\|\tilde{y}(t) - y(t)\right\|_{\infty},\end{aligned} \tag{4.10}$$

where $\Omega = \frac{M_1L_1}{\alpha!} + \frac{M_2L_2}{(1-\beta+\alpha)!}\frac{1}{|1-\beta|}$.

If $\Omega < 1$, then, by contraction mapping theorem, the fractional-order Fredholm-Volterra-Hammerstein integro-differential equation in Equation (4.2) has a unique solution in $C[0, 1]$.

4.2 The Proposed Numerical Method

In this section, we develop a numerical method to solve the system of fractional-order Fredholm-Volterra-Hammerstein integro-differential equations.

Let us consider Equation (4.1):

$$\lambda_1\left({}_C D_{0,x}^{\alpha_i} y_i(x)\right) = \lambda_2 g_i(x) + \lambda_3 \sum_{k=1}^{n}\left(\int_0^1 k_{ik}(x,t)F_{ik}(t, y_1(t), \ldots\ldots, y_n(t))dt\right) + \lambda_4 \sum_{j=1}^{n}\left(\int_0^x \frac{h_{ij}(x,t)}{(x-t)^{\beta}} G_{ij}(t, y_1(t), \ldots\ldots, y_n(t))dt\right). \tag{4.11}$$

Performing fractional integration on both sides of Equation (4.11) results in

$$\lambda_1 y_i(x) = \lambda_1 Y_0 + \lambda_2 J^{\alpha_i}(g_i(x)) + \lambda_3 J^{\alpha_i}\left(\sum_{k=1}^{n}\left(\int_0^1 k_{ik}(x,t)F_{ik}(t, y_1(t), \ldots\ldots, y_n(t))dt\right)\right) + \lambda_4\Gamma(1-\beta)\sum_{j=1}^{n}\left(\frac{1}{\Gamma(1+\alpha_i-\beta)}\int_0^x \frac{h_{ij}(x,t)}{(x-t)^{\beta-\alpha_i}} G_{ij}(t, y_1(t), \ldots\ldots, y_n(t))dt\right), \tag{4.12}$$

where $Y0 = \sum_{k=0}^{K}\left(\frac{x^k}{\Gamma(k+1)} y_i^{(k)}(0)\right)$, $K = Q_i - 1$, $Q_i - 1 < \alpha_i \leq Q_i$.

By using Equations (1.65) and (1.66), we can have the approximation for $y_i(x)$, $g_i(x)$, $F_{ik}(x, y_1(x), \ldots, y_n(x))$ and $G_{ij}(x, y_1(x), \ldots, y_n(x))$ as follows:

$$\begin{aligned} &y_i(x) \approx C_i^T T1_m(x) + D_i^T T2_m(x),\ g_i(x) \approx C_{i0}^T T1_m(x) + D_{i0}^T T2_m(x), \\ &Y0 \approx C_Y^T T1_m(x) + D_Y^T T2_m(x), \end{aligned} \tag{4.13}$$

$$\begin{aligned} F_{ik}(x, y_1(x), \ldots, y_n(x)) &\approx C_{ik}^T T1_m(x) + D_{ik}^T T2_m(x), \\ G_{ij}(x, y_1(x), \ldots, y_n(x)) &\approx C_{ij}^T T1_m(x) + D_{ij}^T T2_m(x). \end{aligned} \tag{4.14}$$

The kernel $k_{ik}(x,t)$ is estimated in the TF domain using Equation (1.78),

$$k_{ik}(x,t) \approx T1_m^T(x)\left(F_1^{ik} T1_m(t) + F_2^{ik} T2_m(t)\right) + T2_m^T(x)\left(F_3^{ik} T1_m(t) + F_4^{ik} T2_m(t)\right). \tag{4.15}$$

The product $k_{ik}(x,t)F_{ik}(t, y_1(t), y_2(t), \ldots, y_n(t))$ can be expressed in the TF domain as

$$\begin{aligned} & k_{ik}(x,t)F_{ik}(t, y_1(t), y_2(t), \ldots, y_n(t)) \\ & \approx \left(T1_m^T(x)\left(F_1^{ik} T1_m(t) + F_2^{ik} T2_m(t)\right) + T2_m^T(x)\left(F_3^{ik} T1_m(t) + F_4^{ik} T2_m(t)\right)\right) \\ & \quad \times \left(T1_m^T(t)C_{ik} + T2_m^T(t)D_{ik}\right), \\ & = T1_m^T(x)F_1^{ik} T1_m(t) T1_m^T(t) C_{ik} + T1_m^T(x) F_2^{ik} T2_m(t) T1_m^T(t) C_{ik} \\ & \quad + T1_m^T(x) F_1^{ik} T1(t) T2_m^T(t) D_{ik} + T1_m^T(x) F_2^{ik} T2_m(t) T2_m^T(t) D_{ik} \\ & \quad + T2_m^T(x) F_3^{ik} T1_m(t) T1_m^T(t) C_{ik} + T2_m^T(x) F_4^{ik} T2_m(t) T1_m^T(t) C_{ik} \\ & \quad + T2_m^T(x) F_3^{ik} T1_m(t) T2_m^T(t) D_{ik} + T2_m^T(x) F_4^{ik} T2_m(t) T2_m^T(t) D_{ik}. \end{aligned} \tag{4.16}$$

Employing Equations (1.69) to (1.71) in Equation (4.16),

$$\begin{aligned} & k_{ik}(x,t)F_{ik}(t, y_1(t), y_2(t), \ldots, y_n(t)) \\ & \quad \approx T1_m^T(x) F_1^{ik} diag(T1_m(t)) C_{ik} + T1_m^T(x) F_2^{ik} O C_{ik} + T1_m^T(x) F_1^{ik} O D_{ik} \\ & \quad + T1_m^T(x) F_2^{ik} diag(T2_m(t)) D_{ik} + T2_m^T(x) F_3^{ik} diag(T1_m(t)) C_{ik} \\ & \quad + T2_m^T(x) F_4^{ik} O C_{ik} + \ T2_m^T(x) F_3^{ik} O D_{ik} + T2_m^T(x) F_4^{ik} diag(T2_m(t)) D_{ik}. \end{aligned} \tag{4.17}$$

Equation (4.17) can also be written as

$$\begin{aligned} & k_{ik}(x,t)F_{ik}(t, y_1(t), y_2(t), \ldots, y_n(t)) \\ & \quad \approx T1_m^T(x) F_1^{ik} diag(C_{ik}) T1_m(t) + T1_m^T(x) F_2^{ik} diag(D_{ik}) T2_m(t) \\ & \quad + T2_m^T(x) F_3^{ik} diag(C_{ik}) T1_m(t) + \ T2_m^T(x) F_4^{ik} diag(D_{ik}) T2_m(t). \end{aligned} \tag{4.18}$$

The TF approximation for one-fold definite integral of (4.18) is

$$\int_0^1 (k_{ik}(x,t)F_{ik}(t,y_1(t),y_2(t),\ldots,y_n(t)))dt$$

$$\approx (T1_m^T(x)F_1^{ik}diag(C_{ik}) + T2_m^T(x)F_3^{ik}diag(C_{ik}))\int_0^1 T1_m(t)dt$$
$$+ \left(T1_m^T(x)F_2^{ik}diag(D_{ik}) + T2_m^T(x)F_4^{ik}diag(D_{ik})\right)\int_0^1 T2_m(t)dt \tag{4.19}$$
$$= A_{ik}T1_m(x) + B_{ik}T2_m(x),$$

where $A_{ik} = a[1 \quad 1 \quad \cdots \quad 1]_{1\times m}$, $B_{ik} = a[1 \quad 1 \quad \cdots \quad 1]_{1\times m}$, $a = B(\text{end}) - B(1)$, $B = [b1 \quad b2(\text{end})]$, $b1 = \tilde{X}_1 + \tilde{X}_2$, $b_2 = \tilde{X}_3 + \tilde{X}_4$, $\tilde{X}_1 = diag\left(F_1^{ik}diag(C_{ik})P_1\right)$, $\tilde{X}_2 = diag\left(F_2^{ik}diag(D_{ik})P_1\right)$, $\tilde{X}_3 = diag\left(F_3^{ik}diag\,(C_{ik})P_2\right)$, $\tilde{X}_4 = diag\left(F_4^{ik}diag(D_{ik})P_2\right)$.

Carrying out fractional integration J^{α_i}on Equation (4.19),

$$\frac{1}{\Gamma(\alpha_i)}\int_0^x (x-t)^{\alpha_i-1}(A_{ik}T1(t) + B_{ik}T2(t))dt = Y_1T1_m(x) + Y_2T2_m(x), \tag{4.20}$$

where $Y_1 = \left(A_{ik}P_1^{\alpha_i} + B_{ik}P_3^{\alpha_i}\right)$, $Y_2 = \left(A_{ik}P_2^{\alpha_i} + B_{ik}P_4^{\alpha_i}\right)$.

Similar to Equation (4.18), the TF estimate for $h_{ij}(x,t)G_{ij}(t,y_1(t), y_2(t),\ldots,y_n(t))$ is obtained as

$$\begin{aligned} &h_{ij}(x,t)G_{ij}(t,y_1(t),y_2(t),\ldots,y_n(t)) \\ &\approx T1_m^T(x)G_1^{ij}diag(C_{ij})T1_m(t) + T1_m^T(x)G_2^{ij}diag(D_{ij})T2_m(t) \\ &\quad + T2_m^T(x)G_3^{ij}diag(C_{ij})T1_m(t) + T2_m^T(x)G_4^{ij}diag(D_{ij})T2_m(t). \end{aligned} \tag{4.21}$$

From Equation (4.12),

$$\begin{aligned} &\frac{1}{\Gamma(1+\alpha_i-\beta)}\int_0^x \frac{h_{ij}(x,t)}{(x-t)^{\beta-\alpha_i}}G_{ij}(t,y_1(t),\ldots\ldots,y_n(t))dt \\ &= \frac{1}{\Gamma(1+\alpha_i-\beta)}\int_0^x \frac{(AT1_m(t)+BT2_m(t))}{(x-t)^{\beta-\alpha_i}}dt, \\ &= A\left(P_1^{\gamma}T1_m(x) + P_2^{\gamma}T2_m(x)\right) + B\left(P_3^{\gamma}T1_m(x) + P_4^{\gamma}T2_m(x)\right), \\ &= (AP_1^{\gamma} + BP_3^{\gamma})T1_m(x) + (AP_2^{\gamma} + BP_4^{\gamma})T2_m(x), \end{aligned} \tag{4.22}$$

where $A=(T1_m^T(x)G_1^{ij}diag(C_{ij})+T2_m^T(x)G_3^{ij}diag(C_{ij}))$, $B=(T1_m^T(x)G_2^{ij}$ $diag(D_{ij})+T2_m^T(x)G_4^{ij}diag(D_{ij}))$, $\gamma=1+\alpha_i-\beta$.

Equation (4.22) can be written as

$$\frac{1}{\Gamma(1+\alpha_i-\beta)}\int_0^x \frac{h_{ij}(x,t)}{(x-t)^{\beta-\alpha_i}}G_{ij}(t,y_1(t),\ldots\ldots,y_n(t))dt=\tilde{Y}_1T1_m(x)+\tilde{Y}_2T2_m(x), \quad (4.23)$$

where

$\tilde{Y}_1=diag\left(G_1^{ij}diag(C_{ij})P_1^{\gamma}+G_2^{ij}diag(D_{ij})P_3^{\gamma}\right)$,

$\tilde{Y}_2=diag\left(G_3^{ij}diag(C_{ij})P_2^{\gamma}+G_4^{ij}diag(D_{ij})P_4^{\gamma}\right)$.

Equation (4.12) becomes

$$\begin{aligned}\lambda_1(C_i^TT1_m(x)+D_i^TT2_m(x))&=\lambda_1(C_Y^TT1_m(x)+D_Y^TT2_m(x))\\&+\lambda_2(Z_1T1_m(x)+Z_2T2_m(x))\\&+\lambda_3\sum_{k=1}^{n}(Y_1T1_m(x)+Y_2T2_m(x))+\lambda_4\Gamma(1-\beta)\sum_{j=1}^{n}\left(\tilde{Y}_1T1_m(x)+\tilde{Y}_2T2_m(x)\right),\end{aligned} \quad (4.24)$$

where $Z_1=C_{i0}^TP_1^{\alpha_i}+D_{i0}^TP_3^{\alpha_i}$, $Z_2=C_{i0}^TP_2^{\alpha_i}+D_{i0}^TP_4^{\alpha_i}$.

Equating the coefficients of LHTF and RHTF,

$$\lambda_1C_i^T=\left(\lambda_1C_Y^T+\lambda_2Z_1+\lambda_3\sum_{k=1}^{n}Y_1+\lambda_4\Gamma(1-\beta)\sum_{j=1}^{n}\tilde{Y}_1\right), \quad (4.25)$$

$$\lambda_1D_i^T=\left(\lambda_1D_Y^T+\lambda_2Z_2+\lambda_3\sum_{k=1}^{n}Y_2+\lambda_4\Gamma(1-\beta)\sum_{j=1}^{n}\tilde{Y}_2\right). \quad (4.26)$$

The solution of the system of nonlinear algebraic equations is the approximate solution of Equation (4.11).

4.3 Convergence Analysis

We now split the given interval into an m number of subintervals of equal width as follows.

$$[0, h), [h, 2h), [2h, 3h), \ldots\ldots\ldots, [jh, (j+1)h), \ldots\ldots, [(m-1)h, mh = 1]. \quad (4.27)$$

We choose an arbitrary subinterval $x, t \in [jh, (j+1)h)$, $j = 0, 1, 2, \ldots\ldots$, $m-1$. On the selected subinterval, we prove that the piecewise linear numerical solution, $\tilde{y}_i(x)$, obtained by the proposed numerical method approaches the exact solution, $y_i(x)$, of the system of fractional-order Fredholm-Volterra-Hammerstein integro-differential equations in the limit of step size, h, tends to zero.

Let $Y(x) = [y_1(x) \ y_2(x) \ \cdots \ \cdots \ y_n(x)]$ and $\tilde{Y}(x) = [\tilde{y}_1(x) \ \tilde{y}_2(x) \ \cdots \ \cdots \ \tilde{y}_n(x)]$ be the original and approximate solution, respectively, of Equation (4.1).

Let us suppose that the nonlinear functions $F_{ik}(x, y_1(x), \ldots\ldots, y_n(x))$ and $G_{iq}(x, y_1(x), \ldots\ldots, y_n(x))$ gratify the following conditions of Lipschitz continuity.

$$\begin{aligned} \left\| F_{ik}(x, Y(x)) - F_{ik}\left(x, \tilde{Y}(x)\right) \right\| &\le L_1 \left\| Y(x) - \tilde{Y}(x) \right\|, \\ \left\| G_{iq}(x, Y(x)) - G_{iq}\left(x, \tilde{Y}(x)\right) \right\| &\le L_2 \left\| Y(x) - \tilde{Y}(x) \right\|. \end{aligned} \quad 4.28$$

where L_1 and L_2 are Lipschitz constants, $L_1, L_2 \in (0, 1)$.

First, we derive the error between the original source term $g_i(x)$ and its TF estimate $\tilde{g}_i(x)$ in the jth subinterval $[x_j, x_{j+1}) = [jh, (j+1)h)$.

Approximating $g_i(x)$ in the jth subinterval,

$$\begin{aligned} \tilde{g}_i(x) &= g_i(jh)\left(1 - \left(\frac{x - jh}{h}\right)\right) + g_i((j+1)h)\left(\frac{x - jh}{h}\right), \\ &= g_i(jh) + \left(\frac{g_i((j+1)h) - g_i(jh)}{h}\right)\left(\frac{x - jh}{h}\right), \\ &= g_i(jh) + \left.\frac{dg_i(x)}{dx}\right|_{x=jh}\left(\frac{x - jh}{h}\right). \end{aligned} \quad (4.29)$$

The Taylor series expansion of $g_i(x)$ around the point $x_j = jh$,

$$g_i(x) \approx g_i(jh) + \left.\frac{dg_i(x)}{dx}\right|_{x=jh}(x - jh) + \left.\frac{d^2 g_i(x)}{dx^2}\right|_{x=jh}\frac{(x - jh)^2}{2} + \left.\frac{d^3 g_i(x)}{dx^3}\right|_{x=jh}\frac{(x - jh)^3}{3} + \cdots\cdots \quad (4.30)$$

Truncating the above series to the first three terms,

$$g_i(x) \approx g_i(jh) + \frac{dg_i(x)}{dx}\bigg|_{x=jh}(x - jh) + \frac{d^2g_i(x)}{dx^2}\bigg|_{x=jh}\frac{(x - jh)^2}{2}. \tag{4.31}$$

Subtracting Equation (4.29) from Equation (4.31),

$$g_i(x) - \tilde{g}_i(x) = \frac{d^2g_i(x)}{dx^2}\bigg|_{x=jh}\frac{(x - jh)^2}{2}. \tag{4.32}$$

Because $x \in [jh, (j+1)h)$, the term $(x - jh)$ is always less than h. So we rewrite Equation (4.32) as follows:

$$g_i(x) - \tilde{g}_i(x) = \frac{d^2g_i(x)}{dx^2}\bigg|_{x=jh}\frac{h^2}{2}. \tag{4.33}$$

Taking the norm,

$$\left\|g_i(x) - \tilde{g}_i(x)\right\| \leq M_1\frac{h^2}{2}, \; where\, M_1 = \left\|\frac{d^2g_i(x)}{dx^2}\bigg|_{x=jh}\right\|, \; M_1 \; is \; a \; positive \; number. \tag{4.34}$$

In the same fashion, the norm of the error between Y_0 and $\tilde{Y}_0$ can be derived as

$$\left\|Y_0 - \tilde{Y}_0\right\| \leq M_2\frac{h^2}{2}, \; where\, M_2 = \left\|\frac{d^2Y_0(x)}{dx^2}\bigg|_{x=jh}\right\|, \; M_2 \; is \; a \; positive \; number. \tag{4.35}$$

We now derive the expression for the error between the kernel $k_{ik}(x,t)$ and its TF approximation in the jth subinterval.

By using Equation (1.78), we get the following.

$$\begin{aligned} k_{ik}(x,t) \approx\, & k_{ik}(jh, jh)T1_j(x)T1_j(t) + k_{ik}(jh, (j+1)h)T1_j(x)T2_j(t) \\ & + k_{ik}((j+1)h, jh)T2_j(x)T1_j(t) + k_{ik}((j+1)h, (j+1)h)\; T2_j(x)T2_j(t)\cdot \end{aligned} \tag{4.36}$$

By using Equations (1.65) and (1.66),

$$\begin{aligned}\tilde{k}_{ik}(x,t) \approx k_{ik}(jh,jh)\left(1-\left(\frac{x-jh}{h}\right)\right)\left(1-\left(\frac{t-jh}{h}\right)\right) \\ + k_{ik}(jh,(j+1)h)\left(1-\left(\frac{x-jh}{h}\right)\right)\left(\frac{t-jh}{h}\right) \\ + k_{ik}((j+1)h,jh)\left(\frac{x-jh}{h}\right)\left(1-\left(\frac{t-jh}{h}\right)\right) \\ + k_{ik}((j+1)h,(j+1)h)\left(\frac{x-jh}{h}\right)\left(\frac{t-jh}{h}\right).\end{aligned} \tag{4.37}$$

Performing multiplication in the right-hand side of Equation (4.37) and rearranging the resulting equation gives

$$\begin{aligned}\tilde{k}_{ik}(x,t) \approx k_{ik}(jh,jh) + \left(\frac{k_{ik}((j+1)h,jh)-k_{ik}(jh,jh)}{h}\right)(x-jh) \\ + \left(\frac{k_{ik}(jh,(j+1)h)-k_{ik}(jh,jh)}{h}\right)(t-jh) \\ + \left(\frac{k_{ik}((j+1)h,(j+1)h)-k_{ik}(jh,(j+1)h)-k_{ik}((j+1)h,jh)+k_{ik}(jh,jh)}{h^2}\right) \\ (x-jh)(t-jh).\end{aligned} \tag{4.38}$$

Use of forward difference formula in Equation (4.38) yields

$$\begin{aligned}\tilde{k}_{ik}(x,t) \approx k_{ik}(jh,jh) + \left(\frac{dk_{ik}(x,t)}{dx}\right)_{x=jh,t=jh}(x-jh) + \left(\frac{dk_{ik}(x,t)}{dt}\right)_{x=jh,t=jh}(t-jh) \\ + \left(\frac{d^2k_{ik}(x,t)}{dxdt}\right)_{x=jh,t=jh}(x-jh)(t-jh).\end{aligned} \tag{4.39}$$

Similar to Equation (4.31), we now expand the actual kernel $k_{ik}(x,t)$ by Taylor series around the point (jh,jh), and consider the first six terms of the resulting series as shown in the next equation:

$$k_{ik}(x,t) \approx k_{ik}(jh,jh) + \left(\frac{dk_{ik}(x,t)}{dx}\right)_{x=jh,t=jh}(x-jh) + \left(\frac{dk_{ik}(x,t)}{dt}\right)_{x=jh,t=jh}(t-jh)$$
$$+ \left(\frac{d^2k_{ik}(x,t)}{dx^2}\right)_{x=jh,t=jh}\frac{(x-jh)^2}{2} + \left(\frac{d^2k_{ik}(x,t)}{dt^2}\right)_{x=jh,t=jh}\frac{(t-jh)^2}{2}$$
$$+ \left(\frac{d^2k_{ik}(x,t)}{dxdt}\right)_{x=jh,t=jh}(x-jh)(t-jh). \tag{4.40}$$

From Equations (4.39) and (4.40),

$$k_{ik}(x,t) - \tilde{k}_{ik}(x,t) = \left(\frac{d^2k_{ik}(x,t)}{dx^2}\right)_{x=jh,t=jh}\frac{(x-jh)^2}{2} + \left(\frac{d^2k_{ik}(x,t)}{dt^2}\right)_{x=jh,t=jh}\frac{(t-jh)^2}{2}. \tag{4.41}$$

Assuming the following,

$$M_3 = \left\|\left(\frac{d^2k_{ik}(x,t)}{dx^2}\right)_{x=jh,t=jh}\right\|, \; M_4 = \left\|\left(\frac{d^2k_{ik}(x,t)}{dt^2}\right)_{x=jh,t=jh}\right\|, \tag{4.42}$$

where M_3 and M_4 are positive numbers.

Using Equation (4.42) in Equation (4.41),

$$\left\|k_{ik}(x,t) - \tilde{k}_{ik}(x,t)\right\| \leq M_3\frac{(x-jh)^2}{2} + M_4\frac{(t-jh)^2}{2}. \tag{4.43}$$

As $x, t \in [jh, (j+1)h)$, the terms; $(x - ih)$ and $(t - ih)$ are always less than h.

So,

$$\left\|k_{ik}(x,t) - \tilde{k}_{ik}(x,t)\right\| \leq M_3\frac{h^2}{2} + M_4\frac{h^2}{2}. \tag{4.44}$$

Likewise, the error between $h_{iq}(x,t)$ and its TF approximation on the jth subinterval is derived as

$$\left\|h_{iq}(x,t) - \tilde{h}_{iq}(x,t)\right\| \leq M_5\frac{h^2}{2} + M_6\frac{h^2}{2}, \tag{4.45}$$

where $M_5 = \left\| \left(\frac{d^2 h_{iq}(x,t)}{dx^2} \right)_{x=jh,t=jh} \right\|$, $M_6 = \left\| \left(\frac{d^2 h_{iq}(x,t)}{dt^2} \right)_{x=jh,t=jh} \right\|$.

In the TF domain, Equation (4.12) is rewritten as

$$\begin{aligned}
\lambda_1 \tilde{y}_i(x) &= \lambda_1 \tilde{Y}_0 + \lambda_2 J^{\alpha_i}\left(\tilde{g}_i(x)\right) \\
&+ \lambda_3 \sum_{k=1}^{n} \left(\frac{1}{\Gamma(\alpha_i)} \int_0^{\tau} (\tau - x)^{\alpha_i - 1} \left(\int_0^1 \tilde{k}_{ik}(x,t) F_{ik}\left(t, \tilde{y}_1(t), \ldots\ldots, \tilde{y}_n(t)\right) dt \right) dx \right) \\
&+ \lambda_4 \Gamma(1-\beta) \sum_{q=1}^{n} \left(\frac{1}{\Gamma(1+\alpha_i-\beta)} \int_0^x \frac{\tilde{h}_{iq}(x,t)}{(x-t)^{\beta-\alpha_i}} G_{iq}\left(t, \tilde{y}_1(t), \ldots\ldots, \tilde{y}_n(t)\right) dt \right).
\end{aligned} \tag{4.46}$$

We now define the error between $y_i(x)$ and $\tilde{y}_i(x)$ on the jth subinterval,

$$\varepsilon = \left\| y_i(x) - \tilde{y}_i(x) \right\|, \; x \in [jh, (j+1)h). \tag{4.47}$$

Multiplying Equation (4.47) by λ_1,

$$\begin{aligned}
\lambda_1 \varepsilon &= \lambda_1 \left\| y_i(x) - \tilde{y}_i(x) \right\| = |\lambda_1| \left\| Y_0 - \tilde{Y}_0 \right\| + \left| \lambda_2 \left\| \frac{1}{\Gamma(\alpha_i)} \int_{jh}^{(j+1)h} (x-t)^{\alpha_i - 1} \left\| g_i(t) - \tilde{g}_i(t) \right\| dt \right\| \right| \\
&+ \left| \lambda_3 \sum_{k=1}^{n} \left(\frac{1}{\Gamma(\alpha_i)} \int_{jh}^{(j+1)h} ((j+1)h - x)^{\alpha_i - 1} \left(\sum_{p=0}^{m-1} \int_{ph}^{(p+1)h} \left\| k_{ik}(x,t) - \tilde{k}_{ik}(x,t) \right\| \left\| F_{ik}(y_1(t), \ldots, y_n(t)) \right. \right. \right. \right. \\
&\qquad \left. \left. \left. \left. - F_{ik}\left(\tilde{y}_1(t), \ldots, \tilde{y}_n(t)\right) \right\| dt \right) dx \right) \right| \\
&+ \left| \lambda_4 \Gamma(1-\beta) \sum_{q=1}^{n} \left(\frac{1}{\Gamma(1+\alpha_i-\beta)} \int_{jh}^{(j+1)h} \frac{\left\| h_{iq}(x,t) - \tilde{h}_{iq}(x,t) \right\|}{((j+1)h - t)^{\beta - \alpha_i}} \left\| G_{iq}(y_1(t), \ldots, y_n(t)) \right. \right. \right. \\
&\qquad \left. \left. \left. - G_{iq}\left(\tilde{y}_1(t), \ldots, \tilde{y}_n(t) \right) \right\| dt \right) \right|.
\end{aligned} \tag{4.48}$$

Using Equations (4.34), (4.35), (4.44), and (4.45) in Equation (4.48),

$$
\begin{aligned}
\lambda_1\varepsilon = \lambda_1\left\|y_i(x)-\tilde{y}_i(x)\right\| \le & \left|\lambda_1\left(M_2\frac{h^2}{2}\right)\right| + \left|\lambda_2\left\|\frac{1}{\Gamma(\alpha_i)}\int\limits_{jh}^{(j+1)h}(x-t)^{\alpha_i-1}\left(M_1\frac{h^2}{2}\right)dt\right\|\right| \\
& + \left|\lambda_3\sum_{k=1}^{n}\left(\frac{1}{\Gamma(\alpha_i)}\int\limits_{jh}^{(j+1)h}((j+1)h-x)^{\alpha_i-1}\left(\sum_{p=0}^{m-1}\int\limits_{ph}^{(p+1)h}\left(M_3\frac{h^2}{2}+M_4\frac{h^2}{2}\right)L_1\left\|Y(t)-\tilde{Y}(t)\right\|dt\right)dx\right)\right| \\
& + \left|\lambda_4\Gamma(1-\beta)\sum_{q=1}^{n}\left(\frac{1}{\Gamma(1+\alpha_i-\beta)}\int\limits_{jh}^{(j+1)h}\frac{\left(M_5\frac{h^2}{2}+M_6\frac{h^2}{2}\right)}{((j+1)h-t)^{\beta-\alpha_i}}L_2\left\|Y(t)-\tilde{Y}(t)\right\|dt\right)\right| = \left|\lambda_1\left(M_2\frac{h^2}{2}\right)\right| \\
& + \left|\lambda_2\left(M_1\frac{h^2}{2}\right)\frac{h^{\alpha_i}}{\Gamma(\alpha_i+1)}\left((j+1)^{\alpha_i}-j^{\alpha_i}\right)\right| \\
& + \left|\lambda_3L_1\left(M_3\frac{h^2}{2}+M_4\frac{h^2}{2}\right)\sum_{k=1}^{n}\left(\frac{1}{\Gamma(\alpha_i)}\int\limits_{jh}^{(j+1)h}((j+1)h-x)^{\alpha_i-1}\left(\sum_{p=0}^{m-1}\int\limits_{ph}^{(p+1)h}\left\|Y(t)-\tilde{Y}(t)\right\|dt\right)dx\right)\right| \\
& + \left|\lambda_4\Gamma(1-\beta)L_2\left(M_5\frac{h^2}{2}+M_6\frac{h^2}{2}\right)\sum_{q=1}^{n}\left(\frac{1}{\Gamma(1+\alpha_i-\beta)}\int\limits_{jh}^{(j+1)h}\frac{1}{((j+1)h-t)^{\beta-\alpha_i}}\left\|Y(t)-\tilde{Y}(t)\right\|dt\right)\right|, \\
= & \left|\lambda_1\left(M_2\frac{h^2}{2}\right)\right| + \left|\lambda_2\left(M_1\frac{h^2}{2}\right)\frac{h^{\alpha_i}}{\Gamma(\alpha_i+1)}\left((j+1)^{\alpha_i}-j^{\alpha_i}\right)\right| \\
& + \left|\lambda_3L_1\left(M_3\frac{h^2}{2}+M_4\frac{h^2}{2}\right)\sum_{k=1}^{n}\left(\frac{1}{\Gamma(\alpha_i)}\int\limits_{jh}^{(j+1)h}((j+1)h-x)^{\alpha_i-1}\left(\sum_{p=0}^{m-1}\int\limits_{ph}^{(p+1)h}\sum_{r=1}^{n}\left\|y_r(t)-\tilde{y}_r(t)\right\|dt\right)dx\right)\right| \\
& + \left|\lambda_4\Gamma(1-\beta)L_2\left(M_5\frac{h^2}{2}+M_6\frac{h^2}{2}\right)\sum_{q=1}^{n}\left(\frac{1}{\Gamma(1+\alpha_i-\beta)}\int\limits_{jh}^{(j+1)h}\frac{1}{((j+1)h-t)^{\beta-\alpha_i}}\sum_{r=1}^{n}\left\|y_r(t)-\tilde{y}_r(t)\right\|dt\right)\right|, \\
= & \left|\lambda_1\left(M_2\frac{h^2}{2}\right)\right| + \left|\lambda_2\left(M_1\frac{h^2}{2}\right)\frac{h^{\alpha_i}}{\Gamma(\alpha_i+1)}\left((j+1)^{\alpha_i}-j^{\alpha_i}\right)\right| \\
& + \left|\lambda_3L_1\left(M_3\frac{h^2}{2}+M_4\frac{h^2}{2}\right)(n\varepsilon mh)\sum_{k=1}^{n}\left(\frac{1}{\Gamma(\alpha_i)}\int\limits_{jh}^{(j+1)h}((j+1)h-x)^{\alpha_i-1}dx\right)\right| \\
& + \left|\lambda_4\Gamma(1-\beta)L_2\left(M_5\frac{h^2}{2}+M_6\frac{h^2}{2}\right)n\varepsilon\sum_{q=1}^{n}\left(\frac{1}{\Gamma(1+\alpha_i-\beta)}\int\limits_{jh}^{(j+1)h}\frac{1}{((j+1)h-t)^{\beta-\alpha_i}}dt\right)\right|, \\
= & \left|\lambda_1\left(M_2\frac{h^2}{2}\right)\right| + \left|\lambda_2\left(M_1\frac{h^2}{2}\right)\frac{h^{\alpha_i}}{\Gamma(\alpha_i+1)}\left((j+1)^{\alpha_i}-j^{\alpha_i}\right)\right| \\
& + \left|\lambda_3L_1\left(M_3\frac{h^2}{2}+M_4\frac{h^2}{2}\right)(n\varepsilon mh)\sum_{k=1}^{n}\left(\frac{h^{\alpha_i}}{\Gamma(\alpha_i+1)}\left((j+1)^{\alpha_i}-j^{\alpha_i}\right)\right)\right| \\
& + \left|\lambda_4\Gamma(1-\beta)L_2\left(M_5\frac{h^2}{2}+M_6\frac{h^2}{2}\right)n\varepsilon\sum_{q=1}^{n}\left(\frac{h^{1+\alpha_i-\beta}}{\Gamma(1+\alpha_i-\beta+1)}\left((j+1)^{1+\alpha_i-\beta}-j^{1+\alpha_i-\beta}\right)\right)\right|, \\
= & \left|\frac{\lambda_1M_2h^2}{2}\right| + \left|\frac{\lambda_2M_1h^{2+\alpha_i}}{2\Gamma(\alpha_i+1)}\left((j+1)^{\alpha_i}-j^{\alpha_i}\right)\right| + \left|\lambda_3L_1(M_3+M_4)\varepsilon n^2\frac{h^{2+\alpha_i}}{2\Gamma(\alpha_i+1)}\left((j+1)^{\alpha_i}-j^{\alpha_i}\right)\right| \\
& + \left|\lambda_4\Gamma(1-\beta)L_2(M_5+M_6)n^2\varepsilon\frac{h^{3+\alpha_i-\beta}}{2\Gamma(2+\alpha_i-\beta)}\left((j+1)^{1+\alpha_i-\beta}-j^{1+\alpha_i-\beta}\right)\right|.
\end{aligned}
\tag{4.49}
$$

As the step size, h, approaches zero,

$$\lim_{h \to 0} \lambda_1 \varepsilon = 0. \tag{4.50}$$

Thus, we have theoretically proved that the proposed TFs-based numerical method converges the approximate solution to the exact solution of the system of fractional-order Fredholm-Volterra-Hammerstein integro-differential equations in Equation (4.12).

4.4 Numerical Experiments

In this section, the effectiveness of the proposed TFs-based numerical method is demonstrated through case studies. Table 4.1 provides the step size and computation time needed by the proposed method.

Case study 4.1: Fractional-order Fredholm-Hammerstein integro-differential equation

Consider the following fractional-order Fredholm-Hammerstein integro-differential equation

$$D^{5/3}y(x) - \int_0^1 (x+t)^2 (f(t))^3 dt = \frac{6}{\Gamma(1/3)} \sqrt[3]{x} - \frac{x^2}{7} - \frac{x}{4} - \frac{1}{9},\ x \in [0,1], \tag{4.51}$$

with the exact solution $f(x) = x^2$.

The absolute errors produced by the proposed method are compared in Table 4.2 with those reported in [1]. It can be noticed that TF approximate solution is more accurate.

Case study 4.2: Fractional-order Volterra-Fredholm integro-differential equation

The fractional-order Volterra-Fredholm integro-differential equation with weak singularity is

TABLE 4.1

CPU usage required by our method

Case study	m	CT (seconds)
4.1	1500	364.685779
4.2	600	18.379803
	800	41.060861

TABLE 4.2

Absolute errors for Case study 4.1

	$\varepsilon(x) = \lvert y(x) - \tilde{y}(x)\rvert$	
x	**Method in Ref. [1]**	**Proposed method**
0.2	2.3540e-004	1.5145e-05
0.4	1.9681e-003	2.7330e-05
0.6	2.5632e-004	4.1346e-05
0.8	3.5128e-004	4.8516e-05
1	1.3321e-003	7.9939e-05

$$D^{0.15}y(x) = \frac{1}{4}\int_0^x \frac{y(t)}{(x-t)^{1/2}}dt + \frac{1}{7}\int_0^1 e^{x+t}y(t)dt + g(x),\ x \in [0,1], \tag{4.52}$$

where $g(x) = \frac{\Gamma(3)}{\Gamma(2.85)}x^{1.85} - \frac{\Gamma(2)}{\Gamma(1.85)}x^{0.85} - \sqrt{\pi}\frac{\Gamma(3)x^{2.5}}{4\Gamma(3.5)} + \sqrt{\pi}\frac{\Gamma(2)x^{1.5}}{4\Gamma(2.5)} - \frac{(e^{x+1}-3e^{x})}{7}$.
The given problem has the actual solution $y(x) = x(x-1)$.

Table 4.3 compares the approximate solution obtained by the proposed method with the numerical solution computed by employing Legendre wavelets method (LWM) and Legendre wavelets collocation method (LWCM) [2], second Chebyshev wavelets method (SCW) [3] and Cos and Sin (CAS) wavelets method [4], respectively. It is seen from Table 4.3 that the proposed numerical method outperforms LWM, LWCM, SCW, and CAS.

Case study 4.3: Fractional-order population growth model

Consider the fractional population growth model of a species within a closed system. The model is given by the following nonlinear fractional Volterra integro-differential equation.

$$D^{\mu}p(t) = ap(t) - bp^2(t) - cp(t)\int_0^t p(x)dx,\ p(0) = 0.1,\ t \in [0,5], \tag{4.53}$$

where $p(t)$ is the population of identical individuals at time t, μ is a constant describing the order of the fractional-order derivative, $a > 0$ is the birth rate coefficient, $b > 0$ is the crowding coefficient and c is the toxicity coefficient denoting the behavior of the population evolution before its level drops to zero in the long run. If $c = 0$, Equation (4.53) turns to be the famous logistic equation. The individual death rate is proportional to the integral term of Equation (4.53) and so the population death rate due to toxicity must include factor μ. As the system is closed,

TABLE 4.3

Comparison between approximate and exact solution (Case study 4.2)

x	SCW in Ref. [3]	CAS in Ref. [4]	Proposed method	ES
0	-3.4910e-03	9.0900e-04	2.717e-13	0
1/6	−1.3912e-01	−1.3345e-01	−0.1388864	−1.3889e-01
2/6	−2.2230e-01	−2.1534e-01	−0.2222194	−2.2222e-01
3/6	−2.4999e-01	−2.4266e-01	−0.2499968	−2.5000e-01
4/6	−2.2216e-01	−2.1514e-01	−0.2222184	−2.2222e-01
5/6	−1.3878e-01	−1.3269e-01	−0.1388844	−1.3889e-01
x	**LWM in Ref. [2]**	**LWCM in Ref. [2]**	**Proposed method**	**ES**
0	0.0000	0.0002	2.705e-13	0.0000
1/8	−0.1090	−0.1088	−0.109373	−0.1094
2/8	−0.1869	−0.1861	−0.187498	−0.1875
3/8	−0.2338	−0.2320	−0.234373	−0.2344
4/8	−0.2498	−0.2497	−0.249998	−0.2500
5/8	−0.2337	−0.2332	−0.234372	−0.2344
6/8	−0.1869	−0.1862	−0.187497	−0.1875
7/8	−0.1084	−0.1081	−0.109372	−0.1093

ES: exact solution

the presence of the toxic term always causes the population level to decrease to zero ultimately. The relative size of the sensitivity to toxins determines the manner in which the population evolves before its extinction.

Using the following nondimensional variables for time and population in Equation (4.53), we get the nondimensional population growth model as given in Equation (4.54).

$$t = \frac{tc}{b},\ u = \frac{pb}{a}, \tag{4.54}$$

$$kD^{\mu}u(t) = u(t) - \mu^2(t) - u(t)\int_0^t u(x)dx, \tag{4.55}$$

where $k = c/(ab)$ is dimensionless parameter.

The closed form expression for the maximum value of $u(t)$ is derived in [5] as given below.

TABLE 4.4
Maximum value of $u(t)$

	Approximation u_{max}				
k	VIM in Ref. [6]	HPM in Ref. [7]	Proposed method	Exact $u_{\max}$	CT (seconds)
0.04	0.86124018	0.86124017	0.87372009	0.87371998	97.320960
0.02	0.90383805	0.90383805	0.92342749	0.92342717	113.360808
0.1	0.76511308	0.76511308	0.76974134	0.76974149	84.860805
0.2	0.65791230	0.65791230	0.65905008	0.65905038	88.108040
0.5	0.48528234	0.48528234	0.48519016	0.48519029	84.348369

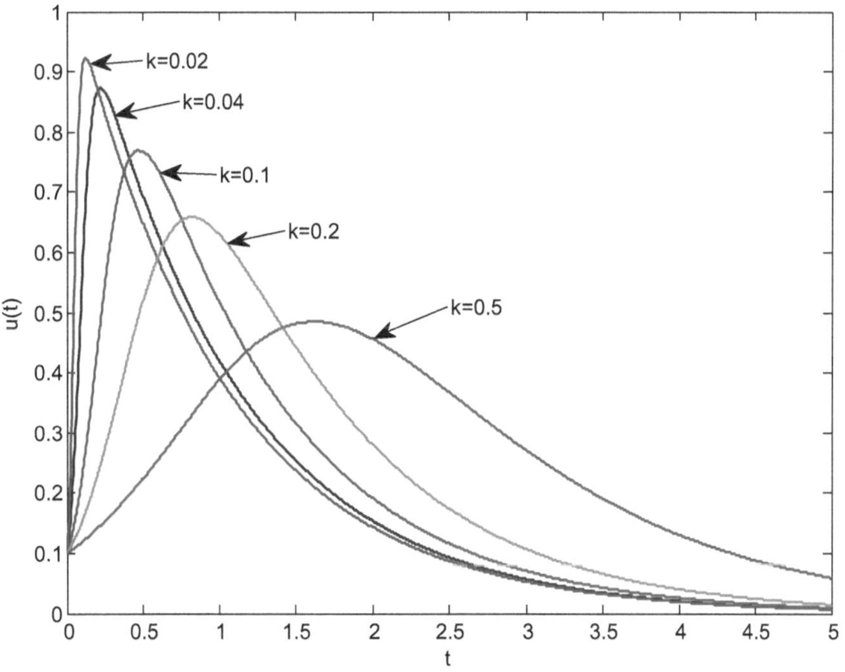

FIGURE 4.1
Approximated population profiles for different k

$$u_{\max} = 1 + k \ln\left(\frac{k}{1 + k - u(0)}\right), \; u(0) = 0.1. \tag{4.56}$$

The step size of 0.005 is used in simulations. The maximum value of $u(t)$ obtained for different values of k and $\mu = 1$ by the proposed method is compared (Table 4.4) with those obtained by the Variational Iteration Method (VIM) in [6], and by the homotopy perturbation method (HPM) in [7]. The proposed TFs-based numerical method yields more accurate approximations than the semianalytical techniques; VIM and HPM. The approximated dynamic behavior of the population $u(t)$ for various values of dimensionless parameter k is displayed in Figure 4.1, and the effect of fractional-order μ on the population of identical individuals is depicted in Figure 4.2.

Case study 4.4: Fractional-order integro-differential equations in anomalous diffusion process

The fractional-order integro-differential equations often arise in anomalous diffusion process which is characterized by comparatively slow decay of

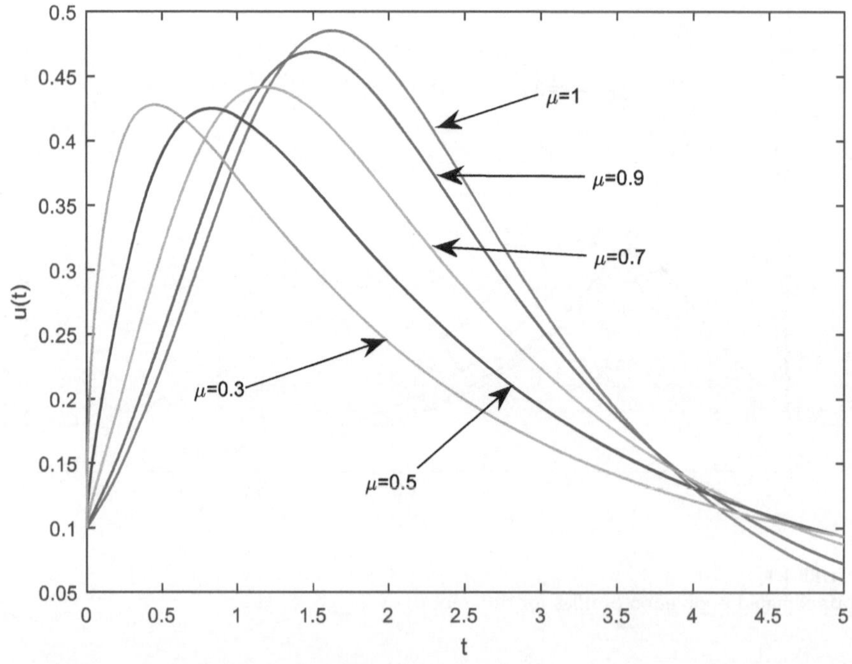

FIGURE 4.2
Approximated population profiles for various μ and $k = 0.5$

autocorrelation function that differs it from the classical Brownian motion [8].

Let us consider the following Abel-Volterra integro-differential equation directly related to the anomalous diffusion and to the particle motion under the joint action of an external force excitation and Basset-type viscous damping.

$$\frac{dv(t)}{dt} = f(t) + \lambda \int_0^t \frac{1}{\sqrt{t-\tau}} v(\tau)d\tau,\ v(0) = 0, \tag{4.57}$$

where $f(t) = \frac{1}{1+t} - \left[-4\sqrt{t} + 2\sqrt{1+t}\ln\left(1 + 2\sqrt{\frac{t}{1+t}} + 2t\left(1 + \sqrt{\frac{t}{1+t}}\right)\right)\right]$.

The exact solution is $v(t) = \ln(1+t)$.

The TF solution obtained with step size of 0.001 is compared with the actual solution in Table 4.5.

The Abel-Volterra fractional-order integro-differential equation given below is representative of a large class of viscoelastic media, which

TABLE 4.5

TF and exact solution

t	Equation (4.57)		Equation (4.58)	
	Exact solution	**TF solution**	**Exact solution**	**TF solution**
0	0	–4.512731e-14	1	0.9999999999
0.1	0.0953101798	0.0953101416	1.1051709180	1.1051721761
0.2	0.1823215567	0.1823215049	1.2214027581	1.2214053104
0.3	0.2623642644	0.2623641997	1.3498588075	1.3498626736
0.4	0.3364722366	0.3364721580	1.4918246976	1.4918299079
0.5	0.4054651081	0.4054650138	1.6487212707	1.6487278706
0.6	0.4700036292	0.4700035171	1.8221188003	1.8221268529
0.7	0.5306282510	0.5306281184	2.0137527074	2.0137622959
0.8	0.5877866649	0.5877865086	2.2255409284	2.2255521589
0.9	0.6418538861	0.6418537027	2.4596031111	2.4596161150
1	0.6931471805	0.6931469657	2.7182818284	2.7182967655
		Performance Indices		
L1	N/A	3.658569e-06	N/A	2.5561617e-04
L2	N/A	2.147767e-07	N/A	1.4937042e-05
CT	N/A	50.190007	N/A	31.992863

particularly attracts an attention with respect to geophysical wave propagation and power-law attenuation.

$$\frac{d^2x(t)}{dt^2} = f(t) + \lambda \int_0^t \frac{1}{\sqrt{t-\tau}} x(\tau)d\tau,\ x(0) = 1,\ x'(0) = 1, \tag{4.58}$$

where $f(t) = e^t\left[1 - \sqrt{\Pi} erf\left(\sqrt{t}\right)\right]$.

The true solution is $x(t) = \exp(t)$.

The step size of 0.001 is used to compute the TF approximate solution, which is given in Table 4.5 along with actual solution. The TF solution accuracy can be improved by considering a smaller step size.

4.5 MATLAB® Codes for Numerical Experiments

This section provides the source codes developed in MATLAB. An initial guess x0 is required to be given as input to all programs.

Program 4.1

```
%%%%%%%%%%%%%%%%%%%%%%%%%%%%%%% MATLAB code for Case study 4.1 %
  %%%%%%%%%%%%%
function [ee1]= Casestudy_4_1(x0)
t0=0;T=1;aa=100;m=aa*10;h=(T-t0)/m;t=[t0:h:T];
I=ones(1,length(t));
tic
CC1=(6/gamma(1/3))*(t.^(1/3))-(1/7)*t.*t-0.25*t-(1/9)*I;
  C01=CC1(1:end-1);
D01=CC1(2:end);CC1=t.*t;C014=CC1(1:end-1);D014=CC1
  (2:end);CC1=I;
C015=CC1(1:end-1);D015=CC1(2:end);CC1=t;C016=CC1
  (1:end-1);D016=CC1(2:end);
[P1alph1,P2alph1]=TOF1(t0,T,m,1);[P3alph1,P4alph1]=
  TOF12(t0,T,m,1);
[P1alph2,P2alph2]=TOF1(t0,T,m,5/3);[P3alph2,P4alph2]=
  TOF12(t0,T,m,5/3);
x=fsolve(@Problem_Fun,x0);
function f=Problem_Fun(x)
for i=1:1:length(t)
  C(i)=x(i);
end
C1=C(1:end-1);D1=C(2:end);CC1=(C.^3);C011=CC1(1:end-1);
  D011=CC1(2:end);
CC1=t.*t.*(C.^3);C012=CC1(1:end-1);D012=CC1(2:end);
CC1=t.*(C.^3);
```

```
C013=CC1(1:end-1);D013=CC1(2:end);c1=((C011)*P1alph1+
  (D011)*P3alph1);
c2=((C011)*P2alph1+(D011)*P4alph1);c3=[c1 c2(end)];
c41=(c3(end)-c3(1));c1=((C012)*P1alph1+(D012)*P3alph1);
c2=((C012)*P2alph1+(D012)*P4alph1);c3=[c1 c2(end)];
c42=(c3(end)-c3(1));c1=((C013)*P1alph1+(D013)*P3alph1);
c2=((C013)*P2alph1+(D013)*P4alph1);c3=[c1 c2(end)];
c43=(c3(end)-c3(1));C02=c41*C014+c42*C015+2*c43*C016;
D02=c41*D014+c42*D015+2*c43*D016;
f1=((C01)*P1alph2+(D01)*P3alph2)+((C02)*P1alph2+
  (D02)*P3alph2)-C1;
f2=((C01)*P2alph2+(D01)*P4alph2)+((C02)*P2alph2+
  (D02)*P4alph2)-D1;
f=[f1 f2];
end
toc
exact=t.*t;  % exact solution
ee1=abs(x-exact);  % absolute error
end
% Algorithm ends
```

Program 4.2

```
%%%%%%%%%%%%%%%%%%%%%%%%%% MATLAB code for Case study 4.2 %
%%%%%%%%%%%%%
function [ee1]= Casestudy_4_2(x0)
t0=0;T=1;
% tt=[0 1/6 2/6 3/6 4/6 5/6 6/6];
% h=1/600;
% t1=[tt(1):h:tt(2)];
% t2=[tt(2):h:tt(3)];
% t3=[tt(3):h:tt(4)];
% t4=[tt(4):h:tt(5)];
% t5=[tt(5):h:tt(6)];
% t6=[tt(6):h:tt(7)];
% t=[t1 t2(2:end) t3(2:end) t4(2:end) t5(2:end) t6(2:end)];
tt=[0 1/8 2/8 3/8 4/8 5/8 6/8 7/8 8/8];h=1/800;
t1=[tt(1):h:tt(2)];t2=[tt(2):h:tt(3)];t3=[tt(3):h:
  tt(4)];t4=[tt(4):h:tt(5)];
t5=[tt(5):h:tt(6)];t6=[tt(6):h:tt(7)];t7=[tt(7):h:
  tt(8)];t8=[tt(8):h:tt(9)];
t=[t1 t2(2:end) t3(2:end) t4(2:end) t5(2:end) t6(2:end)
  t7(2:end) t8(2:end)];
m=length(t)-1;alpha=0.15;I=ones(1,length(t));
  tic
CC1=(gamma(3)/gamma(2.85))*(t.^(1.85))-(gamma(2)/
  gamma(1.85))*(t.^(0.85))-(sqrt(pi)/2)*(1/gamma(3.5))*
  (t.^2.5)+0.25*sqrt(pi)*(1/gamma(2.5))*(t.^1.5)-(1/7)*
  (exp(t+I)-3*exp(t));
```

```
C01=CC1(1:end-1);D01=CC1(2:end);CC1=exp(t);C03=CC1
  (1:end-1);D03=CC1(2:end);t=7/8;[P1alph1,P2alph1]=
  TOF1(t0,T,m,1);[P3alph1,P4alph1]=TOF12(t0,T,m,1);
[P1alph2,P2alph2]=TOF1(t0,T,m,alpha);[P3alph2,P4alph2]=
 TOF12(t0,T,m,alpha);
[P1alph3,P2alph3]=TOF1(t0,T,m,alpha+0.5);
[P3alph3,P4alph3]=TOF12(t0,T,m,alpha+0.5);
x=fsolve(@Problem_Fun,x0);
function f=Problem_Fun(x)
for i=1:1:length(t)
  C(i)=x(i);
end
C1=C(1:end-1);D1=C(2:end);CC1=exp(t).*C;C02=CC1
  (1:end-1);D02=CC1(2:end);
c1=((C02)*P1alph1+(D02)*P3alph1);c2=((C02)*P2alph1+
  (D02)*P4alph1);
c3=[c1 c2(end)];c4=(c3(end)-c3(1));
f1=0.25*gamma(0.5)*((C1)*P1alph3+(D1)*P3alph3)+c4*(1/7)*
  ((C03)*P1alph2+(D03)*P3alph2)+((C01)*P1alph2+(D01)*
  P3alph2)-C1;
f2=0.25*gamma(0.5)*((C1)*P2alph3+(D1)*P4alph3)+c4*(1/7)*
  ((C03)*P2alph2+(D03)*P4alph2)+((C01)*P2alph2+(D01)*
  P4alph2)-D1;f=[f1 f2];
end
toc
exact=t.*(t-1); % Exact Solution
ee1=abs(x-exact);  % absolute error
end
```

Program 4.3

```
%%%%%%%%%%%%%%%%%%%%%%%%% MATLAB code for Case study 4.3 %
%%%%%%%%%%%%%
function [umax,error,x]= Casestudy_4_3(x0)
  t0=0;T=5;aa=100;m=aa*10;h=(T-t0)/m;t=[t0:h:T];
kk1=0.02;mu0=0.1;
FO=1; % fractional order
tic
CC1=mu0*ones(1,length(t));C01=CC1(1:end-1);D01=
  CC1(2:end);
[P1alph1,P2alph1]=TOF1(t0,T,m,1);[P3alph1,P4alph1]=
  TOF12(t0,T,m,1);
[P1alph2,P2alph2]=TOF1(t0,T,m,FO);[P3alph2,P4alph2]=
  TOF12(t0,T,m,FO);
x=fsolve(@Problem_Fun,x0);
function f=Problem_Fun(x)
for i=1:1:length(t)
  C(i)=x(i);
end
```

```
C1=C(1:end-1);D1=C(2:end);C02=((C1)*P1alph1+(D1)*
  P3alph1);
D02=((C1)*P2alph1+(D1)*P4alph1);CC=[C02 D02(end)];CC1=
  (1/kk)*C.*CC;
C03=CC1(1:end-1);D03=CC1(2:end);CC2=(1/kk)*(C-C.*C);
C04=CC2(1:end-1);
D04=CC2(2:end);
f1=C01+((C04)*P1alph2+(D04)*P3alph2)-((C03)*P1alph2+
  (D03)*P3alph2)-C1;
f2=D01+((C04)*P2alph2+(D04)*P4alph2)-((C03)*P2alph2+
  (D03)*P4alph2)-D1;
f=[f1 f2];
end
toc
[umax,Indx]=max(x);
% approximated umax uex=1+kk1*log(kk1/(1+kk1-mu0));
% exact umax error=abs(uex-umax);
end
```

Program 4.4

```
%%%%%%%%%% MATLAB code for Case study 4.4 (Equation (4.57))
%%%%%%%%%%%
function [x1,exact1,norm2,norminf]= Casestudy_4_4_
  PartI(x0)t0=0;T=1;aa=100;m=aa*10;h=(T-t0)/m;t=[t0:h:T];
I=ones(1,length(t));
tic
aa1=I+2*sqrt((t./(I+t)))+2*t.*[1+sqrt(t./(I+t))];
CC1=(I./(I+t))-[-4*sqrt(t)+2*sqrt(t+I).*log(aa1)];
C01=CC1(1:end-1);
D01=CC1(2:end);[P1alph1,P2alph1]=TOF1(t0,T,m,1);
[P3alph1,P4alph1]=TOF12(t0,T,m,1);[P1alph2,P2alph2]=
  TOF1(t0,T,m,1.5);
[P3alph2,P4alph2]=TOF12(t0,T,m,1.5);x=fsolve
  (@Proble_Fun,x0);
function f=Problem_Fun(x)
for i=1:1:length(t)
  C(i)=x(i);
end
C1=C(1:end-1);D1=C(2:end);
f1=((C01)*P1alph1+(D01)*P3alph1)+gamma(1/2)*((C1)*
  P1alph2+(D1)*P3alph2)-C1;
f2=((C01)*P2alph1+(D01)*P4alph1)+gamma(1/2)*((C1)*
  P2alph2+(D1)*P4alph2)-D1;
f=[f1 f2];
end
toc
exact=log(1+t);error=abs(exact-x);
for i=1:11
  b1=aa*(i-1)+1;x1(i)=x(b1);exact1(i)=exact(b1);
```

```
end
x1=x1';exact1=exact1';norm2=norm(error,2);
norminf=norm(error,inf);
end
```

Program 4.6

```
%%%%%%%%%% MATLAB code for Case study 4.4 (Equation (4.58))
%%%%%%%%%%%
function [x1,exact1,norm2,norminf]= Casestudy_4_4_PartII
  (x0)t0=0;T=1;aa=100;m=aa*10;h=(T-t0)/m;t=[t0:h:T];
I=ones(1,length(t));
tic
CC1=exp(t).*(I-sqrt(pi)*erf(sqrt(t)));C02=CC1(1:end-1);
D02=CC1(2:end);
CC2=I+t;C01=CC2(1:end-1);D01=CC2(2:end);[P1alph1,
  P2alph1]=TOF1(t0,T,m,2);
[P3alph1,P4alph1]=TOF12(t0,T,m,2);[P1alph2,P2alph2]=
  TOF1(t0,T,m,2.5);
[P3alph2,P4alph2]=TOF12(t0,T,m,2.5);x=fsolve
  (@Problem_Fun,x0);
function f=Problem_Fun(x)
for i=1:1:length(t)
  C(i)=x(i);
end
C1=C(1:end-1);D1=C(2:end);
f1=C01+((C02)*P1alph1+(D02)*P3alph1)+gamma(1/2)*
  ((C1)*P1alph2+(D1)*P3alph2)-C1;
f2=D01+((C02)*P2alph1+(D02)*P4alph1)+gamma(1/2)*
  ((C1)*P2alph2+(D1)*P4alph2)-D1;
f=[f1 f2];
end
toc
exact=exp(t);error=abs(exact-x);
for i=1:11
  b1=aa*(i-1)+1;x1(i)=x(b1);exact1(i)=exact(b1);
end
x1=x1';exact1=exact1';norm2=norm(error,2);norminf=
  norm(error,inf);
end
```

References

[1] J. Hou, B. Qin, C. Yang (2012). Numerical solution of nonlinear Fredholm integrodifferential equations of fractional order by using hybrid functions and the collocation method. *J. Appl. Math.*, vol. 2012, Article ID 687030, 11 pages.

[2] M. Yi, L. Wang, J. Huang (2016). Legendre wavelets method for the numerical solution of fractional integro-differential equations with weakly singular kernel. *Appl. Math. Model.*, vol. 40, pp. 3422–3437.
[3] Y. Wang, L. Zhu (2016). SCW method for solving the fractional integro-differential equations with a weakly singular kernel. *Appl. Math. Comput.*, vol. 275, pp. 72–80.
[4] M.X. Yi, J. Huang (2015). CAS wavelet method for solving the fractional integro-differential equation with a weakly singular kernel. *Int. J. Comput. Math.*, vol. 92(8), 1715–1728.
[5] K. TeBeest (1997). Numerical and analytical solutions of Volterra's population model. *SIAM Rev.*, vol. 39, pp. 484–493.
[6] Y. Alwesabi, A.A. Dahawi, Y.S. Daniel, A.H.M. Murid (2014). Analytical solution of Volterra's population model using Variation Iteration Method (VIM). In *Proceedings of 1st International Conference of Recent Trends in Information and Communication Technologies*.
[7] S.T. Mohyud-Din, A. Yildirim, Y. Gulkanat (2010). Analytical solution of Volterra's population model. *J. King Saud Univ.*, vol. 22, pp. 247–250.
[8] V.K. Singh, E.B. Postnikov (2013). Operational matrix approach for solution of integro-differential equations arising in theory of anomalous relaxation processes in vicinity of singular point. *Appl. Math. Model.*, vol. 37, pp. 6609–6616.

5

Numerical Method for Simulation of Physical Processes Represented by Stiff and Nonstiff Fractional-Order Differential Equations, and Differential-Algebraic Equations

This chapter deals with computing the numerical solution of stiff and nonstiff fractional-order ordinary differential and differential-algebraic equations. It has been proved that numerous physical processes in varied applied areas of science and engineering such as electrochemistry, physics, geology, astrophysics, seismic wave analysis, sound wave propagation, psychology and life sciences, biology, and so on can be better described by the mathematical models involving fractional differential and/or differential-algebraic equations than the integer order models [1–9]. Gaining insights into the aforementioned physical processes having inherent fractional-order description demands the exact solutions of the respective fractional-order models. Nonetheless, it is extremely difficult (even impossible in some circumstances) to find analytical solution of those fractional-order differential equations. This drawback limits the practical applications of the subject (fractional calculus). To enable the subject to be successfully applied in the said application areas, some analytical methods [10] were proposed but their range of applicability is very narrow, therefore, several numerical methods [11–16], each having its own advantages and disadvantages, were devised. It is well known that no analytical or numerical method can solve all classes (nonstiff and stiff) of fractional-order differential equations. Some real-world processes are inherently very complex and highly nonlinear, especially biological processes, that pose computational challenges which the available semianalytical or numerical methods may not meet. Significant efforts continue to develop more efficient and reliable numerical methods which are able to precisely approximate the solution of fractional-order differential and differential-algebraic equations. From this standpoint, we propose a numerical method to solve system of fractional-order ordinary differential equations of the following form:

$$
{}_0^C D_t^{\alpha} y_i(t) = \sum_{k=1}^{\lambda} b_k {}_0^C D_t^{\beta_k} y_i(t) + F(t, y_1(t), y_2(t), \ldots, y_N(t)) + f_i(t), \qquad n-1 \le \alpha n,\ n \in Z^+, \tag{5.1}
$$

where $i = 1, 2, \ldots, N$ (N is an integer), $y_i(t)$ is the ith unknown function, $F_i(t, y_1(t), y_2(t), y_3(t), \ldots, y_N(t))$ can be linear or nonlinear, $f_i(t)$ is the given function, and ${}_0^C D_t^{\alpha}$ is the Caputo fractional-order derivative.

Our numerical technique is based on triangular orthogonal functions presented in Chapter 1. The proposed numerical method will be tested on linear and nonlinear fractional-order differential equations. We shall also try to apply the proposed method on a special class of fractional-order differential equations (FDEs), that is, fractional-order differential-algebraic equations (FDAEs). In [17–20], the semianalytical techniques, the homotopy analysis method, the Adomian decomposition method, the variational iteration method, the fractional differential transform method, and the iterative decomposition method (also the new iterative method) were extended to solve linear and nonlinear FDAEs. Jaradat et al. [21] proposed a hybrid of the Laplace transform and the homotopy analysis method, and Ding and Jiang [22] employed waveform relaxation method for the approximate solution of FDAEs. In the works cited earlier, no attempt was made to simulate physical process models involving stiff fractional-order (or integer-order) differential equations or stiff fractional-order (or integer-order) differential-algebraic equations. It is one of our objectives to find stable and accurate approximate solutions to stiff FDEs and FDAEs in the desired interval.

5.1 Existence and Uniqueness of Solution

It is proved in this section that the fractional-order nonlinear differential equation conferred in Equation (5.1) has a unique solution under certain conditions.

For $N = 1$, Equation (5.1) becomes

$$ {}_0^C D_t^{\alpha} y(t) = \sum_{k=1}^{\lambda} b_k {}_0^C D_t^{\beta_k} y(t) + F(t, y(t)) + f(t),\ n-1 \leq \alpha < n,\ n \in Z^+. \tag{5.2} $$

The initial conditions are $y^{(s)}(0) = a_s$, $s = 0, 1 \ldots, n-1$.

Here $t \in [0, T]$, $y(t) : [0, T] \to [0, \infty)$, $F(t, y(t)) : [0, \infty) \to [0, \infty)$, $f(t) : [0, T] \to [0, \infty)$.

An equivalent integral form of Equation (5.2) is given by

$$ y(t) = Y_1(t) + \sum_{k=1}^{\lambda} b_k {}_0 J_t^{\alpha-\beta_k} y(t) + {}_0 J_t^{\alpha} F(t, y(t)) + g(t), \tag{5.3} $$

where $Y_1(t) = \sum_{p=0}^{n-1} \frac{t^p}{p!} y^{(p)}(0)$, $g(t) = {}_0 J_t^{\alpha} f(t) - {}_0 J_t^{\alpha-\beta_k} Y_2(t)$, $Y_2(t) = \sum_{k=1}^{\lfloor \beta_k \rfloor} b_k \frac{t^k}{k!} y^{(k)}(0)$.

Any function satisfying Equation (5.3) also satisfies the above integral equation.

Let us define Banach space as

$$C = [0, T] \times [0, \infty), \tag{5.4}$$

with metric $d : [0, \infty) \times [0, \infty) \to [0, \infty)$ defined as the distance function, $d(y(t), \tilde{y}(t)) = \|y(t) - \tilde{y}(t)\|$.

We assume that the function F satisfies the following condition of Lipschitz continuity.

$$d(F(t, y(t)), F(t, \tilde{y}(t))) \leq Ld(y(t), \tilde{y}(t)), \tag{5.5}$$

where L is a Lipschitz constant, $L \in (0, 1)$.

By considering the following successive approximations, we prove that Equation (5.3) has a unique solution.

$$\phi_0 = Y_1(t), \tag{5.6}$$

$$\phi_{k+1}(t) = \phi_0 + \sum_{k=1}^{\lambda} b_{k0} J_t^{\alpha-\beta_k} \phi_k + {}_0J_t^{\alpha} F(t, \phi_k) + g(t),\ k = 0, 1, 2, 3, \ldots. \tag{5.7}$$

Let us define the Picard operator on Banach Space C as

$$\Gamma\phi = \phi_0 + \sum_{k=1}^{\lambda} b_{k0} J_t^{\alpha-\beta_k} \phi_k + {}_0J_t^{\alpha} F(t, \phi_k) + g(t). \tag{5.8}$$

We now show that the Picard operator is a contraction on Banach space. Let t be such that $\|(\Gamma\phi_2 - \Gamma\phi_1)t\| = \|\Gamma\phi_2(t) - \Gamma\phi_1(t)\|$.

$$\begin{aligned}
&\|(\Gamma\phi_2(t) - \Gamma\phi_1(t))\| \\
&\leq \left| \frac{1}{\Gamma(\alpha)} \int_0^t |t-\tau|^{\alpha-1} \|F(\tau, \phi_2(\tau)) - F(\tau, \phi_1(\tau))\| d\tau + \sum_{k=1}^{\lambda} b_{k0} J_t^{\alpha-\beta_k} (\|\varphi_2(t) - \varphi_1(t)\|) \right|, \\
&\leq \left| \frac{1}{\Gamma(\alpha)} \int_0^t |t-\tau|^{\alpha-1} L \|\varphi_2(t) - \varphi_1(t)\| d\tau + \sum_{k=1}^{\lambda} b_k \frac{1}{\Gamma(\alpha-\beta_k)} \int_0^t |t-\tau|^{\alpha-\beta_k-1} \|\varphi_2(t) - \varphi_1(t)\| d\tau \right|, \\
&\leq \left| \frac{L\|\varphi_2(t) - \varphi_1(t)\|}{\Gamma(\alpha)} \int_0^t |t-\tau|^{\alpha-1} d\tau + \sum_{k=1}^{\lambda} b_k \frac{\|\varphi_2(t) - \varphi_1(t)\|}{\Gamma(\alpha-\beta_k)} \int_0^t |t-\tau|^{\alpha-\beta_k-1} d\tau \right|, \\
&\leq \left| \frac{L\|\varphi_2(t) - \varphi_1(t)\|}{\Gamma(\alpha+1)} |t|^{\alpha} + \sum_{k=1}^{\lambda} b_k \frac{\|\varphi_2(t) - \varphi_1(t)\|}{\Gamma(\alpha-\beta_k+1)} |t|^{\alpha-\beta_k+1} \right|, \\
&\leq \Omega \|\varphi_2(t) - \varphi_1(t)\|,
\end{aligned} \tag{5.9}$$

where $\Omega = \left| \frac{L}{\Gamma(\alpha+1)} |t|^{\alpha} + \sum_{k=1}^{\lambda} b_k \frac{1}{\Gamma(\alpha-\beta_k+1)} |t|^{\alpha-\beta_k+1} \right|$.

If $\Omega < 1$, by contraction mapping theorem, Equation (5.2) has a unique solution on $[0, T]$.

5.2 The Proposed Numerical Method

Recalling Equation (5.3),

$$y(t) = Y_1(t) + \sum_{k=1}^{\lambda} b_k {}_0J_t^{\alpha-\beta_k} y(t) + {}_0J_t^{\alpha} F(t, y(t)) + g(t). \tag{5.10}$$

Using Equations (1.68) and (1.73),

$$y(t) \approx C^T T1_m(t) + D^T T2_m(t), \tag{5.11}$$

$$Y_1(t) \approx C_1^T T1_m(t) + D_1^T T2_m(t), \tag{5.12}$$

$$g(t) \approx C_2^T T1_m(t) + D_2^T T2_m(t), \tag{5.13}$$

$$F(t, y(t)) \approx C_3^T T1_m(t) + D_3^T T2_m(t). \tag{5.14}$$

Substituting Equations (5.11) to (5.14) and replacing the Riemann-Liouville fractional-order integral with its TF estimate (Equation (1.97)) in Equation (5.10),

$$\begin{aligned} &C^T T1_m(t) + D^T T2_m(t) \\ &= C_1^T T1_m(t) + D_1^T T2_m(t) + (C_3^T P_1^{\alpha} + D_3^T P_3^{\alpha}) T1_m(t) + (C_3^T P_2^{\alpha} + D_3^T P_4^{\alpha}) T2_m(t) \\ &\quad + \sum_{k=1}^{\lambda} b_k \left(\left(C^T P_1^{\alpha-\beta_k} + D^T P_3^{\alpha-\beta_k} \right) T1_m(t) + \left(C^T P_2^{\alpha-\beta_k} + D^T P_4^{\alpha-\beta_k} \right) T2_m(t) \right) \\ &\quad + C_2^T T1_m(t) + D_2^T T2_m(t). \end{aligned} \tag{5.15}$$

Equating the coefficients of LHTF and RHTF vector,

$$C^T = C_1^T + +(C_3^T P_1^{\alpha} + D_3^T P_3^{\alpha}) + \sum_{k=1}^{\lambda} b_k \left(C^T P_1^{\alpha-\beta_k} + D^T P_3^{\alpha-\beta_k} \right) + C_2^T, \tag{5.16}$$

$$D^T = D_1^T + \left(C_3^T P_{\frac{\alpha}{2}} + D_3^T P_{\frac{\alpha}{4}}\right) + \sum_{k=1}^{\lambda} b_k\left(C^T P_{\frac{\alpha-\beta_k}{2}} + D^T P_{\frac{\alpha-\beta_k}{4}}\right) + D_2^T. \quad (5.17)$$

This system of algebraic equations can be solved by using any algebraic equation solver. From Equations (5.11), (5.16), and (5.17), we can have the TF approximate solution to nonlinear fractional-order ordinary differential equation in Equation (5.1).

5.3 Convergence Analysis

In this section, we show that the TF approximate solution converges to the exact solution of fractional-order nonlinear ordinary differential equations in Equation (5.1).

Let $\tilde{y}(t)$ be the TF estimate for the exact solution, $y(t)$, of the fractional-order nonlinear ordinary differential equation in Equation (5.1).

The error between the exact solution and TF approximate solution is defined on the jth subinterval $[jh, (j+1)h)$ as

$$\varepsilon_j = y(t) - \tilde{y}(t),\ t \in [jh, (j+1)h),\ j = 0, 1, \ldots, m-1. \quad (5.18)$$

Using Equation (1.68),

$$\varepsilon_j = y(t) - \left[y(jh) + \tfrac{y((j+1)h)-y(jh)}{h}(t-jh)\right] = y(t) - \left[y(jh) + \tfrac{dy(t)}{dt}\Big|_{t=jh}(t-jh)\right]. \quad (5.19)$$

The Taylor series expansion of $y(t)$ around the center jh is

$$y(t) = y(jh) + \tfrac{dy(t)}{dt}\Big|_{t=jh}(t-jh) + \tfrac{d^2y(t)}{dt^2}\Big|_{t=jh}\tfrac{(t-jh)^2}{2!} + \sum_{q=0}^{\infty}\left(\tfrac{d^qy(t)}{dt^q}\Big|_{t=jh}\tfrac{(t-jh)^q}{q!}\right). \quad (5.20)$$

Considering the second-order Taylor series approximation for $y(t)$ and employing it in Equation (5.19),

$$\varepsilon_j = \tfrac{d^2y(t)}{dt^2}\Big|_{t=jh}\tfrac{(t-jh)^2}{2!} = y''(jh)\tfrac{(t-jh)^2}{2!}. \quad (5.21)$$

Let us make the following assumption

$$M = \max\{|y''(0), y''(h), \ldots, y''(mh)|\}. \quad (5.22)$$

We now calculate $\left\|\varepsilon_j\right\|_1$ on the jth subinterval,

$$\left\|\varepsilon_j\right\|_1 = \int_{jh}^{(j+1)h} \left|\varepsilon_j\right| dt = \int_{jh}^{(j+1)h} \left|y''(jh)\right| \frac{(t-jh)^2}{2} dt = M \int_{jh}^{(j+1)h} \frac{(t-jh)^2}{2} dt = \frac{Mh^3}{6}. \quad (5.23)$$

Let ε_m be the sum of errors, ε_j,

$$\varepsilon_m = \sum_{j=0}^{m-1} \varepsilon_j. \quad (5.24)$$

Calculating $\|\varepsilon_m\|_1$,

$$\|\varepsilon_m\|_1 = \int_0^T |\varepsilon_m| dt = \int_0^T \left(\sum_{j=0}^{m-1} \left|\varepsilon_j\right|\right) dt = \sum_{j=0}^{m-1} \left(\int_0^T \left|\varepsilon_j\right| dt\right) = \sum_{j=0}^{m-1} \left\|\varepsilon_j\right\|_1 = \frac{mTMh^3}{6} = \frac{TM}{6m^2}. \quad (5.25)$$

Taking limit,

$$\lim_{m\to\infty} \left\|\varepsilon_j\right\|_1 = \lim_{m\to\infty} \frac{M}{m^2} = 0. \quad (5.26)$$

Therefore,

$$\lim_{m\to\infty} \varepsilon_m = 0\cdot \quad (5.27)$$

5.4 Numerical Experiments

In this section, the proposed numerical method is applied to different types of FDEs and FDAEs to show its effectiveness and applicability.

5.4.1 Investigation of Validity and Accuracy

Example 5.1: Simple linear multiorder fractional differential equation (FDE)

Consider the following linear multiorder FDE:

$${}_0^C D_t^{\alpha} y(t) + {}_0^C D_t^{\beta} y(t) + y(t) = f(t),\ y(0) = 0,\ y'(0) = 0 \quad (5.28)$$

The exact solution is $y(t) = t^3\cdot$

We consider the following two cases.

Case 1

$$\alpha = 2, \beta = 0.5, f(t) = t^3 + 6t + \left({}^{3.2}\!/\!_{\Gamma(0.5)}\right)t^{2.5}. \tag{5.29}$$

The multiorder FDE in Equation (5.28) with the above parameters is solved using the proposed numerical method with a step size of 0.1. The absolute error between the exact solution and the piecewise linear TF solution is computed. The ∞-norm of the absolute error is compared in Table 5.1 with the maximal absolute error produced in [23–25]. It is evident in Tables 5.1 and 5.2 that our numerical method is able to give more accurate approximate solution even with small number of subintervals $m = 10$, that is, $h = 0.1$ than those achieved in [23–25].

Case 2

$$\alpha = 2, \beta = 0.75, f(t) = t^3 + 6t + \frac{8.533333333}{\Gamma(0.25)}t^{2.25}. \tag{5.30}$$

The maximal absolute error attained by proposed numerical method is compared (Table 5.3) with that got in [23] and [24]. In this case, too, the proposed numerical method provides superior result with the step size of 0.1. The time elapsed during the computation of the TF solution of Equation (5.28) (case 2) is noted and tabulated in Table 5.2. The linear multiorder FDE in Equation (5.28) is also solved using the numerical method based on block pulse functions (BPF) using the same step size that is used in the proposed method.

TABLE 5.1

Error analysis for Example 5.1

Method	$\|\|\varepsilon(t)\|\|_\infty$
Proposed method	1.18e-14
BPFs based method	0.16331
Method in Ref. [24]	1.86e-09
Method 1a in Ref. [23]	2.96e-04
Method 1b in Ref. [23]	2.71e-04
Method 2 in Ref. [23]	1.79e-05
Method 3 in Ref. [23]	2.96e-04
Method 1a(2) in Ref. [23]	8.14e-07
Method 3(2) in Ref. [23]	8.14e-07
Method in [25]	8.55e–10

TABLE 5.2
Computational time needed by TFs

		CT (seconds)	
Example	h	**Case 1**	**Case 2**
5.1	1/10	0.237018	0.289367
5.2	1/500	11.222759	12.064899
5.3	1/500	11.377852	11.250266

TABLE 5.3
Error analysis for Example 5.1

Method	h	$\|\varepsilon(t)\|_\infty$
Proposed method	1/10	5.79e-12
BPFs based method	1/10	0.163105
Method in Ref. [24]	1/512	1.86e-09
Method 1a in Ref. [23]	1/512	3.54e-03
Method 1b in Ref. [23]	1/512	5.93e-05
Method 2 in Ref. [23]	1/512	1.18e-04
Method 3 in Ref. [23]	1/512	5.43e-04
Method 1a(2) in Ref. [23]	1/512	3.10e-06
Method 3(2) in Ref. [23]	1/512	5.07e-06

Example 5.2: Complex linear high-order FDE

The linear multiorder FDE is

$$a\,{}_0^C D_t^{\alpha} x(t) + b\,{}_0^C D_t^{\alpha_2} x(t) + c\,{}_0^C D_t^{\alpha_1} x(t) + e x(t) = \frac{2b}{\Gamma(3-\alpha_2)} t^{2-\alpha_2} + \frac{2c}{\Gamma(3-\alpha_1)} t^{2-\alpha_1} - A1,$$
$$A1 = \frac{c}{\Gamma(2-\alpha_1)} t^{1-\alpha_1} + e(t^2 - t), x(0) = 0, x'(0) = -1, x''(0) = 2, x'''(0) = 0. \tag{5.31}$$

The exact solution for Equation (5.31) is $x(t) = t^2 - t$.

The given multiorder FDE is solved using the proposed method as well as the BPFs-based method for the following two cases:

Case 1

$$a = 1,\ b = 1,\ c = 1,\ e = 1,\ \alpha_1 = 0.77,\ \alpha_2 = 1.44,\ \alpha = 3.91. \tag{5.32}$$

Case 2

$$a = 1,\ b = 1,\ c = 0.5,\ e = 0.5,\ \alpha_1 = \sqrt{2}/_{20},\ \alpha_2 = \sqrt{2},\ \alpha = \sqrt{11}. \qquad (5.33)$$

In [26], the linear multiorder FDE in Equation (5.31) is solved using ET (Euler's method with product trapezoidal quadrature formula) and ER (Euler's method with product rectangle rule) methods using the step size of 0.001. Comparing with the results of [26], our numerical method provides a better result (Tables 5.2 and 5.4) with a smaller number of subintervals.

Example 5.3: Complex linear low-order FDE

The linear multiorder FDE is

$$aD^2x(t) + bDx(t) + c\,{}_0^C D_t^{\alpha_2} x(t) + e\,{}_0^C D_t^{\alpha_1} x(t) + kx(t) = f(t),\ x(0) = 1,\ x'(0) = 0, \qquad (5.34)$$

where $f(t) = a + bt + \frac{c}{\Gamma(3-\alpha_2)} t^{2-\alpha_2} + \frac{e}{\Gamma(3-\alpha_1)} t^{2-\alpha_1} + k(1 + 0.5t^2)$.

The given linear nonhomogenous multiorder FDE has the analytical solution $x(t) = 1 + 0.5t^2$.

We consider the following two cases:

Case 1

$$a = 1,\ b = 3,\ c = 2,\ e = 1,\ k = 5,\ \alpha_1 = 0.0159,\ \alpha_2 = 0.1379. \qquad (5.35)$$

Case 2

$$a = 0.2,\ b = 1,\ c = 1,\ e = 0.5,\ k = 2,\ \alpha_1 = 0.00196,\ \alpha_2 = 0.07621. \qquad (5.36)$$

The accuracy of the TFs approximate solution obtained in both cases (Table 5.5; and the respective elapsed times are given in Table 5.2) is

TABLE 5.4

Error analysis for Example 5.2

		$\|\varepsilon(t)\|_\infty$	
Method	h	**Case 1**	**Case 2**
PM	**1/500**	**2.6650e-05**	**7.006e-06**
BBM	1/500	0.02110912	0.0063250
ET	1/1000	0.00099806	0.0009958
ER	1/1000	0.00095385	0.0009805

PM: proposed method, BBM: BPFs-based method

TABLE 5.5

Error analysis for Example 5.3

		$\|\varepsilon(t)\|_{\infty}$	
Method	h	**Case 1**	**Case 2**
PM	**1/500**	**1.8413e-07**	**1.96584e-07**
BBM	$1/500^{a}, 1/1000^{b}$	0.00100066	0.00501666
PECE	$1/1000$	0.000409626	0.000437974

a: case 1, b: case 2

better than the accuracy of the numerical solution acquired by Predict Evaluate-Correct Evaluate method (PECE) in [27] and BBM.

Example 5.4: Nonlinear multiorder FDE

The nonlinear multiorder FDE is

$$a_0 D_t^{\alpha} x(t) + b_0^C D_t^{\alpha_2} x(t) + c_0^C D_t^{\alpha_1} x(t) + e(x(t))^3 = f(t), x(0) = 0, x'(0) = 0, x''(0) = 0, \tag{5.37}$$

where $f_2(t) = \left(2at^{3-\alpha}/\Gamma(4-\alpha)\right) + \left(2b/\Gamma(4-\alpha_2)\right)t^{3-\alpha_2} + \left(2c/\Gamma(4-\alpha_1)\right)t^{3-\alpha_1} + e\left(t^3/3\right)^3$.

The exact solution for the following two cases is $x(t) = t^3/3$.

Case 1

$$a = 1,\ b = 2,\ c = 0.5,\ e = 1,\ \alpha_1 = 0.00196,\ \alpha_2 = 0.07621,\ \alpha = 2. \tag{5.38}$$

Case 2

$$a = 1,\ b = 0.1,\ c = 0.2,\ e = 0.3,\ \alpha_1 = \sqrt{5}/5,\ \alpha_2 = \sqrt{2}/2,\ \alpha = 2. \tag{5.39}$$

Using the step size of 1/10, the nonlinear nonhomogenous multiorder FDE in Equation (5.37) is solved for the above two cases. Table 5.6 presents the maximal absolute errors gained via our method and those obtained by BBM, ET, and ER methods in [26], through ADM and modified PNM in [28] and via a numerical method (NM) transforming the fractional-order differential equations into a system of first-order ordinary differential equations in [29]. In terms of accuracy and computation time (Table 5.7), the performance of our numerical method is far superior to that of ET, ER, PNM, ADM, and NM.

Example 5.5: Linear multiorder FDE with variable coefficients [27]

The linear multiorder FDE with variable coefficients is

TABLE 5.6
Error analysis for Example 5.4

		$\|\varepsilon(t)\|_\infty$	
Method	h	**Case 1**	**Case 2**
PM	**1/10**	**4.607425e-15**	**2.997602e-15**
BBM	1/10	0.0544348769	0.0549991496
ET	1/1000	0.0008924007	0.0009717941
ER	1/1000	0.0007891357	0.0009438396
PNM	1/2000	0.000399235	0.000388881
ADM*	N/A	0.000150218	5.74351e-06
NM	1/2000	9.39e-05	2.6866e-04

*The ADM series solution is truncated to first four terms.

TABLE 5.7
CPU time

	CT (seconds)	
Method	**Case 1**	**Case 2**
PM	**0.244999**	**0.313121**
PNM	894.999	952.907
ADM	478.577	505.954
NM	5.9380	5.9060

$$\begin{aligned} aD^2x(t)+b(t)Dx(t)+c(t)\,{}_0^CD_t^{\alpha_2}x(t)+e(t)\,{}_0^CD_t^{\alpha_1}x(t)+k(t)x(t)&=f(t),\\ x(0)=2,\ x'(0)=0,& \end{aligned} \tag{5.40}$$

where $f(t)=-a-b(t)t-\frac{c(t)t^{2-\alpha_2}}{\Gamma(3-\alpha_2)}-\frac{e(t)t^{2-\alpha_1}}{\Gamma(3-\alpha_1)}+k(t)\left(2-0.5t^2\right)$.

The given problem possesses the closed form solution $x(t)=2-0.5t^2$.

Case 1

$$a=0.1,\ b(t)=t,\ c(t)=1+t,\ e(t)=t^2,\ k(t)=(1+t)^2,\ \alpha_1=0.781,\ \alpha_2=0.891. \tag{5.41}$$

Case 2

$$a=5,\ b(t)=\sqrt{t},\ c(t)=t^2-t,\ e(t)=3t,\ k(t)=t^3-t,\ \alpha_1=\sqrt{7}/_{70},\ \alpha_2=\sqrt{13}/_{13}. \tag{5.42}$$

As per Tables 5.8 and 5.9, TF solutions are better than ET and ER solutions. The successful implementation of proposed method on different forms of multi-order FDEs makes sure that the method has a broad range of applicability.

Example 5.6: Linear fractional-order differential-algebraic equation (FDAEs) [22]

The fractional-order linear differential-algebraic equations is

$$ {}_0^C D_t^{0.5} x_1(t) + 2x_1(t) - \tfrac{\Gamma(3.5)}{2} x_2(t) + x_3(t) = 2t^{2.5} + \sin t, \tag{5.43}$$

$$ {}_0^C D_t^{0.5} x_2(t) + x_2(t) + x_3(t) = \tfrac{2}{\Gamma(2.5)} t^{1.5} + t^2 + \sin t,\ t \in [0,1], \tag{5.44}$$

$$ 2x_1(t) + x_2(t) - x_3(t) = 2t^{2.5} + t^2 - \sin t,\ x_1(0) = 0,\ x_2(0) = 0,\ x_3(0) = 0. \tag{5.45}$$

The exact solution of this problem is $x_1(t) = t^{2.5}$, $x_2(t) = t^2$, $x_3(t) = \sin t$.

The proposed TFs-based numerical method as well as the numerical-based BPFs are applied to linear fractional-order differential-algebraic equations. As shown in Tables 5.10 and 5.11, the TF solution is more accurate than the BPF solution. The step size of 0.001 is used in TFs-based and BPFs-based numerical method.

Example 5.7: Nonlinear FDAEs [22]

The fractional-order nonlinear differential-algebraic equations is

$$ {}_0^C D_t^{0.5} x_1(t) + x_1(t)x_2(t) - x_3(t) = \tfrac{6}{\Gamma(3.5)} t^{2.5} + 2t^4 + t^7 - e^t - \sin t,\ t \in [0,1], \tag{5.46}$$

$$ \begin{aligned} {}_0^C D_t^{0.5} x_2(t) - \frac{\Gamma(5)}{\Gamma(4.5)} t^{0.5} x_1(t) + 2x_2(t) + x_1(t)x_3(t) \\ = \frac{2}{\Gamma(1.5)} t^{0.5} + 4t + 2t^4 + t^3 e^t + t^4 \sin t, \end{aligned} \tag{5.47}$$

$$ x_1^2(t) - x_2(t)t^2 + x_3(t) = e^t + t\sin t - 2t^3,\ x_1(0) = 0,\ x_2(0) = 0,\ x_3(0) = 1. \tag{5.48}$$

TABLE 5.8

CPU time for Example 5.5

	CT (seconds)	
h	**Case 1**	**Case 2**
1/500	21.105298	29.008036

TABLE 5.9

Error analysis of Example 5.5

		$\|\varepsilon(t)\|_\infty$	
Method	*h*	**Case 1**	**Case 2**
PM	**1/500**	**1.3473590e-07**	**4.00000e-06**
ET in Ref. [26]	1/1000	0.0003154278	0.00050354
ER in Ref. [26]	1/1000	0.0000278985	0.000484466

We have the analytical solution, $x_1(t) = t^3$, $x_2(t) = 2t + t^4$, $x_3(t) = e^t + t \sin t$, for the given fractional-order nonlinear differential-algebraic equation.

Using the step size of 0.001, the TF approximate and the BPF approximate solution are obtained (Tables 5.12 and 5.13). Our numerical method is more accurate and faster than the BPFs-based numerical method.

Example 5.8: System of nonlinear FDEs [30]

The system of nonlinear fractional-order differential equations is

$$Du_1(t) + \frac{\Gamma(7/3)}{2}\,{}_0^C D_t^{2/3} u_1(t) = 1 + 2t + t^{4/3} + u_1^2(t) - u_2(t),\; u_1(0) = 0, \tag{5.49}$$

$$Du_2(t) + \frac{\Gamma(7/3)}{2}\,{}_0^C D_t^{2/3} u_2(t) = 4tu_1(t) + 12\frac{\Gamma(7/3)}{\Gamma(13/3)}t^{10/3},\; u_2(0) = 1. \tag{5.50}$$

The exact solution is $u_1(t) = t^2$, $u_2(t) = t^4 + 1$.

The results of our method and the BPFs-based numerical method are given in Table 5.14. In this example, too, our method gave a better approximate solution than the BPFs-based numerical method.

5.4.2 Application to Physical Processes Described by FDEs and FDAEs

Application 5.1: Bagley-Torvik equation [26]

Consider the Bagley-Torvik equation

$$aD^2y(t) = -b\,{}_0^C D_t^{3/2} y(t) - cy(t) + c(1+t), t \in [0, 1], \tag{5.51}$$

subject to initial conditions $y(0) = y('0) = 1$.

The exact solution is $y(t) = 1 + \text{t}$.

The given fractional model is solved using our method and the BPFs-based numerical method (Table 5.15). The step size of 0.1 is used in both methods. The approximate solution acquired via our method is far better than the approximate solution obtained by BPFs.

TABLE 5.10
Absolute errors for Example 5.6 (proposed method)

t	$\lvert\varepsilon_1(t)\rvert$	$\lvert\varepsilon_2(t)\rvert$	$\lvert\varepsilon_3(t)\rvert$
0	5.59846479e-13	2.65839586e-11	2.77036516e-11
0.1	5.14698659e-08	5.94931441e-08	1.72432875e-07
0.2	5.58746472e-08	4.68328368e-08	1.58582131e-07
0.3	5.75969211e-08	3.34914296e-08	1.48685272e-07
0.4	5.84764880e-08	2.43165152e-08	1.41269491e-07
0.5	5.89900187e-08	1.74737168e-08	1.35453754e-07
0.6	5.93153753e-08	1.21032707e-08	1.30734021e-07
0.7	5.95332362e-08	7.73656422e-09	1.26803036e-07
0.8	5.96850269e-08	4.09211553e-09	1.23462169e-07
0.9	5.97939315e-08	9.87627868e-10	1.20575490e-07
1	5.94273601e-08	2.17449280e-09	1.16680227e-07
Performance Indices			
L1	5.98690821e-08	1.27273517e-07	1.82621664e-07
L2	1.80533111e-06	1.24189960e-06	4.48758816e-06
CT	119.409179	-	-

Application 5.2: Two-point Bagley-Torvik equation [31]

Consider the two-point fractional Bagley-Torvik equation

$$D^2u(t) + \theta\, {}_0^C D_t^{\alpha} u(t) = g(t) + \theta t^{1-\alpha} p(t), \tag{5.52}$$

where $p(t) = \left(\frac{120}{\Gamma(6-\alpha)}t^4 - \frac{348}{5\Gamma(5-\alpha)}t^3 + \frac{456}{25\Gamma(4-\alpha)}t^2 - \frac{339}{125\Gamma(3-\alpha)}t + \frac{27}{125\Gamma(2-\alpha)}\right)$,

$$g(t) = 20t^3 - \tfrac{174}{5}t^2 + \tfrac{456}{25}t - \tfrac{339}{125},$$

$$\theta = 1.$$

The actual solution is $u(t) = t^5 - \frac{29}{10}t^4 + \frac{76}{25}t^3 - \frac{339}{250}t^2 + \frac{27}{125}t$.

Using the step size of 0.0025, the TF as well as BPF approximations are obtained. Table 5.16 compares the absolute errors of our method with those achieved by using discrete spline function [31], Bessel collocation method [32], and Haar wavelets [33]. The approximate solution obtained by our method is more accurate than the BPF solution and that achieved in [31–33].

Application 5.3: Plant-herbivore model [34]

The following model describes the interaction between a plant and an insect.

TABLE 5.11
Absolute errors for Example 5.6 (BPFs-based method)

t	$\lvert\varepsilon_1(t)\rvert$	$\lvert\varepsilon_2(t)\rvert$	$\lvert\varepsilon_3(t)\rvert$
0	1.44506105e-08	4.47039961e-07	4.75941182e-07
0.1	3.97779987e-05	0.000100403399	0.000179959397
0.2	0.00011213904	0.000200380439	0.000424658523
0.3	0.00020579608	0.000300366994	0.000711959155
0.4	0.00031668169	0.000400357769	0.001033721149
0.5	0.00044244281	0.000500350897	0.001385236518
0.6	0.00058149106	0.000600345507	0.001763327637
0.7	0.00073266008	0.000700341128	0.002165661292
0.8	0.00089504599	0.000800337475	0.002590429468
0.9	0.00106792152	0.000900334365	0.003036177419
1	0.00125068496	0.001000331675	0.003501701601
		Performance Indices	
L1	0.00125068496	0.001000331675	0.003501701601
L2	0.01979868638	0.018280585316	0.057749553896
CT	1240.218884	-	-

$$\frac{dP(t)}{dt} = aP(t)(1 - bP(t)) + cdF(t)A(t) - a_1P(t)L(t),$$

$$\frac{dF(t)}{dt} = b_1P(t) - c_1F(t) - dF(t)A(t), \tag{5.53}$$

$$\frac{dL(t)}{dt} = d_1F(t)A(t) + a_2A(t) - b_2a_1P(t)L(t) - c_2L(t),$$

$$\frac{dA(t)}{dt} = b_2a_1P(t)L(t) - d_2A(t),$$

where P is the density of plant species, L is the density of larval species of insect, A is the density of adult species of insect, F is the density of flower species involving in pollination, a is the plant intrinsic growth rate, b is the plant intraspecific self-regulation coefficient, d denotes the pollination rate, a_1 is the herbivory rate, b_1 is the flower production rate, c_1 is the flower decay rate, c_2 and d_2 are larva and adult mortality rates, c is the plant pollination efficiency ratio, d_1 indicates the adult consumption ratio, and a_2 represents the reproduction rate resulting from the pollination of the other plant species. In comparison to the life cycles of plant species and insects, flowers last for a short time; hence, equating $d/dtF(t)$ to zero.

TABLE 5.12

Absolute errors for Example 5.7 (proposed method)

t	$\lvert\varepsilon_1(t)\rvert$	$\lvert\varepsilon_2(t)\rvert$	$\lvert\varepsilon_3(t)\rvert$
0	2.51942599e-18	5.03921158e-18	0
0.1	2.55205650e-08	1.04930407e-05	1.04981448e-07
0.2	3.41979098e-08	5.32826485e-06	2.13677762e-07
0.3	4.01577795e-08	3.42840129e-06	3.10724638e-07
0.4	4.82906511e-08	2.43279225e-06	3.95427966e-07
0.5	5.96869647e-08	1.80951699e-06	4.67300992e-07
0.6	7.31841570e-08	1.37061290e-06	5.25036208e-07
0.7	8.63288122e-08	1.03264951e-06	5.652198349e-07
0.8	9.63067177e-08	7.53423248e-07	5.808089671e-07
0.9	1.00792959e-07	5.11112959e-07	5.609576430e-07
1	9.85600492e-08	2.97129017e-07	4.942491260e-07
		Performance Indices	
L1	1.00901019e-07	2.88615905e-04	5.808089671e-07
L2	2.15841644e-06	4.946965648e-04	1.380591851e-05
CT	205.238986	-	-

The three-dimensional integer order system is

$$\begin{aligned}\frac{dP(t)}{dt} &= aP(t)(1-bP(t)) + cdF(t)A(t) - a_1P(t)L(t),\\ \frac{dL(t)}{dt} &= d_1F(t)A(t) + a_2A(t) - b_2a_1P(t)L(t) - c_2L(t),\\ \frac{dA(t)}{dt} &= b_2a_1P(t)L(t) - d_2A(t),\end{aligned} \tag{5.54}$$

where $F = (b_1P)/(c_1 + dA)$.

Using the transformations; $x = bP(t)$, $y(t) = bL(t)$, $z(t) = bA(t)$, and $t = a\tau$ in the above system of FDEs, we get the following dimensionless system

$$\begin{aligned}\frac{dx(t)}{dt} &= x(t)(1-x(t)) + \frac{\lambda cx(t)z(t)}{\eta+z(t)} - \beta x(t)y(t),\\ \frac{dy(t)}{dt} &= \frac{\lambda d_1x(t)z(t)}{\eta+z(t)} + \varphi z(t) - b_2\beta x(t)y(t) - \mu y(t),\\ \frac{dz(t)}{dt} &= \beta x(t)y(t) - vz(t),\end{aligned} \tag{5.55}$$

where $\lambda = b_1/a$, $\eta = c_1b/d$, $\beta = a_1/ab$, $\mu = c_2/a$, $v = d_2/a$, $\phi = a_2/a$.

TABLE 5.13

Absolute errors for Example 5.7 (BPFs-based method)

t	$\lvert\varepsilon_1(t)\rvert$	$\lvert\varepsilon_2(t)\rvert$	$\lvert\varepsilon_3(t)\rvert$
0	1.18855302e-05	0.00112604239	1.41265776e-10
0.1	0.02129700059	0.00262332946	0.000384734938
0.2	0.05145247798	0.00474510933	0.001634313469
0.3	0.07818380754	0.00684779299	0.001274480785
0.4	0.09557802934	0.00943783873	0.004608882259
0.5	0.10081373834	0.01259392375	0.018188505686
0.6	0.09404909872	0.01538856947	0.037323862689
0.7	0.07806719530	0.01623829173	0.055416371950
0.8	0.05703975074	0.01409884906	0.064178434994
0.9	0.03497202870	0.00965269481	0.057584857856
1	0.01473878278	0.00553067986	0.034791013710
	Performance Indices		
L1	0.10085737013	0.01629342365	0.064271380550
L2	2.19268792586	0.33688287495	1.1420279955513
CT	8185.474037	-	-

We take the following parameters.

$k = 0.005,\ c = 0.8,\ \lambda = 2.7,\ \eta = 1.1,\ d_1 = 0.7,\ b_2 = 1.4,\ \phi = 2.7,\ \mu = 4,$ $\nu = 1.5, \beta = 10.48, x(0) = 0.48, y(0) = 0.1, z(0) = 0.32.$

As the model has no closed form solution, the numerical solution obtained by Runge-Kutta fourth order method is treated as the true solution. The TF and BPF approximate solution are obtained using the step size of 0.001 and $\alpha_1 = \alpha_2 = \alpha_3 = 1$, compared with the RK solution in Tables 5.17 and 5.18. The computational time needed by our method is larger than the BPFs-based numerical method, but the TF solution accuracy is better than the BPF solution accuracy.

Application 5.4: Financial model [35]

The integer order financial model is

$$\frac{dX(t)}{dt} = Z(t) + (Y(t) - a)X(t),$$

$$\frac{dY(t)}{dt} = 1 - bY(t) - X^2(t), \tag{5.56}$$

TABLE 5.14

Absolute errors for Example 5.8

	Proposed method		BPFs-based method	
t	$\lvert\varepsilon_1(t)\rvert$	$\lvert\varepsilon_2(t)\rvert$	$\lvert\varepsilon_1(t)\rvert$	$\lvert\varepsilon_2(t)\rvert$
0	1.97478254e-14	7.52731210e-14	4.96598639e-07	1.68753899e-14
0.1	2.45119556e-10	8.53095238e-09	0.00010047490	1.45475210e-06
0.2	1.81546478e-09	3.26520570e-08	0.00020063726	1.17351393e-05
0.3	5.82955465e-09	7.11255772e-08	0.00030133165	3.99317724e-05
0.4	1.33471358e-08	1.22838204e-07	0.00040314440	9.53724556e-05
0.5	2.54705739e-08	1.86432343e-07	0.00050688911	0.00018765972
0.6	4.34496283e-08	2.60103194e-07	0.00061364512	0.00032676858
0.7	5.88136607e-08	3.41399182e-07	0.00072483462	0.00052319834
0.8	1.03553516e-07	4.26972799e-07	0.00084235436	0.00078819417
0.9	1.50387221e-07	5.12228619e-07	0.00096878960	0.00113407025
1	2.13169026e-07	5.90795238e-07	0.00110776197	0.00157469941
		Performance Indices		
L1	2.13169026e-07	5.90795238e-07	0.00110776197	0.00157469941
L2	2.47422193e-06	9.24140159e-06	0.01924401935	0.01859375815
CT	198.029121	-	1385.236520	-

$$\frac{dZ(t)}{dt} = -X(t) - cZ(t),$$

where X is the interest rate, Y is the investment demand, and Z is the price index.

The following parameter values are considered for simulations.
$a = 3,\ b = 0.1,\ c = 0.1,\ X(0) = 2,\ Y(0) = 3,\ Z(0) = 2.$

The TF as well BPF solutions (for $q_1 = q_2 = q_3 = 1$) are compared with the numerical solution obtained by the Runge-Kutta (RK) fourth-order method and the error analysis is carried out (Tables 5.19 and 5.20).

Application 5.5: Epidemiological model for computer viruses [36]

A computer virus is a type of deliberately harmful software that is capable of interfering with the correct operation of a computer. Computer viruses first were developed in the 1980s. At first, they could not cause major damage and their spread was somewhat insignificant. However, as computer technology advanced, computer viruses became strong and were able to access personal data, causing huge damage to individuals as well as corporations. An epidemiological model for computer viruses helps us to

TABLE 5.15

Absolute errors for Application 5.1

t	Proposed method	BPFs-based method
0	4.99600361e-15	0.049999998326
0.1	5.66133814e-16	0.049999994820
0.2	4.44089209e-16	0.050000003638
0.3	4.44089209e-16	0.050000005709
0.4	4.44089209e-16	0.049999992387
0.5	0	0.049999992808
0.6	4.44089209e-16	0.049999994397
0.7	0	0.050000004613
0.8	0	0.049999980541
0.9	0	0.050000100636
1	5.66133814e-15	0.050000002457
	Performance Indices	
L1	5.66133814e-15	0.050000100636
L2	8.40036148e-15	0.165831260725
CT	0.243149	1.374632

understand the nature of viruses and the effectiveness of an antivirus program installed in the virus-infected computer. Let us consider a network of computers, some of which are equipped with antivirus programs. The total population of the network is divided into four groups: the population of noninfected computers subjected to possible infection ($S(t)$), the population of noninfected computers equipped with antivirus protection ($A(t)$), the population of infected computers ($I(t)$), and the population of computers removed from the network whether or not they are infected ($R(t)$).

Let us consider the following epidemiological model for computer viruses:

$$
\begin{aligned}
\tfrac{dS(t)}{dt} &= N - \alpha_{SA}S(t)A(t) - \beta S(t)I(t) - \mu S + \sigma R(t),\ S(0) = s_1,\\
\tfrac{dI(t)}{dt} &= \beta S(t)I(t) - \alpha_{IA}A(t)I(t) - \delta I(t) - \mu I,\ I(0) = s_2,\\
\tfrac{dR(t)}{dt} &= \delta I(t) - \sigma R(t) - \mu R,\ R(0) = 0,\\
\tfrac{dA(t)}{dt} &= \alpha_{SA}S(t)A(t) + \alpha_{IA}A(t)I(t) - \mu A,\ A(0) = 0,
\end{aligned}
\tag{5.57}
$$

where N is influx rate, representing the incorporation of new computers to the network, μ is proportion coefficient for the mortality rate, not due to the virus.

TABLE 5.16

Absolute error obtained via various methods

t	Proposed method	BPFs-based method	Method in Ref. [31]	Method in Ref. [32]	Method in Ref. [33]
0.1	3.12e-07	1.76e-05	2.00E-03	1.08E −2	3.59E −3
0.2	7.19e-06	3.28e-05	1.74E-03	8.96E −3	1.58E −3
0.3	1.18e-06	3.8e-05	1.53E-03	3.78E −3	1.79E −3
0.4	1.68e-06	1.65e-05	1.32E-03	1.44E −7	1.63E −3
0.5	2.18e-06	2.26e-06	1.11E-03	1.01E −3	1.16E −3
0.6	2.66e-06	1.06e-05	8.92E-04	5.62E −8	5.84E −4
0.7	3.10e-06	5.02e-05	5.53E-04	1.26E −3	1.27E −4
0.8	3.47e-06	5.15e-06	4.19E-04	1.28E −3	1.19E −4
0.9	3.76e-05	9.34e-05	2.02E-04	2.07E −8	5.54E −4
			Performance Indices		
L1	3.93e-06	1.53e-04	NV	NV	NV
L2	5.50e-05	8.53e-04	NV	NV	NV
CT	18.1706	780.0160	NV	NV	NV

NV: no value was reported in the cited paper.

The population S is infected with a rate that is related to the probability of susceptible computers to establish effective communications with infected ones. Therefore, this rate is proportional to the product SI, with the proportion factor represented by β. Conversion of susceptible to antidotal is proportional to the product SA and is controlled by α_{SA}, that is, an operational parameter defined by the antivirus distribution strategy of the network administration. Infected computers can be fixed by using antivirus programs converted to antidotal ones with a rate proportional to AI, with a proportion factor given by α_{IA}, or become useless and removed with a rate controlled by δ. Removed computers can be restored and converted into susceptible ones with a proportion factor σ. This model represents the dynamics of the propagation of the infection of a known virus and, consequently, the conversion of antidotal into infected is not considered. Therefore, by using this model, a vaccination strategy can be defined, providing an economical use of the antivirus programs. It is assumed that there is no introduction of new computers into the network during the propagation of the considered virus; thus, the influx rate equals zero. As the machine obsolescence time is larger than the time of the virus action, we choose $\mu = 0$.

TABLE 5.17

Error analysis using proposed method (Application 5.3)

t	$\lvert\varepsilon_1(t)\rvert$	$\lvert\varepsilon_2(t)\rvert$	$\lvert\varepsilon_3(t)\rvert$
0	0	1.38777878e-17	0
0.1	3.23105959e-06	2.09320915e-06	4.51336719e-06
0.2	5.48745216e-06	8.47001430e-06	2.34350857e-06
0.3	1.06205560e-05	1.19880103e-05	5.23094067e-06
0.4	5.02459631e-06	5.89062521e-06	4.46885358e-07
0.5	3.20243931e-06	1.79977494e-06	2.41487815e-06
0.6	3.32703890e-06	1.11683195e-06	2.42458928e-06
0.7	3.98172481e-06	1.25019422e-06	1.87335445e-06
0.8	3.57645815e-06	3.86049652e-07	2.22579264e-06
0.9	3.44223434e-06	4.65178760e-08	2.26999390e-06
1	3.53897039e-06	1.09333213e-07	2.05618787e-06
		Performance Indices	
L1	1.95586198e-05	2.252561948e-05	1.919557934e-05
L2	2.36551932e-04	2.505201021e-04	1.743052576e-04
CT	445.005851	-	-

The modified epidemiological model is defined as

$$\begin{aligned} \tfrac{dS(t)}{dt} &= -\alpha_{SA}S(t)A(t) - \beta S(t)I(t) + \sigma R(t),\ t \in [0, 25], \\ \tfrac{dI(t)}{dt} &= \beta S(t)I(t) - \alpha_{IA}A(t)I(t) - \delta I(t), \\ \tfrac{dR(t)}{dt} &= \delta I(t) - \sigma R(t), \\ \tfrac{dA(t)}{dt} &= \alpha_{SA}S(t)A(t) + \alpha_{IA}A(t)I(t). \end{aligned} \tag{5.58}$$

Here, the total population of the network $T = S + I + R + A$.

We take the parameters and the initial conditions
$\alpha_{SA} = -0.025$, $\alpha_{IA} = 0.25$ $\beta = 0.1$, $\sigma = 0.8$, $\delta = 9$, $S(0) = 5$, $I(0) = 95$, $R(0) = 0$, $A(0) = 0$.

The modified computer virus model does not have closed-form solution. Therefore, the model is solved by the RK fourth-order method and the obtained numerical solution is regarded as the original solution to verify the accuracy of the proposed method. The TF approximate solutions are computed using $h = 0.001$ and compared with the RK solution. The 2-norm and infinity-norm of the error between the TF solution and RK solution are

TABLE 5.18

Error analysis using the BPFs-based method (Application 5.3)

t	$\lvert\varepsilon_1(t)\rvert$	$\lvert\varepsilon_2(t)\rvert$	$\lvert\varepsilon_3(t)\rvert$
0	7.83215707e-05	0.000456346062	1.04052476e-05
0.1	0.000242083185	0.000219158071	0.000141698913
0.2	0.000299909610	9.678399770e-05	0.000187940576
0.3	0.000305427251	3.275949251e-05	0.000187630982
0.4	0.000286425974	3.340013340e-07	0.000170472445
0.5	0.000252334711	9.404925780e-06	0.000144373399
0.6	0.000214013413	8.322386129e-06	0.000117514578
0.7	0.000178870651	5.103796118e-06	9.507775661e-05
0.8	0.000149612039	3.459996848e-06	7.840391680e-05
0.9	0.000124591712	2.215292446e-06	5.529466577e-05
1	0.000103204433	1.238522762e-06	5.481470295e-05
		Performance Indices	
L1	3.088314330e-04	4.563460629e-04	1.917106045e-04
L2	0.0071595684073	0.0037477447933	0.00416724113783
CT	232.442715	-	-

computed and tabulated in Table 5.21. It is observed from Table 5.21 and Figure 5.1 that the TF solution is in good accordance with RK solution.

Application 5.6: Chemical Akzo Nobel problem [37]

The Akzo Nobel problem, which was originally formulated in Akzo Nobel Central Research in Arnhem (The Netherlands), represents a chemical reaction involving two species FLB and ZHU. The carbon dioxide is continuously added to this mixture. The reaction scheme is shown in Figure 5.2. Among the resulting species, ZLA is important. Because the process is commercially developed, the names of the species are not real.

The rate equations are given by

$$\begin{aligned} r_1 &= k_1[FLB]^4[CO_2]^{0.5}, \\ r_2 &= k_2[FLBT][ZHU], \\ r_3 &= \frac{k_2}{K}[FLB][ZLA], \\ r_4 &= k_3[FLB][ZHU]^2, \end{aligned} \tag{5.59}$$

TABLE 5.19

Error analysis using the TFs-based method (Application 5.4)

t	$\lvert\varepsilon_1(t)\rvert$	$\lvert\varepsilon_2(t)\rvert$	$\lvert\varepsilon_3(t)\rvert$
0	7.19424519e-14	1.89626092e-13	3.01536573e-13
0.1	4.14954892e-07	9.22758105e-07	4.55879557e-07
0.2	9.60013700e-07	5.22141731e-07	2.00811352e-07
0.3	2.32845009e-06	1.31772787e-06	3.72357321e-07
0.4	2.60179146e-06	3.90890409e-06	8.52900355e-07
0.5	1.72120989e-06	5.45617641e-06	1.01756653e-06
0.6	5.74332778e-07	5.77373732e-06	9.55424525e-07
0.7	5.73354707e-08	5.47423183e-06	8.16811670e-07
0.8	4.80299399e-07	5.00003613e-06	5.78203439e-07
0.9	7.03412389e-07	4.51689911e-06	5.57192959e-07
1	8.29346539e-07	4.03473545e-06	4.93329621e-07
		Performance Indices	
L1	5.12333190e-06	5.77358248e-06	1.34387287e-06
L2	5.82858729e-05	1.41159582e-04	2.31464957e-05
CT	412.896871	-	-

$$r_3 = \frac{k_2}{K}[FLB][ZLA],$$

$$r_5 = k_4[FLB.ZHU]^2[CO_2]^{0.5}.$$

The last reaction in Figure 5.2 describes an equilibrium $K_S = [FLB.ZHU]/[FLB][ZHU]$.

The inflow of carbon dioxide per volume unit is denoted by $F_{in} = klA((p(CO_2)/H) - [CO_2])$, where klA is the mass transfer coefficient, $H$is the Henry constant, $p(CO_2)$ is the partial carbon dioxide pressure, and $p(CO_2)$ is assumed to be independent of $[CO_2]$.

The chemical process commences with by mixing 0.444 mol/liter FLB with 0.007 mol/liter ZHU. The concentration of carbon dioxide at the beginning is 0.00123 mol/liter. Initially, no other species are present. The concentrations [FLB], [CO_2], [FLBT], [ZHU], [ZLA], and [FLB.ZHU] are indicated by $y_1(t)$, $y_2(t)$, $y_3(t)$, $y_4(t)$, $y_5(t)$, $y_6(t)$, respectively.

The following mathematical model describes the rate of change of the concentrations: [FLB], [CO_2], [FLBT], [ZHU], [ZLA], and [FLB.ZHU].

TABLE 5.20

Error analysis using the BPFs-based method (Application 5.4)

t	$\lvert\varepsilon_1(t)\rvert$	$\lvert\varepsilon_2(t)\rvert$	$\lvert\varepsilon_3(t)\rvert$
0	0.000995998264	0.003000492246	0.001999498250
0.1	0.000167837041	0.002939995104	0.001863613535
0.2	0.000571797487	0.002570883934	0.001665116046
0.3	0.001065466926	0.002007143844	0.001426511777
0.4	0.001285179830	0.001411188057	0.001177014921
0.5	0.001296002475	0.000897544636	0.000940922793
0.6	0.001186608315	0.000509393361	0.000732633828
0.7	0.001027337095	0.000241397764	0.000557436471
0.8	0.000860643680	5.840174566e-05	0.000414704643
0.9	0.000707149207	3.688748398e-05	0.000300913766
1	0.000574387435	9.685819512e-05	0.000211643362
		Performance Indices	
L1	0.001311621131	0.003016375068	0.0019994982500
L2	0.030114898394	0.051958271391	0.0369756360541
CT	240.586883	-	-

$$\begin{aligned}
\tfrac{dy_1(t)}{dt} &= -2k_1y_1^4(t)y_2^{0.5}(t) + k_2y_3(t)y_4(t) - \tfrac{k_2}{K}y_1(t)y_5(t) - k_3y_1(t)y_4^2(t),\\
\frac{dy_2(t)}{dt} &= -\,0.5k_1y_1^4(t)y_2^{0.5}(t) - k_3y_1(t)y_4^2(t) - 0.5k_4y_6^2(t)y_2^{0.5}(t)\\
&\quad + klA\Big(\big(p(CO_2)/H\big) - y_2(t)\Big),\\
\tfrac{dy_3(t)}{dt} &= k_1y_1^4(t)y_2^{0.5}(t) - k_2y_3(t)y_4(t) + \tfrac{k_2}{K}y_1(t)y_5(t),\\
\tfrac{dy_4(t)}{dt} &= -k_2y_3(t)y_4(t) + \tfrac{k_2}{K}y_1(t)y_5(t) - 2k_3y_1(t)y_4^2(t),\\
\tfrac{dy_5(t)}{dt} &= k_2y_3(t)y_4(t) - \tfrac{k_2}{K}y_1(t)y_5(t) + k_4y_6^2(t)y_2^{0.5}(t),\\
0 &= K_Sy_1(t)y_4(t) - y_6(t).
\end{aligned} \tag{5.60}$$

The values of parameters are $k_1 = 18.7$, $k_2 = 0.58$, $k_3 = 0.09$, $k_4 = 0.42$, $K_S = 115.83$, $K = 34.4$, $klA = 3.3$, $H = 737$, $p(CO_2) = 0.9$,

and the initial values are $y_1(0) = 0.444$, $y_2(0) = 0.00123$, $y_3(0) = 0$, $y_4(0) = 0.007$, $y_5(0) = 0$, $y_6(0) = K_S \times 0.444 \times 0.007$.

TABLE 5.21

Performance of proposed method for $t \in [0, 25]$

PI	S	I	R	A
L1	0.011467407964	0.006687565561	0.008356660683	0.014269569653
L2	0.412920423681	0.059747679693	0.242294169284	0.579628584324
CT	493.487355	-	-	-

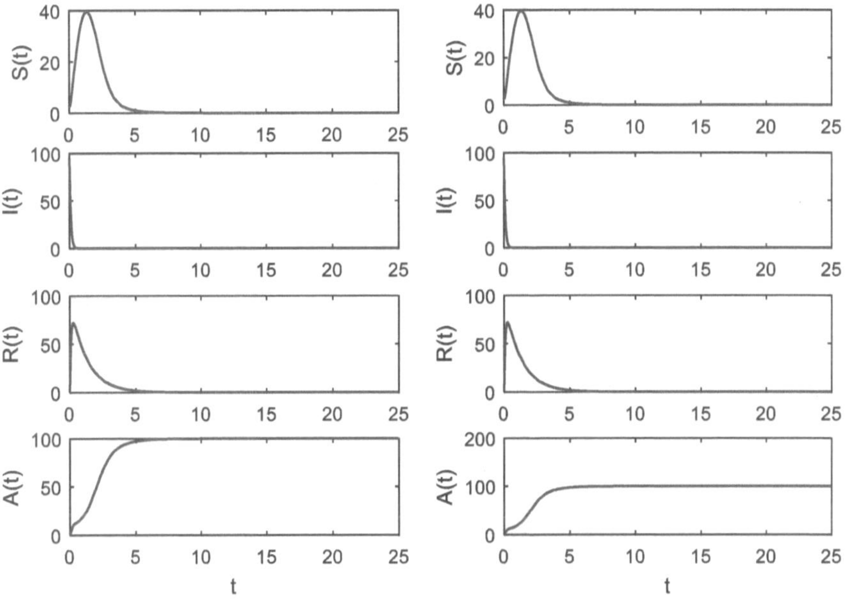

FIGURE 5.1
Numerical solution of modified epidemiological model. First column: TF solution, second column: RK solution

The chemical Akzo nobel problem (CANP) is solved in the interval [0, 1] using the Modified Rosenbrock method of order 2 (MRM2) [38]. The numerical solution obtained by MRM2 is treated as the exact solution. The TF solution obtained in the interval [0, 1] is compared with the MRM2 solution in Tables 5.22 and 5.23. The semianalytical techniques such as the Adomian decomposition method (ADM), the homotopy analysis method (HAM), and the fractional differential transform method with Adomian polynomials (FDTM) [39] are also employed to solve CANP. The 2-norm and maximum norm of the error between the MRM2 solution and each semianalytical solution are calculated and tabulated in Table 5.24. It is seen from Table 5.24

$$2\,\mathrm{FLB} + \frac{1}{2}\mathrm{CO_2} \xrightarrow{k_1} \mathrm{FLBT} + \mathrm{H_2O}$$

$$\mathrm{ZLA} + \mathrm{FLB} \underset{k_2}{\overset{k_2/K}{\rightleftharpoons}} \mathrm{FLBT} + \mathrm{ZHU}$$

$$\mathrm{FLB} + 2\,\mathrm{ZHU} + \mathrm{CO_2} \xrightarrow{k_3} \mathrm{LB} + \mathrm{nitrate}$$

$$\mathrm{FLB.ZHU} + \frac{1}{2}\mathrm{CO_2} \xrightarrow{k_4} \mathrm{ZLA} + \mathrm{H_2O}$$

$$\mathrm{FLB} + \mathrm{ZHU} \rightleftharpoons \mathrm{FLB.ZHU}$$

FIGURE 5.2
Reaction scheme for chemical Akzo Nobel problem

and Figures 5.3, 5.4, and 5.5 that ADM, HAM, and FDTM solution are unrealistic and do not show the true dynamic behavior of CANP. The proposed method is tested further to verify whether it can simulate real process models in the larger time interval [0, 180]. MRM2 solution is also obtained in the same interval and compared with the TF solution (Figures 5.6 and 5.7 and the last row of Table 5.24). In contrast to ADM, HAM, and FDTM, the proposed method has a larger region of convergence.

Application 5.7: Robertson's system

$$\begin{aligned} \tfrac{dy_1(t)}{dt} &= -0.04y_1(t) + 10^4 y_2(t)y_3(t),\ y_1(0) = 1, \\ \tfrac{dy_2(t)}{dt} &= 0.04y_1(t) - 10^4 y_2(t)y_3(t) - 3\times 10^7 y_2^2(t),\ y_2(0) = 0, \\ \tfrac{dy_3(t)}{dt} &= 3\times 10^7 y_2^2(t),\ y_3(0) = 0,\ t\in[0,40]. \end{aligned} \tag{5.61}$$

As can be seen in Figures 5.8 and 5.9 and Table 5.25, in this application as well the proposed method shows superior performance over ADM, HAM, and FDTM.

Application 5.8: High Irradiance Responses (HIRES) of photo morphogenesis [37]

The HIRES problem, which arises from plant physiology, explains how light is involved in morphogenesis. To be exact, it describes the "High Irradiance Responses (HIRES)" of photo morphogenesis on the basis of phytochrome, through a chemical reaction involving eight reactants. Figure 5.10 shows the reaction scheme. Pr and Pfr describe the red and far-red absorbing form of phytochrome, respectively, which are bound by the receptors X and X′ partially influenced by the enzyme *E*.

TABLE 5.22

Performance of the proposed method for Application 5.6

t	$\lvert\varepsilon_1(t)\rvert$	$\lvert\varepsilon_2(t)\rvert$	$\lvert\varepsilon_3(t)\rvert$
0	0	2.1684043449e-19	1.5717114790e-22
0.1	3.4477018323e-06	1.1247220217e-06	1.7239698605e-06
0.2	3.0849887631e-06	9.3953188501e-07	1.5425248716e-06
0.3	1.2889304020e-06	3.6505670270e-07	5.4441117923e-07
0.4	5.7237129852e-07	1.6913472824e-07	3.3610850134e-07
0.5	2.9769270287e-07	4.8633961203e-08	1.4876438306e-07
0.6	1.8362119652e-07	1.1191507633e-08	9.1735711884e-08
0.7	1.6515841955e-07	4.4433070489e-09	8.2515015531e-08
0.8	1.6468386032e-07	3.4802451494e-09	8.2288754279e-08
0.9	1.6685184306e-07	3.3666960940e-09	8.3383434022e-08
1	1.6385047485e-07	1.9392357822e-09	8.1885517465e-08
		Performance Indices	
L1	3.8558960131e-06	1.2256535385e-06	1.92804150207e-06
L2	2.1428191849e-05	5.6485457397e-06	1.07143339351e-05
CT	50.912321		

TABLE 5.23

Performance of the proposed method for Application 5.6

t	$\lvert\varepsilon_4(t)\rvert$	$\lvert\varepsilon_5(t)\rvert$	$\lvert\varepsilon_6(t)\rvert$
0	0	2.8985117078e-22	0
0.1	2.3778917463e-10	1.3388257734e-07	2.7830046346e-06
0.2	5.1391471767e-11	1.2224159084e-07	2.4975085350e-06
0.3	1.0711706348e-10	5.3805681956e-08	1.0499658352e-06
0.4	1.5291133376e-10	3.0001686996e-08	5.5245934960e-07
0.5	1.6196419726e-10	1.5214622933e-08	2.4925883823e-07
0.6	1.4722553037e-10	1.0532858804e-08	1.5605386738e-07
0.7	1.2528279282e-10	9.5929770328e-09	1.3993801151e-07
0.8	1.0267711333e-10	9.3624435059e-09	1.3835691992e-07
0.9	8.0713724766e-11	9.2400548546e-09	1.3894788236e-07
1	7.4802600745e-11	9.0055247094e-09	1.3615518479e-07
L1	2.4528410032e-10	1.5075644228e-07	3.1162901695e-06
L2	1.8834033917e-09	8.51185992405e-07	1.7345423854e-05

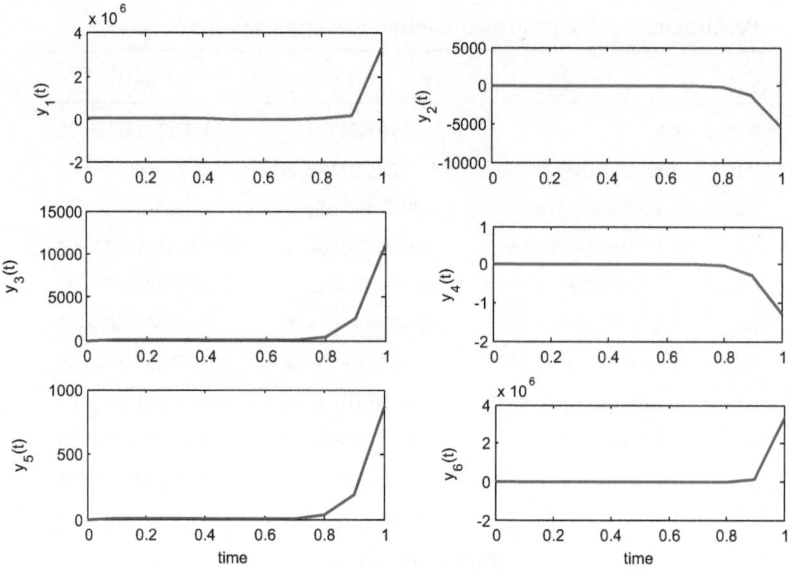

FIGURE 5.3
ADM numerical solution of Chemical Akzo Nobel Problem in the interval [0, 1]

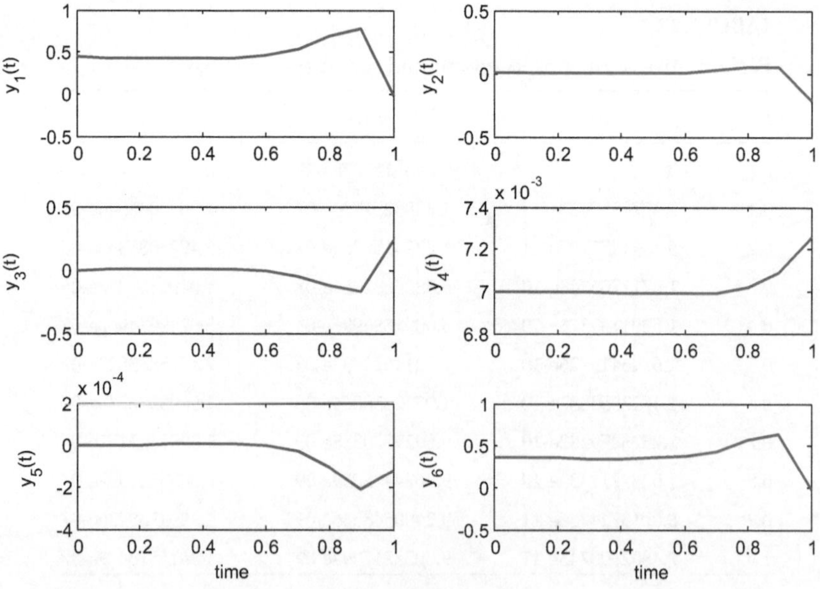

FIGURE 5.4
HAM numerical solution of Chemical Akzo Nobel Problem in the interval [0, 1]

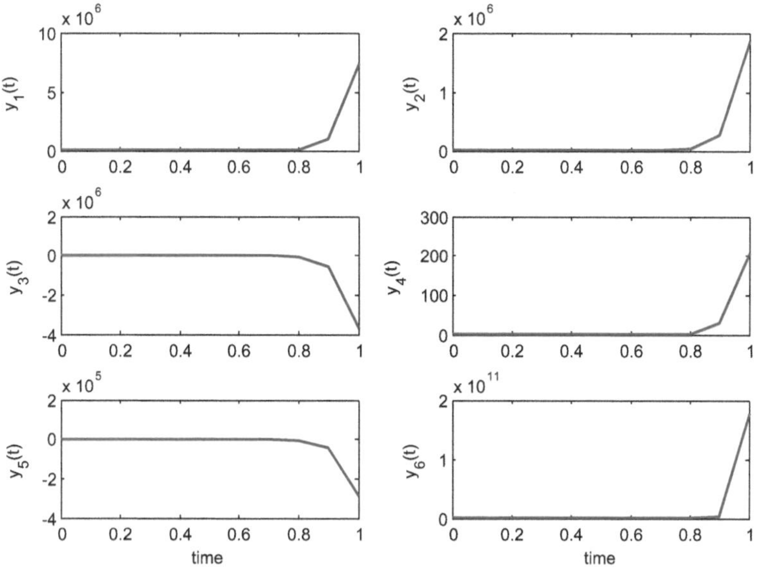

FIGURE 5.5
FDTM numerical solution of Chemical Akzo Nobel Problem in the interval [0, 1]

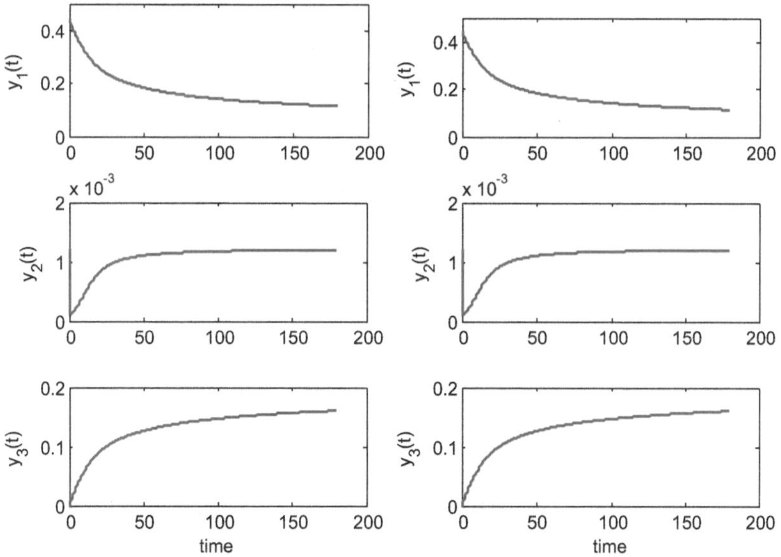

FIGURE 5.6
Numerical solution of Chemical Akzo Nobel Problem. First column: TF solution, second column: MRM2 solution

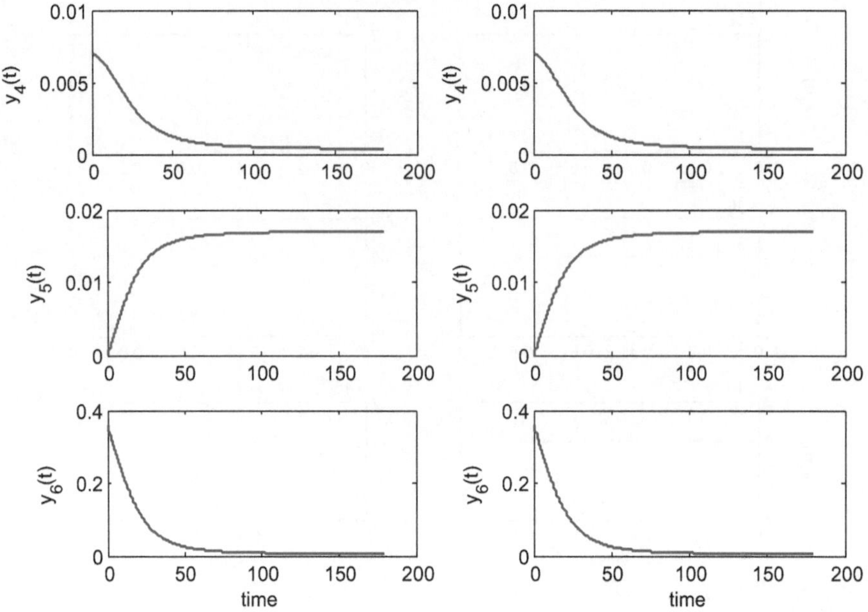

FIGURE 5.7
Numerical solution of Chemical Akzo Nobel Problem. First column: TF solution, second column: MRM2 solution

The values of parameters are $k_1 = 1.71$, $k_2 = 0.43$, $k_3 = 8.32$, $k_4 = 0.69$, $k_5 = 0.035$, $k_6 = 8.32$, $k_+ = 280$, $k_- = 0.69$, $k^* = 0.69$, $O_{k_s} = 0.0007$.

The concentrations of P_r, P_{fr}, P_rX, $P_{fr}X$, P_rX', $P_{fr}X'$, $P_{fr}X'E$ and E are denoted by $y_1(t)$, $y_2(t)$, $y_3(t)$, $y_4(t)$, $y_5(t)$, $y_6(t)$, $y_7(t)$ and $y_8(t)$, respectively.

The following integer-order model describes the rate of change of the concentrations of P_r, P_{fr}, P_rX, $P_{fr}X$, P_rX', $P_{fr}X'$, $P_{fr}X'E$, and E. The reader is referred to [40] for more details on the process and its model parameters.

$$\frac{dy_1(t)}{dt} = -1.7y_1(t) + 0.43y_2(t) + 8.32y_3(t) + 0.0007,$$

$$\frac{dy_2(t)}{dt} = 1.7y_1(t) + 8.75y_2(t),$$

$$\frac{dy_3(t)}{dt} = -10.03y_3(t) + 0.43y_4(t) + 0.035y_5(t), \qquad (5.62)$$

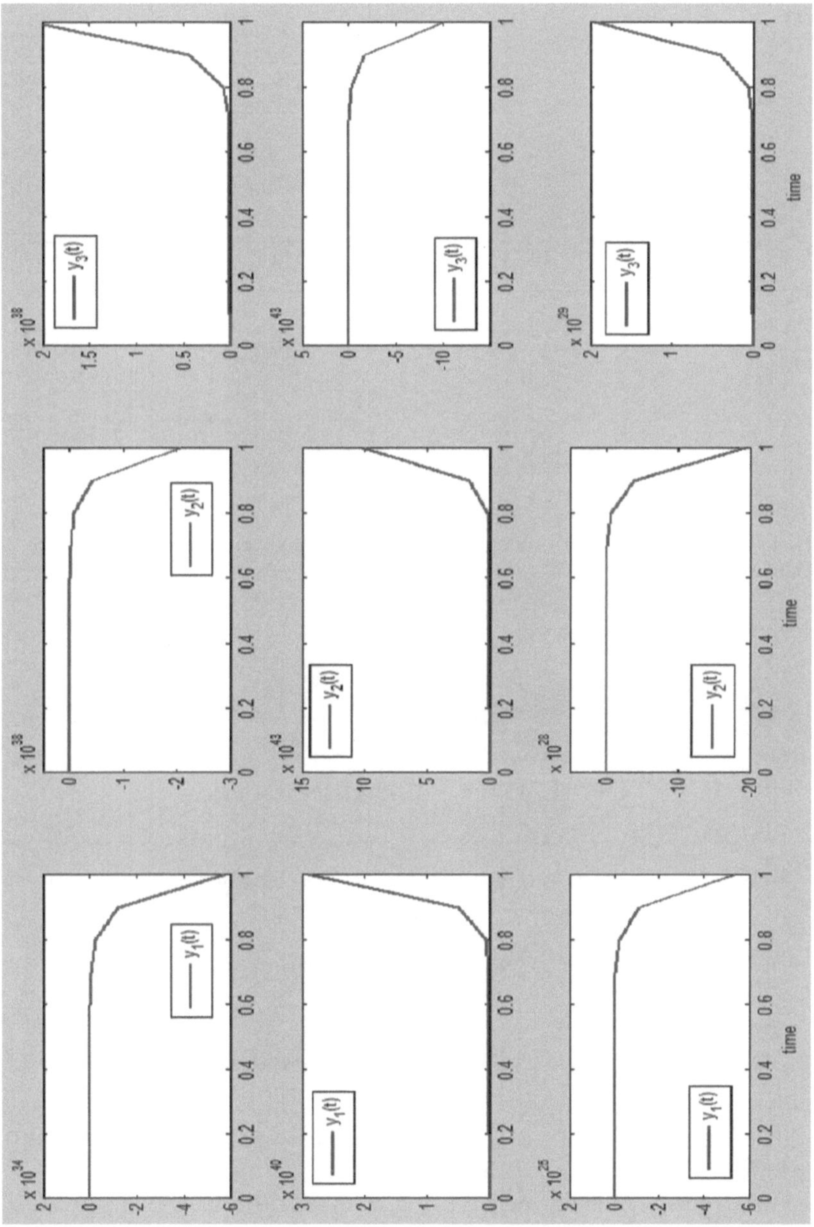

FIGURE 5.8
Numerical solutions of Robertson's system; ADM solution: first row, FDTM: second row, HAM: third row

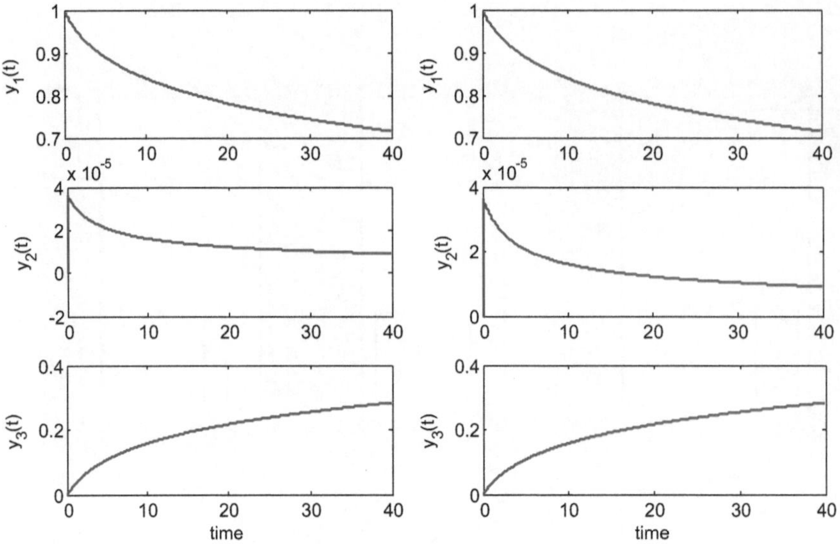

FIGURE 5.9
Numerical solution of Robertson's system in the interval [0, 40]; first column: TF solution, second column: MRM2

$$\frac{dy_4(t)}{dt} = 8.32y_2(t) + 1.71y_3(t) - 1.12y_4(t),$$

$$\frac{dy_5(t)}{dt} = -1.745y_5(t) + 0.43y_6(t) + 0.43y_7(t),$$

$$\frac{dy_6(t)}{dt} = -280y_6(t)y_8(t) + 0.69y_4(t) + 1.71y_5(t) - 0.43y_6(t) + 0.69y_7(t),$$

$$\frac{dy_7(t)}{dt} = 280y_6(t)y_8(t) - 1.81y_7(t),$$

$$\frac{dy_8(t)}{dt} = -280y_6(t)y_8(t) + 1.81y_7(t).$$

The initial values are $y_1(0) = 1$, $y_2(0) = 0$, $y_3(0) = 0$, $y_4(0) = 0$, $y_5(0) = 0$, $y_6(0) = 0$, $y_7(0) = 0$, $y_8(0) = 0.0057$.

The HIRES problem is solved by using MRM2 with the step size of 0.005 on the interval $t \in [0, 1]$, and the obtained numerical solution is considered as the true solution. The Adomian decomposition method and homotopy analysis method are applied to the system of differential equations in

TABLE 5.24
Performance analysis of ADM, HAM, FDTM and proposed method (Application 5.6)

Method	PI	$y_1(t)$	$y_2(t)$	$y_3(t)$	$y_4(t)$	$y_5(t)$	$y_6(t)$
ADM	L1	3.350704099e+06	5.44094005e+03	1.12450423e+04	1.29317518477	8.7551638e+02	3.3507041e+06
	L2	3.354273276e+06	5.57760396e+03	1.15253828e+04	1.32319487466	8.9731134e+02	3.3542733e+06
HAM	L1	0.4526671598716	0.21485211616	0.22647233744	2.76899958e-04	8.2036517e-04	0.36674627961
	L2	0.6408540607026	0.22790588624	0.32048977342	2.97441421e-04	0.00161949086	0.52504069267
FDTM	L1	7.333553241e+06	1.87196305e+06	3.66667176e+06	2.09178964e+02	2.83631139e+05	1.7769206e+11
	L2	7.406615524e+06	1.89065229e+06	3.70320190e+06	2.11178319e+02	2.86455126e+05	1.7772542e+11
TF	L1	3.629139133e-05	1.22565250e-06	1.79166477e-05	2.08848780e-06	3.78125483e-06	4.6698790e-05
	L2	0.00311112468	2.95241461e-05	0.00147823346	1.887671289e-04	5.24939243e-04	0.00444992099

ADM, HAM, and FDTM solution are acquired in the interval [0, 1], whereas the TF solution is computed in the larger interval [0, 180]. PI: performance indices

TABLE 5.25

Performance analysis of ADM, HAM, FDTM and proposed method (Application 5.7)

Method	PI	$y_1(t)$	$y_2(t)$	$y_3(t)$
ADM	L1	5.808303182690760e+34	2.084517126397520e+38	2.085097956715780e+38
	L2	5.933720626256325e+34	2.129527655804214e+38	2.130121027866830e+38
HAM	L1	5.398771578363330e+25	1.937422385060880e+29	1.937962262218710e+29
	L2	5.515234925330266e+25	1.979215584550776e+29	1.979767108043300e+29
FDTM	L1	2.889814455412640e+40	1.013618564980670e+44	1.013907546426210e+44
	L2	2.930456118323780e+40	1.027873854985858e+44	1.028166900597689e+44
TF	L1	9.653000577414250e-05	4.372973039191384e-07	9.653780334198392e-05
	L2	0.025004821626201	1.565169329742837e-05	0.025009776170167

The step size of 2e-04 is used in computing TF solutions in the larger time interval [0, 40].

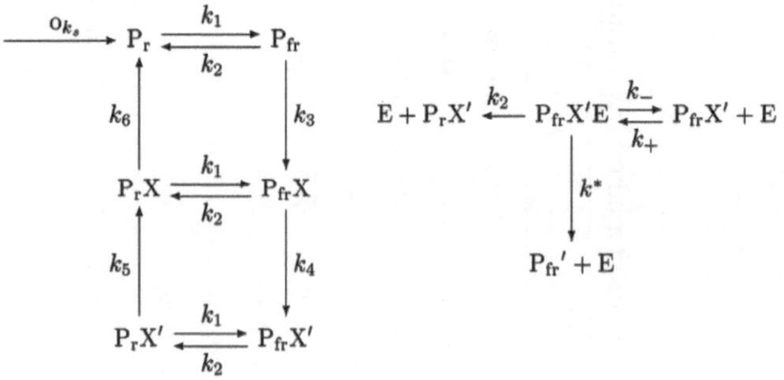

FIGURE 5.10
Reaction scheme for HIRES problem

Equation (5.62), and the obtained solutions are compared with the true solution. Table 5.26 provides the 2-norm and maximum norm of the error between the true solution and each approximate solution. It can be seen in Table 5.26 and Figure 5.11 (ADM solution) and Figure 5.12 (HAM solution) that ADM and HAM solution do not depict the true dynamic behavior of HIRES problem. By contrast, it is evident from Table 5.27 and Figure 5.13 that the TF solution is in good agreement with the true solution, even on the larger time interval [0, 180].

TABLE 5.26

Performance analysis of ADM and HAM (Application 5.8)

Method	PI	$y_1(t)$	$y_2(t)$	$y_3(t)$	$y_4(t)$
ADM	L1	20.07892283287	24.47814766640	20.42073362976	25.9575428319
	L2	20.43982061519	24.95153296116	20.80914808192	25.3949135599
HAM	L1	0.47646418250	0.101124666942	0.01778204779	0.40470360330
	L2	1.11499263793	0.220997833353	0.04155729432	0.91760144621
Method	**PI**	$y_1(t)$	$y_2(t)$	$y_3(t)$	$y_4(t)$
ADM	L1	11.62172248037	1.68078623e+05	1.70356127e+05	1.70356127e+05
	L2	11.84809848186	1.71513671e+05	1.73838444e+05	1.73838444e+05
HAM	L1	0.15079310475	0.545636340787	0.022448102318	0.02244810231
	L2	0.33124183409	1.369733741166	0.033005848487	0.03300584848

True solution, ADM solution, and HAM solution are computed on [0, 1].

TABLE 5.27

Performance analysis of proposed method (Application 5.8)

Method	PI	$y_1(t)$	$y_2(t)$	$y_3(t)$	$y_4(t)$
TF[a]	L1	5.66956153e-05	3.80319248e-05	8.76997546e-06	2.00734048e-04
	L2	0.00153132307	3.82201225e-04	1.72365789e-04	0.00391292889
Method	**PI**	$y_1(t)$	$y_2(t)$	$y_3(t)$	$y_4(t)$
TF[a]	L1	7.23614443e-05	2.15517845e-04	4.98672411e-06	4.98672410e-06
	L2	0.00138324172	0.00427389887	2.69037163e-05	2.69054830e-05

a: both the true solution and TF solution are obtained on the interval [0, 180].

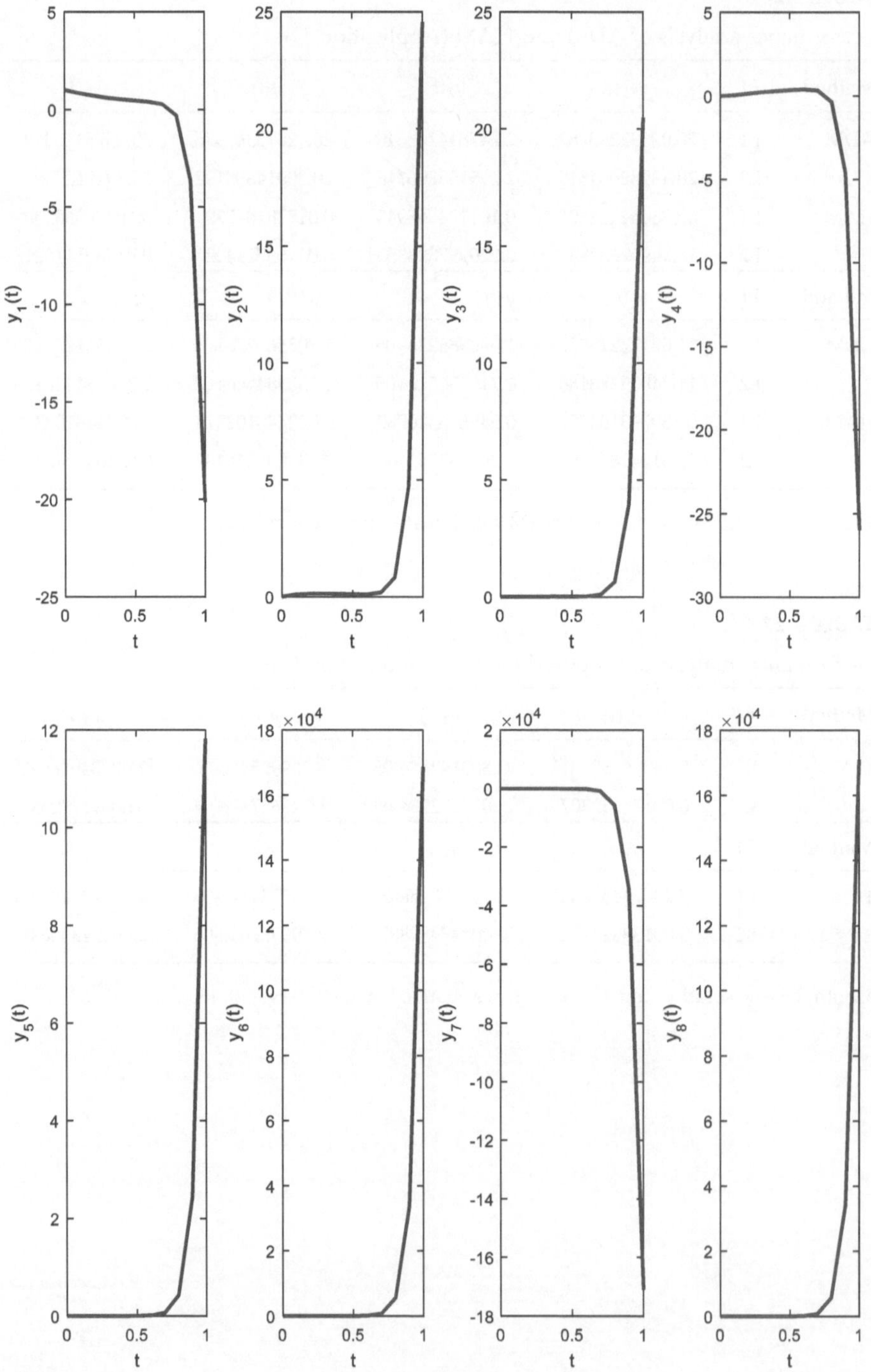

FIGURE 5.11
ADM solution of HIRES problem in the interval [0, 1]

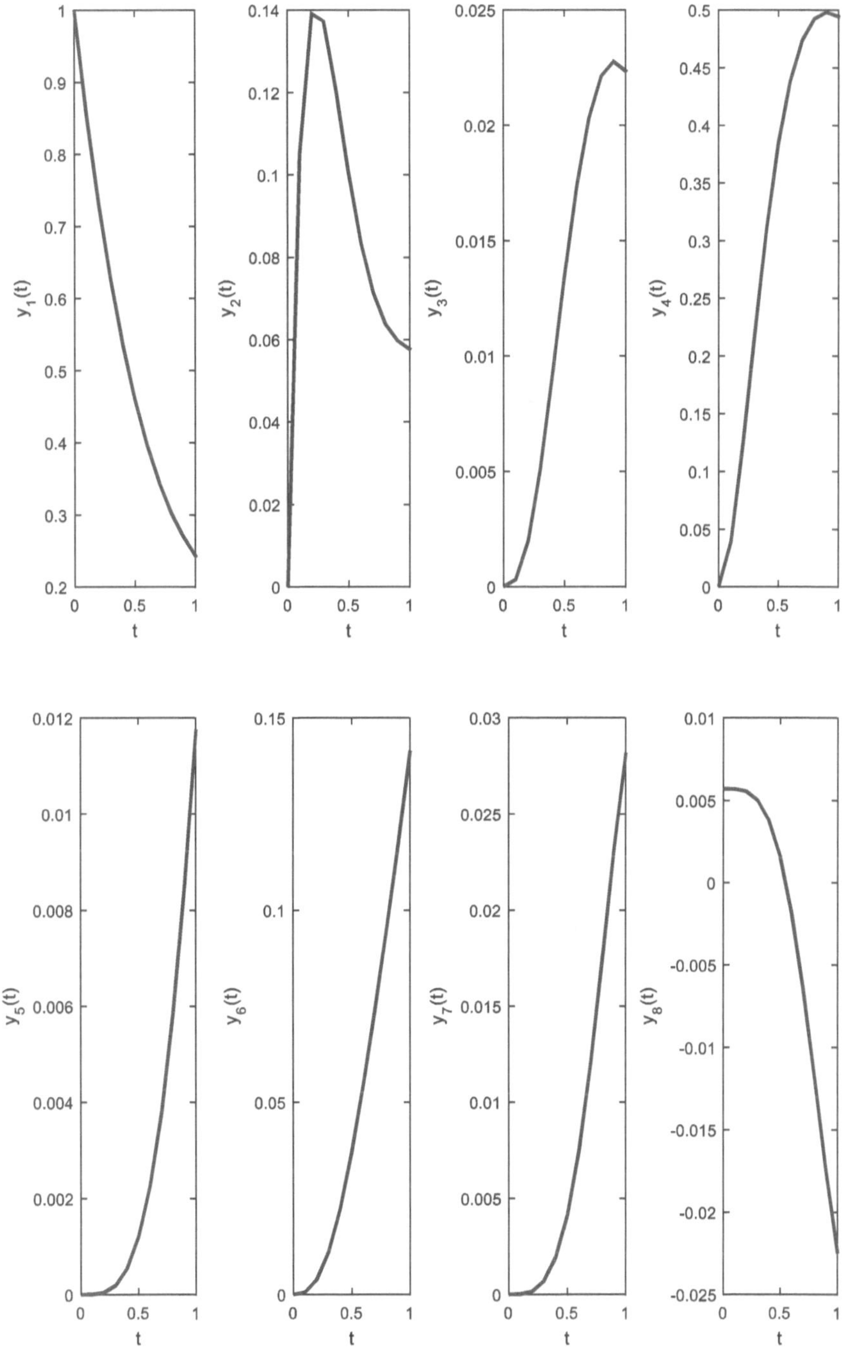

FIGURE 5.12
HAM solution of HIRES problem in the interval [0, 1]

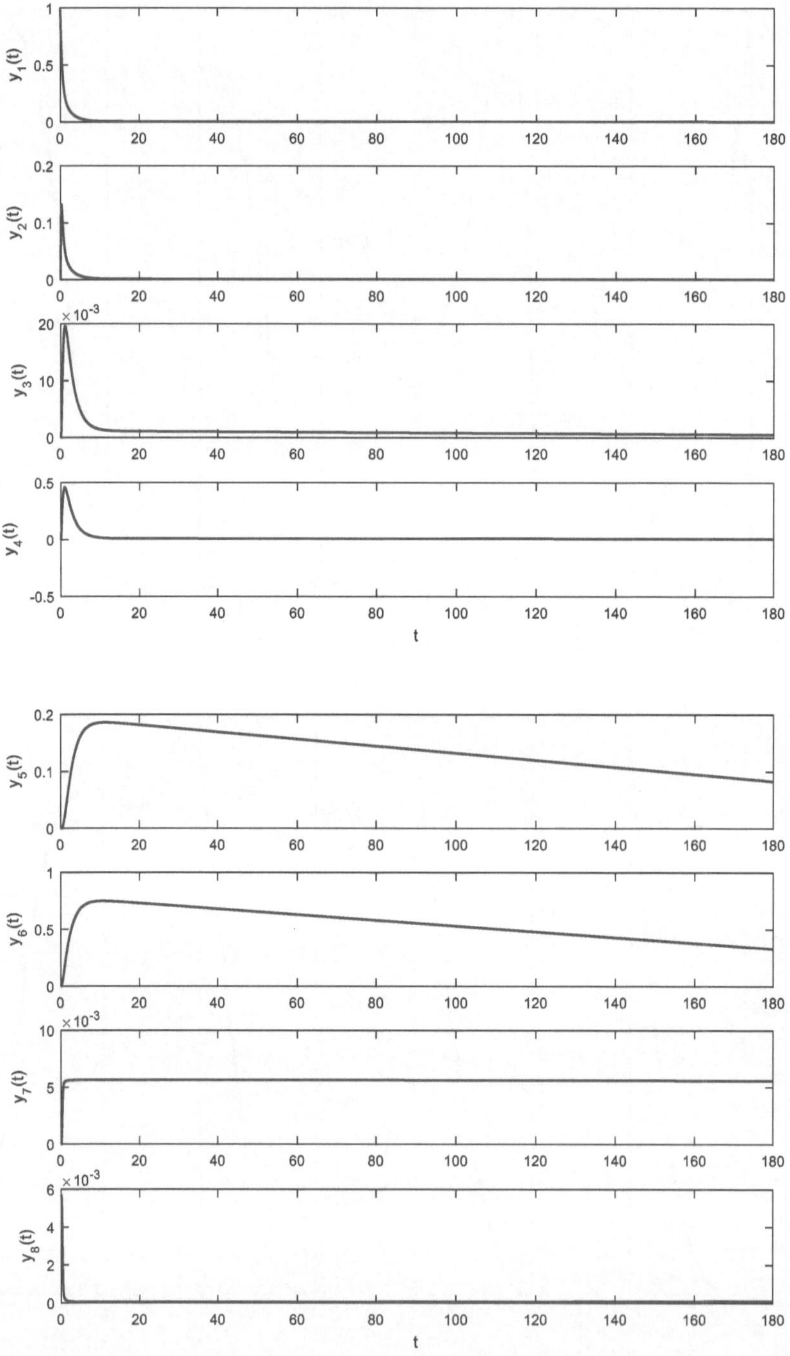

FIGURE 5.13
TF solution of HIRES problem in the interval [0, 180]

5.5 MATLAB® Codes for Numerical Experiments

This section provides MATLAB codes for Application 5.1 to 5.8.

Program 5.1

```
%%%%%%%%%%%%%%%%%%%%%%%%% MATLAB code for Application 5.1
%%%%%%%%%%%%%%%
function [e11,norminf1,normrms1]=Application_5_1
t0=0;T=1;a=1;m=a*10;h=(T-t0)/m;t=[t0:h:T];I=ones(1,
  length(t));bb=0.5;cc=0.5;
tic
C002=I+t;C001=I+t;C01=C001(1:end-1);D01=C001(2:end);
C02=C002(1:end-1);
D02=C002(2:end);[P1alph1,P2alph1]=TOF1(t0,T,m,0.5);
[P3alph1,P4alph1]=TOF12(t0,T,m,0.5);[P1alph2,P2alph2]
=TOF1(t0,T,m,2);
[P3alph2,P4alph2]=TOF12(t0,T,m,2);
x=fsolve(@Problem_Fun,zeros(1,m+1));
function f=Problem_Fun(x)
for i=1:1:length(t)
    C11(i)=x(i);
end
C1=C11(1:end-1);D1=C11(2:end);
f1=C02+bb*(C02*P1alph1+D02*P3alph1)-bb*(C1*P1alph1
  +D1*P3alph1)-cc*(C1*P1alph2+D1*P3alph2)+cc*
  (C01*P1alph2+D01*P3alph2)-C1;
f2=D02+bb*(C02*P2alph1+D02*P4alph1)-bb*(C1*P2alph1
+D1*P4alph1)-cc*(C1*P2alph2
  +D1*P4alph2)+cc*(C01*P2alph2
  +D01*P4alph2)-D1;
f=[f1;f2];
end
toc
exact=1+t; % Exact Solution
e1=abs(exact-x);
for i=0:1:10
    e11(i+1)=e1(i*a+1);t1(i+1)=t(i*a+1);
end
e11=e11';norminf1=norm(e1,inf);normrms1=norm(e1,2);
end
```

Program 5.2

```
%%%%%%%%%%%%%%%%%%%%%%%%% MATLAB code for Application 5.2
%%%%%%%%%%%%%%%
function [e11,norminf1,normrms1]=Application_5_2
t0=0;T=1;a=70;m=a*10;h=(T-t0)/m;t=[t0:h:T];I=ones(1,
  length(t));alpha=0.5;
```

```
theta=1;
ticC002=(t.^(1-alpha)).*((120/gamma(6-alpha))*(t.^4)-
  (348/(5*gamma(5-alpha)))*(t.^3)+(456/(25*gamma(4-
  alpha)))*(t.^2)-(339/
  (125*(gamma(3-alpha))))*t+(27/(125*gamma(2-
alpha)))*I);
C001=20*(t.^3)-(174/5)*(t.^2)+(456/25)*(t)-(339/125)*I;
C01=C001(1:end-1);D01=C001(2:end);C02=C002(1:end-1);
D02=C002(2:end);
C003=t*(27/125);C03=C003(1:end-1);D03=C003(2:end);
[P1alph1,P2alph1]=TOF1(t0,T,m,2-alpha);
[P3alph1,P4alph1]=TOF12(t0,T,m,2-alpha);[P1alph2,
P2alph2]=TOF1(t0,T,m,2);
[P3alph2,P4alph2]=TOF12(t0,T,m,2);
x=fsolve(@Problem_Fun,zeros(1,m+1));
function f=Problem_Fun(x)
for i=1:1:length(t)
   C11(i)=x(i);
end
C1=C11(1:end-1);D1=C11(2:end);
f1=C03-
theta*(C1*P1alph1+D1*P3alph1)+(C01*P1alph2+D01*P3alph2)
+theta*(C02*P1alph2+D02*P3alph2)-C1;
  f2=D03-theta*(C1*P2alph1+D1*P4alph1)+(C01*P2alph2
+D01*P4alph2)+theta*(C02*P2alph2+D02*P4alph2)-D1;
f=[f1;f2];
end
toc
% Exact Solution
exact1=(t.^5)-(29/10)*(t.^4)+(76/25)*(t.^3)-(339/250)*
(t.^2)+(27/125)*t;
e1=abs(exact1-x);
for i=0:1:10
   e11(i+1)=e1(i*a+1);t1(i+1)=t(i*a+1);
end
e11=e11';norminf1=norm(e1,inf);normrms1=norm(e1,2);
end
```

Program 5.3

```
%%%%%%%%%%%%%%%%%%%%%%%% MATLAB code for Application 5.3
%%%%%%%%%%%%%%%
function [e11,e22,e33]=Application_5_3
t0=0;T=1;a=1;m=a*10;h=(T-t0)/m;t=[t0:h:T];alpha1=1;
alpha2=1;alpha3=1;
I=ones(1,length(t));x0=0.48;y0=0.1;z0=0.32;C001=I;
tic
C01=x0*C001(1:end-1);D01=x0*C001(2:end);C02=y0*C001(1:
  end-1);
```

```
D02=y0*C001(2:end);C03=z0*C001(1:end-1);D03=z0*C001
  (2:end);
[P1alph1,P2alph1]=TOF1(t0,T,m,alpha1);[P3alph1,P4alph1]
  =TOF12(t0,T,m,alpha1);
[P1alph2,P2alph2]=TOF1(t0,T,m,alpha2);[P3alph2,P4alph2]
  =TOF12(t0,T,m,alpha2);
[P1alph3,P2alph3]=TOF1(t0,T,m,alpha3);[P3alph3,P4alph3]
  =TOF12(t0,T,m,alpha3);
x=fsolve(@Problem_Fun,zeros(1,3*(m+1)));
function f=Problem_Fun(x)
for i=1:1:length(t)
    C11(i)=x(i);
end
C1=C11(1:end-1);D1=C11(2:end);
for j=length(t)+1:1:2*length(t)
    C12(j-(length(t)))=x(j);
end
C2=C12(1:end-1);D2=C12(2:end);
for j1=2*length(t)+1:1:3*length(t)
    C13(j1-2*(length(t)))=x(j1);
end
C3=C13(1:end-1);D3=C13(2:end);X=C11;Y=C12;Z=C13;
k=0.005;c=0.8;lambda=2.7;eta=1.1;d1=0.7;b2=1.4;phi=2.7;
mu=4;v=1.5;beta=10.48;
CCC1=X.*(I-X)+(I./eta+Z).*lambda*c.*X.*Z-beta*X.*Y;
CC1=CCC1(1:end-1);DD1=CCC1(2:end);
CCC2=(I./(eta+Z)).*lambda.*d1.*X.*Z+phi*Z-b2*beta*X.
*Y-mu*Z;
CC2=CCC2(1:end-1);DD2=CCC2(2:end);
CCC3=beta*X.*Y-v*Z;CC3=CCC3(1:end-1);DD3=CCC3(2:end);
f1=[C01+(CC1*P1alph1+DD1*P3alph1)-C1];f2=[D01
  +(CC1*P2alph1+DD1*P4alph1)-D1];
f3=[C02+(CC2*P1alph2+DD2*P3alph2)-C2];f4=[D02
  +(CC2*P2alph2+DD2*P4alph2)-D2];
f5=[C03+(CC3*P1alph3+DD3*P3alph3)-C3];f6=[D03
  +(CC3*P2alph3+DD3*P4alph3)-D3];
f=[f1;f2;f3;f4;f5;f6];
end
toc
% TF solution
x1=[(x(1:m+1))];x2=x(m+2:2*(m+1));x3=x(2*(m+1)+1:3*
(m+1));
% Exact Solution
[tt,x]=ode45(@problem,t,[x0;y0;z0]);
function dy=problem(t,x)
dy=zeros(3,1);
k=0.005;c=0.8;lambda=2.7;eta=1.1;d1=0.7;b2=1.4;phi=2.7;
mu=4;v=1.5;beta=10.48;
X=x(1);Y=x(2);Z=x(3);
```

```
dy(1)=X*(1-X)+(1/eta+Z)*lambda*c*X*Z-beta*X*Y;
dy(2)=(1/(eta+Z))*lambda*d1*X*Z+phi*Z-b2*beta*X*Y-mu*Z;
dy(3)=beta*X*Y-v*Z;
end
exact1=x(:,1);exact2=x(:,2);exact3=x(:,3);
e1=abs(exact1'-x1);e2=abs(exact2'-x2);e3=abs
  (exact3'-x3);
for i=0:10
    b=a*i+1;e11(i+1)=e1(b);e22(i+1)=e2(b);e33(i+1)
=e3(b);
end
e11=e11';e22=e22';e33=e33';norminf1=norm(e1,inf);nor
  minf2=norm(e2,inf);
norminf34=norm(e3,inf);normrms1=norm(e1,2);normrms2=
  norm(e2,2);
normrms3=norm(e3,2);
end
```

Program 5.4

```
%%%%%%%%%%%%%%%%%%%%%%% MATLAB code for Application 5.4
%%%%%%%%%%%%%%%
function [e11,e22,e33]=Application_5_4
t0=0;T=1;a=1;m=a*10;h=(T-t0)/m;t=[t0:h:T];alpha=1;
q1=alpha;q2=alpha;
q3=alpha;I=ones(1,length(t));x0=2;y0=3;z0=2;C001=I;
tic
C01=x0*C001(1:end-1);D01=x0*C001(2:end);C02=y0*C001
  (1:end-1);
D02=y0*C001(2:end);C03=z0*C001(1:end-1);D03=z0*C001
  (2:end);
[P1alph1,P2alph1]=TOF1(t0,T,m,q1);[P3alph1,P4alph1]
  =TOF12(t0,T,m,q1);
[P1alph2,P2alph2]=TOF1(t0,T,m,q2);[P3alph2,P4alph2]
  =TOF12(t0,T,m,q2);
[P1alph3,P2alph3]=TOF1(t0,T,m,q3);[P3alph3,P4alph3]
  =TOF12(t0,T,m,q3);
x=fsolve(@Problem_Fun,zeros(1,3*(m+1)));
function f=Problem_Fun(x)
for i=1:1:length(t)
    C11(i)=x(i);
end
C1=C11(1:end-1);D1=C11(2:end);
for j=length(t)+1:1:2*length(t)
    C12(j-(length(t)))=x(j);
end
C2=C12(1:end-1);D2=C12(2:end);
for j1=2*length(t)+1:1:3*length(t)
    C13(j1-2*(length(t)))=x(j1);
end
```

```
C3=C13(1:end-1);D3=C13(2:end);X=C11;Y=C12;Z=C13;a=3.0;
  b=1;c=1;
CCC1=Z+(Y-a*I).*X;CC1=CCC1(1:end-1);DD1=CCC1(2:end);
CCC2=I-b*Y-X.*X;CC2=CCC2(1:end-1);DD2=CCC2(2:end);
CCC3=-X-c*Z;CC3=CCC3(1:end-1);DD3=CCC3(2:end);
f1=[C01+(CC1*P1alph1+DD1*P3alph1)-C1];f2=[D01
  +(CC1*P2alph1+DD1*P4alph1)-D1];
f3=[C02+(CC2*P1alph2+DD2*P3alph2)-C2];f4=[D02
  +(CC2*P2alph2+DD2*P4alph2)-D2];
f5=[C03+(CC3*P1alph3+DD3*P3alph3)-C3];f6=[D03
  +(CC3*P2alph3+DD3*P4alph3)-D3];
f=[f1;f2;f3;f4;f5;f6];
end
toc
% TF solution
x1=[(x(1:m+1))];x2=x(m+2:2*(m+1));x3=x(2*(m+1)+1:3*
  (m+1));
% Exact Solution
[tt,x]=ode45(@problem,t,[x0;y0;z0]);
function dy=problem(t,x)
dy=zeros(3,1);a=3.0;b=1;c=1;X=x(1);Y=x(2);Z=x(3);
dy(1)=Z+(Y-a)*X;dy(2)=1-b*Y-X*X;dy(3)=-X-c*Z;
end
exact1=x(:,1);exact2=x(:,2);exact3=x(:,3);
e1=abs(exact1'-x1);e2=abs(exact2'-x2);e3=abs
  (exact3'-x3);
for i=0:10
    b=a*i+1;e11(i+1)=e1(b);e22(i+1)=e2(b);e33(i+1)
=e3(b);
end
e11=e11';e22=e22';e33=e33';norminf1=norm(e1,inf);nor-
minf2=
  norm(e2,inf);
norminf34=norm(e3,inf);normrms1=norm(e1,2);
  normrms2=norm(e2,2);
normrms3=norm(e3,2);
end
```

Program 5.5

```
%%%%%%%%%%%%%%%%%%%%%%%%%%%%%%%% MATLAB code for Application 5.5
%%%%%%%%%%%%%%%%%%%%%
function [e1,e2,e3,e4]=Application_5_5
t0=0;T=25;aa=25;m=aa*10;h=(T-t0)/m;t=[t0:h:T];alpha1=1;
alpha2=1;
alpha3=1;alpha4=1;S0=3;I0=95;R0=1;A0=1;x0=[S0 I0 R0 A0];
beta=0.1;sigma=0.8;delta=9;alphaSA=0.025;alphaIA=0.25;
I=ones(1,length(t));
h1=0.001;t11=[];x11=[];x22=[];x33=[];x44=[];
tic
```

```
for i=1:length(t)-1
t1=[t(i):h1:t(i+1)];m1=length(t1)-1;I1=ones(1,length
  (t1));CC1=S0*I1;
C01=CC1(1:end-1);D01=CC1(2:end);CC2=I0*I1;C02=CC2(1:
  end-1);D02=CC2(2:end);
CC3=R0*I1;C03=CC3(1:end-1);D03=CC3(2:end);CC4=A0*I1;
C04=CC4(1:end-1);
D04=CC4(2:end);
[P1alph1,P2alph1]=TOF1(t(i),t(i+1),m1,alpha1);
[P3alph1,P4alph1]=TOF12(t(i),t(i+1),m1,alpha1);
[P1alph2,P2alph2]=TOF1(t(i),t(i+1),m1,alpha2);
[P3alph2,P4alph2]=TOF12(t(i),t(i+1),m1,alpha2);
[P1alph3,P2alph3]=TOF1(t(i),t(i+1),m1,alpha3);
[P3alph3,P4alph3]=TOF12(t(i),t(i+1),m1,alpha3);
[P1alph4,P2alph4]=TOF1(t(i),t(i+1),m1,alpha4);
[P3alph4,P4alph4]=TOF12(t(i),t(i+1),m1,alpha4);
x=fsolve(@ProblemMs_Fun,zeros(1,4*(m1+1)));
function f=ProblemMs_Fun(x)
t=t1;
for i=1:1:length(t)
   C11(i)=x(i);
end
C1=C11(1:end-1);D1=C11(2:end);
for j=length(t)+1:1:2*length(t)
   C12(j-(length(t)))=x(j);
end
C2=C12(1:end-1);D2=C12(2:end);
for k=2*length(t)+1:1:3*length(t)
   C13(k-(2*length(t)))=x(k);
end
C3=C13(1:end-1);D3=C13(2:end);
for kk=3*length(t)+1:1:4*length(t)
   C14(kk-(3*length(t)))=x(kk);
end
C4=C14(1:end-1);D4=C14(2:end);S=C11;I=C12;R=C13;A=C14;
CCC1=-alphaSA*S.*A-beta*S.*I+sigma*R;CC1=CCC1(1:end-1);
DD1=CCC1(2:end);
CCC2=beta*S.*I-alphaIA*A.*I-delta*I;CC2=CCC2(1:end-1);
DD2=CCC2(2:end);
CCC3=delta*I-sigma*R;CC3=CCC3(1:end-1);DD3=CCC3(2:end);
CCC4=alphaSA*S.*A+alphaIA*A.*I;CC4=CCC4(1:end-1);
DD4=CCC4(2:end);
f1=[C01+(CC1*P1alph1+DD1*P3alph1)-C1];f2=[D01
  +(CC1*P2alph1+DD1*P4alph1)-D1];
f3=[C02+(CC2*P1alph2+DD2*P3alph2)-C2];f4=[D02
  +(CC2*P2alph2+DD2*P4alph2)-D2];
f5=[C03+(CC3*P1alph3+DD3*P3alph3)-C3];f6=[D03
  +(CC3*P2alph3+DD3*P4alph3)-D3];
```

```
f7=[C04+(CC4*P1alph4+DD4*P3alph4)-C4];f8=[D04
  +(CC4*P2alph4+DD4*P4alph4)-D4];
f=[f1 f2 f3 f4 f5 f6 f7 f8];
end
x1=[(x(1:m1+1))];x2=x(m1+2:2*(m1+1));x3=x(2*(m1+1)+1:3*
  (m1+1));
x4=x(3*(m1+1)+1:4*(m1+1));S0=x1(end);I0=x2(end);R0=x3
  (end);A0=x4(end);
if i<m
x11=[x11 x1(1:end-1)];x22=[x22 x2(1:end-1)];x33=[x33 x3
  (1:end-1)];
x44=[x44 x4(1:end-1)];t11=[t11 t1(1:end-1)];
elseif i==m
x11=[x11 x1(1:end)];x22=[x22 x2(1:end)];x33=[x33 x3
  (1:end)];
x44=[x44 x4(1:end)];t11=[t11 t1(1:end)];
end
end
toc
% Exact Solution
[t,y]=ode45(@Problem,[t11],x0);
function dy=Problem(t,x)
dy=zeros(4,1);C11=x(1);C12=x(2);C13=x(3);C14=x(4);
S=C11;I=C12;R=C13;A=C14;
dy(1)=-alphaSA*S.*A-beta*S.*I+sigma*R;dy(2)=beta*S.
  *I-alphaIA*A.*I-delta*I;
dy(3)=delta*I-sigma*R;dy(4)=alphaSA*S.*A+alphaIA*A.
  *I;end
exact1=y(:,1);exact2=y(:,2);exact3=y(:,3);
exact4=y(:,4);
[x11' exact1];[x22' exact2];[x33' exact3];[x44' exact4];
e1=abs(exact1-x11');e2=abs(exact2-x22');e3=abs
(exact3-x33');
e4=abs(exact4-x44');norminf1=norm(e1,inf);norminf2=norm
(e2,inf);
norminf3=norm(e3,inf);norminf4=norm(e4,inf);
normrms1=norm(e1,2);normrms2=norm(e2,2);
  normrms3=norm(e3,2);
normrms4=norm(e4,2);
end
```

Program 5.6

```
%%%%%%%%%%%%%%%%%%%%%%%%%% MATLAB code for Application 5.6
%%%%%%%%%%%%%%%
function [e1,e2,e3,e4,e5,e6]=Application_5_6
t0=0;T=180;m=1800;h=(T-t0)/m;t=[t0:h:T];alpha=1;
x01=0.444;x02=0.00123;
x03=0;x04=0.007;x05=0;h1=0.005;x11=[];x22=[];x33=[];
x44=[];x55=[];t11=[];
```

```
warning('off')
tic
for i=1:length(t)-1
t1=[t(i):h1:t(i+1)];m1=length(t1)-1;
[P1alph,P2alph]=TOF1(t1(1),t1(end),m1,alpha);
[P3alph,P4alph]=TOF12(t1(1),t1(end),m1,alpha);
x=fsolve(@Problem_Fun,zeros(1,5*(m1+1)));
function f=Problem_Fun(x)
t=t1;
for i=1:1:length(t)
   C11(i)=x(i);
end
C1=C11(1:end-1);D1=C11(2:end);
for j=length(t)+1:1:2*length(t)
   C12(j-(length(t)))=x(j);
end
C2=C12(1:end-1);D2=C12(2:end);
for k=2*length(t)+1:1:3*length(t)
   C13(k-(2*length(t)))=x(k);
end
C3=C13(1:end-1);D3=C13(2:end);
for w=3*length(t)+1:1:4*length(t)
   C14(w-(3*length(t)))=x(w);
end
C4=C14(1:end-1);D4=C14(2:end);
for w1=4*length(t)+1:1:5*length(t)
   C15(w1-(4*length(t)))=x(w1);
end
C5=C15(1:end-1);D5=C15(2:end);CC=ones(1,length(t));
k1=18.7;k2=0.58;K=34.4;k3=0.09;k4=0.42;klA=3.3;Po2=0.9;
H=737;Ks=115.83;
C001=x01*ones(1,length(t));C002=x02*ones(1,length(t));
C003=x03*ones(1,length(t));C004=x04*ones(1,length(t));
C005=x05*ones(1,length(t));C01=C001(1:end-1);D01=C001
  (2:end);
C02=C002(1:end-1);D02=C002(2:end);C03=C003(1:end-1);
D03=C003(2:end);
C04=C004(1:end-1);D04=C004(2:end);C05=C005(1:end-1);
D05=C005(2:end);
y1=C11; y2=C12; y3=C13; y4=C14;y5=C15;y6=Ks*y1.*y4;r1=k1*
  (y1.^4).*(y2.^0.5);
r2=k2*y3.*y4;r3=(k2/K)*y1.*y5;r4=k3*y1.*(y4.^2);
r5=k4*y6.*y6.*(y2.^0.5);
Fin=klA*((Po2/H)*CC-y2);CCC1=-2*r1+r2-r3-r4;CC1=CCC1(1:
end-1);DD1=CCC1(2:end);CCC2=-0.5*r1-r4-0.5*r5+Fin;
CC2=CCC2(1:end-1);DD2=CCC2(2:end);CCC3=r1-r2+r3;
CC3=CCC3(1:end-1);
DD3=CCC3(2:end);CCC4=-r2+r3-2*r4;CC4=CCC4(1:end-1);
DD4=CCC4(2:end);
```

```
CCC5=r2-r3+r5;CC5=CCC5(1:end-1);DD5=CCC5(2:end);
f1=[C01+(CC1*P1alph+DD1*P3alph)-C1];f2=[D01+(CC1*P2alph
  +DD1*P4alph)-D1];
f3=[C02+(CC2*P1alph+DD2*P3alph)-C2];f4=[D02+(CC2*P2alph
  +DD2*P4alph)-D2];
f5=[C03+(CC3*P1alph+DD3*P3alph)-C3];f6=[D03+(CC3*P2alph
  +DD3*P4alph)-D3];
f7=[C04+(CC4*P1alph+DD4*P3alph)-C4];f8=[D04+(CC4*P2alph
  +DD4*P4alph)-D4];
f9=[C05+(CC5*P1alph+DD5*P3alph)-C5];f10=[D05
  +(CC5*P2alph+DD5*P4alph)-D5];
f=[f1;f2;f3;f4;f5;f6;f7;f8;f9;f10];
end
x1=[(x(1:m1+1))];x2=x(m1+2:2*(m1+1));x3=x(2*(m1+1)+1:3*
  (m1+1));
x4=x(3*(m1+1)+1:4*(m1+1));x5=x(4*(m1+1)+1:5*(m1+1));
x11=[x11 x1(1:end-1)];
x22=[x22 x2(1:end-1)];x33=[x33 x3(1:end-1)];x44=[x44 x4
  (1:end-1)];
x55=[x55 x5(1:end-1)];x01=x1(end);x02=x2(end);x03=x3
  (end);x04=x4(end);
x05=x5(end);t11=[t11 t1(1:end-1)];
end
x66=115.83*x11.*x44;
toc
% Exact Solution
[t,y]=ode23s(@Problem,[t11],[0.444;0.00123;0;0.007;0]);
function dy=Problem(t,x)
dy=zeros(5,1);k1=18.7;k2=0.58;K=34.4;k3=0.09;k4=0.42;
klA=3.3;Po2=0.9;H=737;
Ks=115.83;y1=x(1);y2=x(2);y3=x(3);y4=x(4);y5=x(5);
y6=Ks*y1*y4;
r1=k1*(y1^4)*(y2^0.5);r2=k2*y3*y4;r3=(k2/K)*y1*y5;
r4=k3*y1*y4^2;
r5=k4*y6*y6*y2^0.5;Fin=klA*((Po2/H)-y2);
dy(1)=-2*r1+r2-r3-r4;dy(2)=-0.5*r1-r4-0.5*r5+Fin;dy(3)
  =r1-r2+r3;
dy(4)=-r2+r3-2*r4;dy(5)=r2-r3+r5;
end
exact1=y(:,1);exact2=y(:,2);exact3=y(:,3);exact4=y
  (:,4);exact5=y(:,5);
exact6=115.83*exact1.*exact4;
E=[exact1 exact2 exact3 exact4 exact5 exact6];
e1=abs(exact1-x11');e2=abs(exact2-x22');e3=abs
  (exact3-x33');
e4=abs(exact4-x44');e5=abs(exact5-x55');e6=abs
  (exact6-x66');
```

```
e1max=max(e1);e2max=max(e2);e3max=max(e3);e4max=max
  (e4);e5max=max(e5);
e6max=max(e6);
end
```

Program 5.7

```
%%%%%%%%%%%%%%%%%%%%%%% MATLAB code for Application 5.7
%%%%%%%%%%%%%%
function [E1,E2,E3]=Application_5_7
t0=0;T=40;m=800;h=(T-t0)/m;t=[t0:h:T];alpha1=1;
alpha2=1;alpha3=1;
x01=1;x02=0;x03=0;h1=0.0002;x11=[];x22=[];x33=[];
t11=[];
warning('off')
tic
for i=1:length(t)-1
t1=[t(i):h1:t(i+1)];m1=length(t1)-1;
[P1alph1,P2alph1]=TOF1(t1(1),t1(end),m1,alpha1);
[P3alph1,P4alph1]=TOF12(t1(1),t1(end),m1,alpha1);
[P1alph2,P2alph2]=TOF1(t1(1),t1(end),m1,alpha2);
[P3alph2,P4alph2]=TOF12(t1(1),t1(end),m1,alpha2);
[P1alph3,P2alph3]=TOF1(t1(1),t1(end),m1,alpha3);
[P3alph3,P4alph3]=TOF12(t1(1),t1(end),m1,alpha3);
x=fsolve(@problem_Fun, zeros(1,3*(m1+1)));
function f=problem_Fun(x)
t=t1;
for i=1:1:length(t)
    C11(i)=x(i);
end
C1=C11(1:end-1);D1=C11(2:end);
for j=length(t)+1:1:2*length(t)
    C12(j-(length(t)))=x(j);
end
C2=C12(1:end-1);D2=C12(2:end);
for k=2*length(t)+1:1:3*length(t)
    C13(k-(2*length(t)))=x(k);
end
C3=C13(1:end-1);D3=C13(2:end);C001=x01*ones(1,
  length(t));
C002=x02*ones(1,length(t));C003=x03*ones(1,length(t));
C01=C001(1:end-1);
C02=C002(1:end-1);C03=C003(1:end-1);D01=C001(2:end);
D02=C002(2:end);
D03=C003(2:end);y1=C11;y2=C12;y3=C13;CC=ones(1,length
  (t));cc1=CC(1:end-1);
CCC1=-0.04*y1+(10^4)*y2.*y3;CC1=CCC1(1:end-1);
CCC2=0.04*y1-(10^4)*y2.*y3-(3*10^7)*y2.*y2;CC2=CCC2(1:
  end-1);
CCC3=(3*10^7)*y2.*y2;CC3=CCC3(1:end-1);DD1=CCC1(2:end);
```

```
DD2=CCC2(2:end);DD3=CCC3(2:end);f1=[C01+(CC1*P1alph1
  +DD1*P3alph1)-C1];
f2=[D01+(CC1*P2alph1+DD1*P4alph1)-D1];f3=[C02
  +(CC2*P1alph2+DD2*P3alph2)-C2];
f4=[D02+(CC2*P2alph2+DD2*P4alph2)-D2];f5=[C03
  +(CC3*P1alph3+DD3*P3alph3)-C3];
f6=[D03+(CC3*P2alph3+DD3*P4alph3)-D3];f=[f1;f2;f3;f4;
  f5;f6];
end
x1=[(x(1:m1+1))];x2=x(m1+2:2*(m1+1));x3=x(2*(m1+1)+1:3*
(m1+1));
x11=[x11  x1(1:end-1)];x22=[x22  x2(1:end-1)];x33=[x33  x3
(1:end-1)];
x01=x1(end);x02=x2(end);x03=x3(end);t11=[t11 t1(1:end-1)];
end
toc
% Exact Solution
[t,y]=ode23s(@Problem,t11,[1;0;0]);
function dy=Problem(t,x)
dy=zeros(3,1);
dy(1)=-0.04*x(1)+(10^4)*x(2)*x(3);
dy(2)=0.04*x(1)-(10^4)*x(2)*x(3)-(3*10^7)*x(2)*x(2);
dy(3)=(3*10^7)*x(2)*x(2);
end
exact1=y(:,1);exact2=y(:,2);exact3=y(:,3);x11=x11';
x22=x22';
x33=x33';E1=abs(exact1-x11);E2=abs(exact2-x22);E3=abs
  (exact3-x33);
end
```

Program 5.8

```
%%%%%%%%%%%%%%%%%%%%%%%%%%%%%%%%%%%%%%%%%%%%%%%%%%%%%%%%% MATLAB code for Application 5.8
%%%%%%%%%%%%%%%%%%%%%%%%%%%%%%%%%%%%%%%%%%%
function [ee1,ee2,ee3,ee4,ee5,ee6,ee7,ee8]=Application_5_8
t0=0;T=180;m=1800;h=(T-t0)/m;t=[t0:h:T];alpha1=1;
alpha2=1;alpha3=1;
alpha4=1;alpha5=1;alpha6=1;alpha7=1;alpha8=1;x01=1;
x02=0;x03=0;x04=0;
x05=0;x06=0;x07=0;x08=0.0057;h1=0.01;x11=[];x22=[];x33=
[];x44=[];x55=[];
x66=[];x77=[];x88=[];t11=[];
warning('off')
tic
for i=1:length(t)-1
t1=[t(i):h1:t(i+1)]m1=length(t1)-1;
[P1alph1,P2alph1]=TOF1(t1(1),t1(end),m1,alpha1);
[P3alph1,P4alph1]=TOF12(t1(1),t1(end),m1,alpha1);
[P1alph2,P2alph2]=TOF1(t1(1),t1(end),m1,alpha2);
[P3alph2,P4alph2]=TOF12(t1(1),t1(end),m1,alpha2);
```

```
[P1alph3,P2alph3]=TOF1(t1(1),t1(end),m1,alpha3);
[P3alph3,P4alph3]=TOF12(t1(1),t1(end),m1,alpha3);
[P1alph4,P2alph4]=TOF1(t1(1),t1(end),m1,alpha4);
[P3alph4,P4alph4]=TOF12(t1(1),t1(end),m1,alpha4);
[P1alph5,P2alph5]=TOF1(t1(1),t1(end),m1,alpha5);
[P3alph5,P4alph5]=TOF12(t1(1),t1(end),m1,alpha5);
[P1alph6,P2alph6]=TOF1(t1(1),t1(end),m1,alpha6);
[P3alph6,P4alph6]=TOF12(t1(1),t1(end),m1,alpha6);
[P1alph7,P2alph7]=TOF1(t1(1),t1(end),m1,alpha7);
[P3alph7,P4alph7]=TOF12(t1(1),t1(end),m1,alpha7);
[P1alph8,P2alph8]=TOF1(t1(1),t1(end),m1,alpha8);
[P3alph8,P4alph8]=TOF12(t1(1),t1(end),m1,alpha8);
x=fsolve(@Problem_Fun,zeros(1,8*(m1+1)));
function f=Problem_Fun(x)
t=t1;
for i=1:1:length(t)
   C11(i)=x(i);
end
C1=C11(1:end-1);D1=C11(2:end);
for j=length(t)+1:1:2*length(t)
   C12(j-(length(t)))=x(j);
end
C2=C12(1:end-1);D2=C12(2:end);
for k=2*length(t)+1:1:3*length(t)
   C13(k-(2*length(t)))=x(k);
end
C3=C13(1:end-1);D3=C13(2:end);
for w=3*length(t)+1:1:4*length(t)
   C14(w-(3*length(t)))=x(w);
end
C4=C14(1:end-1);D4=C14(2:end);
for w1=4*length(t)+1:1:5*length(t)
   C15(w1-(4*length(t)))=x(w1);
end
C5=C15(1:end-1);D5=C15(2:end);
for w2=5*length(t)+1:1:6*length(t)
   C16(w2-(5*length(t)))=x(w2);
end
C6=C16(1:end-1);D6=C16(2:end);
for w3=6*length(t)+1:1:7*length(t)
   C17(w3-(6*length(t)))=x(w3);
end
C7=C17(1:end-1);D7=C17(2:end);
for w4=7*length(t)+1:1:8*length(t)
   C18(w4-(7*length(t)))=x(w4);
end
C8=C18(1:end-1);D8=C18(2:end);
C001=(x01)*ones(1,length(t));C002=(x02)*ones(1,
  length(t));
```

```
C003=(x03)*ones(1,length(t));C004=(x04)*ones(1,
  length(t));
C005=(x05)*ones(1,length(t));C006=(x06)*ones(1,
  length(t));
C007=(x07)*ones(1,length(t));C008=(x08)*ones(1,
  length(t));
C01=C001(1:end-1);C02=C002(1:end-1);C03=C003(1:end-1);
C04=C004(1:end-1);
C05=C005(1:end-1);C06=C006(1:end-1);C07=C007(1:end-1);
C08=C008(1:end-1);
D01=C001(2:end);D02=C002(2:end);D03=C003(2:end);
D04=C004(2:end);
D05=C005(2:end);D06=C006(2:end);D07=C007(2:end);
D08=C008(2:end);
CC=ones(1,length(t));cc1=CC(1:end-1);y1=C11;y2=C12;
y3=C13;y4=C14;y5=C15;
y6=C16;y7=C17;y8=C18;CCC1=-1.71*y1+0.43*y2+8.32*y3
  +0.0007*CC;
CC1=CCC1(1:end-1);CCC2=1.71*y1-8.75*y2;CC2=CCC2(1:
  end-1);
CCC3=-10.03*y3+0.43*y4+0.035*y5;CC3=CCC3(1:end-1);
CCC4=8.32*y2+1.71*y3-1.12*y4;CC4=CCC4(1:end-1);
CCC5=-1.745*y5+0.43*y6+0.43*y7;CC5=CCC5(1:end-1);
CCC6=-280*y6.*y8+0.69*y4+1.71*y5-0.43*y6+0.69*y7;
CC6=CCC6(1:end-1);
CCC7=280*y6.*y8-1.81*y7;CC7=CCC7(1:end-1);CCC8=-280*y6.
  *y8+1.81*y7;
CC8=CCC8(1:end-1);DD1=CCC1(2:end);DD2=CCC2(2:end);
DD3=CCC3(2:end);
DD4=CCC4(2:end);DD5=CCC5(2:end);DD6=CCC6(2:end);
DD7=CCC7(2:end);
DD8=CCC8(2:end);f1=[C01+(CC1*P1alph1+DD1*P3alph1)-C1];
f2=[D01+(CC1*P2alph1+DD1*P4alph1)-D1];f3=[C02
  +(CC2*P1alph2+DD2*P3alph2)-C2];
f4=[D02+(CC2*P2alph2+DD2*P4alph2)-D2];f5=[C03
  +(CC3*P1alph3+DD3*P3alph3)-C3];
f6=[D03+(CC3*P2alph3+DD3*P4alph3)-D3];f7=[C04
  +(CC4*P1alph4+DD4*P3alph4)-C4];
f8=[D04+(CC4*P2alph4+DD4*P4alph4)-D4];f9=[C05
  +(CC5*P1alph5+DD5*P3alph5)-C5];
f10=[D05+(CC5*P2alph5+DD5*P4alph5)-D5];
f11=[C06+(CC6*P1alph6+DD6*P3alph6)-C6];
f12=[D06+(CC6*P2alph6+DD6*P4alph6)-D6];
f13=[C07+(CC7*P1alph7+DD7*P3alph7)-C7];
f14=[D07+(CC7*P2alph7+DD7*P4alph7)-D7];
f15=[C08+(CC8*P1alph8+DD8*P3alph8)-C8];
f16=[D08+(CC8*P2alph8+DD8*P4alph8)-D8];
f=[f1;f2;f3;f4;f5;f6;f7;f8;f9;f10;f11;f12;f13;f14;
f15;f16];
```

```
end
toc
x1=[(x(1:m1+1))];x2=x(m1+2:2*(m1+1));x3=x(2*(m1+1)+1:3*
  (m1+1));
x4=x(3*(m1+1)+1:4*(m1+1));x5=x(4*(m1+1)+1:5*(m1+1));
x6=x(5*(m1+1)+1:6*(m1+1));x7=x(6*(m1+1)+1:7*(m1+1));
x8=x(7*(m1+1)+1:8*(m1+1));
if i<m
x11=[x11 x1(1:end-1)];x22=[x22 x2(1:end-1)];x33=[x33 x3(1:
  end-1)];
x44=[x44 x4(1:end-1)];x55=[x55 x5(1:end-1)];x66=[x66 x6(1:
  end-1)];
x77=[x77 x7(1:end-1)];x88=[x88 x8(1:end-1)];
else
x11=[x11x1(1:end)];x22=[x22 x2(1:end)];x33=[x33 x3(1:end)];
x44=[x44x4(1:end)];x55=[x55 x5(1:end)];x66=[x66 x6(1:end)];
x77=[x77 x7(1:end)];x88=[x88 x8(1:end)];
end
x01=x1(end);x02=x2(end);x03=x3(end);x04=x4(end);x05=x5
(end);x06=x6(end);
x07=x7(end);x08=x8(end);
if i<m
t11=[t11 t1(1:end-1)];
else
t11=[t11 t1(1:end)];
end
end
% Exact Solution
[t,y]=ode23s(@problem,[t11],[1 zeros(1,6) 0.0057]);
function dy=problem(t,x)
dy=zeros(8,1);dy(1)=-1.71*x(1)+0.43*x(2)+8.32*x(3)
+0.0007;
dy(2)=1.71*x(1)-8.75*x(2);dy(3)=-10.03*x(3)+0.43*x(4)
  +0.035*x(5);
dy(4)=8.32*x(2)+1.71*x(3)-1.12*x(4);dy(5)=-1.745*x(5)
  +0.43*x(6)+0.43*x(7);
dy(6)=-280*x(6)*x(8)+0.69*x(4)+1.71*x(5)-0.43*x(6)
  +0.69*x(7);
dy(7)=280*x(6)*x(8)-1.81*x(7);dy(8)=-280*x(6)*x(8)
  +1.81*x(7);
end
exact1=y(:,1);exact2=y(:,2);exact3=y(:,3);exact4=y
  (:,4);exact5=y(:,5);
exact6=y(:,6);exact7=y(:,7);exact8=y(:,8);ee1=abs
  (exact1'-x11);
ee2=abs(exact2'-x22);ee3=abs(exact3'-x33);ee4=abs
  (exact4'-x44);
ee5=abs(exact5'-x55);ee6=abs(exact6'-x66);ee7=abs
  (exact7'-x77);
```

```
ee8=abs(exact8'-x88);norminfT1=norm(ee1,inf);
norminfT2=norm(ee2,inf);
norminfT3=norm(ee3,inf);norminfT4=norm(ee4,inf);
norminfT5=norm(ee5,inf);
norminfT6=norm(ee6,inf);norminfT7=norm(ee7,inf);
norminfT8=norm(ee8,inf);
normrmsT1=norm(ee1,2);normrmsT2=norm(ee2,2);
normrmsT3=norm(ee3,2);
normrmsT4=norm(ee4,2);normrmsT5=norm(ee5,2);
normrmsT6=norm(ee6,2);
normrmsT7=norm(ee7,2);normrmsT8=norm(ee8,2);
end
```

5.6 Concluding Remarks

The following conclusions can be drawn from this chapter:

- The proposed method has a broad range of applicability in sense that it can be applied to various kinds of FDEs and FDAEs.
- The proposed method outperforms the BPFs-based method as well as the numerical methods reported in [23–29]. and [31–33].
- All forms of FDEs and FDAEs can be solved better in the TF domain than in the BPF domain.
- The most popular semianalytical methods, the Adomian decomposition method, the fractional differential transform method with Adomian polynomials, and the homotopy analysis method, fail to simulate physical process models involving nonstiff or stiff FDEs or FDAEs. By contrast, the proposed method does well in dealing with the numerical solution of those real process models. Irrespective of model complexity (i.e., nonlinearity, high dimensionality, stiffness, etc.), the proposed method can offer valid approximate solutions in any desired time interval, which may be quite a bit larger than [0, 1].
- As it is evident from Section 5.4, the proposed method is applicable to integer as well as noninteger order process models consisting of FDEs or FDAEs.
- Semianalytical methods, with which the proposed method is compared, involve fractional-order integration of solutions obtained in the previous step or recursive relations which make them more complicated and required more CPU usage. On the contrary, the proposed method is easily implementable.

- The proposed method is absolutely suitable for analyzing mathematical models describing scientific problems in application areas of the fractional calculus.

References

[1] K.B. Oldham, J. Spanier (1974). *The Fractional Calculus: Theory and Applications of Differentiation and Integration to Arbitrary Order*. New York: Dover Publications.

[2] T.M. Atanackovic, S. Pilipovic, B. Stankovic, D. Zorica (2014). *Fractional Calculus with Applications in Mechanics: Vibrations and Diffusion Processes*. London: Wiley.

[3] R. Herrmann (2011). *Fractional Calculus: An Introduction for Physicists*. Singapore: World Scientific Publishing Co. Pte. Ltd.

[4] G.A. Losa, D. Merlini, T.F. Nonnenmacher, E.R. Weibel (2005). *Fractals in Biology and Medicine*. Basel: Birkhauser.

[5] F. Mainardi (2010). *Fractional Calculus and Waves in Linear Viscoelasticity*. London: Imperial College Press.

[6] A.K. Golmankhneh, X.J. Yang, D. Balean (2015). Einstein field equations within local fractional calculus. *Rom. J. Phys.*, vol. 60, pp. 22–31.

[7] M. Biyajima, T. Mizoguchi, N. Suzuki (2015). A new blackbody radiation law based on fractional calculus and its application to NASA COBE data. *Physica A*, vol. 440, pp. 129–138.

[8] W.M. Ahmad, R. El-Khazali (2007). Fractional-order dynamical models of love. *Chaos Soliton. Fract.*, vol. 33, pp. 1367–1375.

[9] L. Song, S. Xu, J. Yang (2007). Dynamical models of happiness with fractional order. *Commun. Nonlinear Sci. Numer. Simulat.*, vol. 15, pp. 616–628.

[10] I. Podlubny (1999). *Fractional Differential Equations*. New York: Academic Press.

[11] K. Diethelm (1997). An algorithm for the numerical solution of differential equations of fractional order. *Electron. Trans. Numer. Anal.*, vol. 5, pp. 1–5.

[12] Z.M. Odibat, S. Momani (2008). An algorithm for the numerical solution of differential equations of fractional order. *J. Appl. Math. Informatics*, vol. 26, pp. 15–27.

[13] G. Adomian (1994). *Solving Frontier Problems of Physics: The Decomposition Method*. Boston: Kluwer Academic Publishers.

[14] J.H. He (1999a). Variational iteration method – A kind of non-linear analytical technique: Some examples. *Int. J. Non-Linear Mech.*, vol. 34, pp. 699–708.

[15] J.H. He (1999b). Homotopy perturbation technique. *Comput. Method Appl. Mech. Eng.*, vol. 178, pp. 257–262.

[16] A. Arigoklu, I. Ozkol (2007). Solution of fractional differential equations by using differential transform method. *Chaos Soliton Fract.*, vol. 34, pp. 1473–1481.

[17] M. Zurigat, S. Momani, A. Alawneh (2010). Analytical approximate solutions of systems of fractional algebraic–Differential equations by homotopy analysis method. *Comput. Math. Appl.*, vol. 59, pp. 1227–1235.

[18] B. Ibiş, M. Bayram (2011). Numerical comparison of methods for solving fractional differential-algebraic equations (FDAEs). *Comput. Math. Appl.*, vol. 62, pp. 3270–3278.
[19] B. Ibiş, M. Bayram, A.G. Agargun (2011). Applications of fractional differential transform method to fractional differential-algebraic equations. *Europ. J. Pure Appl. Math.*, vol. 4(2), pp. 129–141.
[20] O.S. Odetunde, O.A. Taiwo (2015). An algorithm for the approximation of fractional differential-algebraic equations with Caputo-type derivatives. *J. Appl. Computat. Math.*, vol. 4, pp. 242. DOI:10.4172/2168-9679.1000242.
[21] H.M. Jaradat, M. Zurigat, S. Al-Shara, Q. Katatbeh (2014). Toward a new algorithm for systems of fractional differential equations. *Italian J. Pure Appl. Math.*, vol. 32, pp. 579–594.
[22] X.-L. Ding, Y.-L. Jian (2014). Waveform relaxation method for fractional differential-algebraic equations. *Fract. Calc. Appl. Anal.*, vol. 17(3), pp. 585–604.
[23] N.J. Ford, J.A. Connolly (2009). Systems-based decomposition schemes for the approximate solution of multi-term fractional differential equations. *J. Comput. Appl. Math.*, vol. 229, pp. 382–391.
[24] S.C. Shiralashetti, A.B. Deshi (2016). An efficient Haar wavelet collocation method for the numerical solution of multi-term fractional differential equations. *Nonlinear Dyn.*, vol. 83, pp. 293–303.
[25] E. Hesameddini, A. Rahimi, E. Asadollahifard (2016). On the convergence of a new reliable algorithm for solving multi-order fractional differential equations. *Commun. Nonlinear Sci. Numer. Simulat.*, vol. 34, pp. 154–164.
[26] A.E.M. El-Mesiry, A.M.A. El-Sayed, H.A.A. El-Saka (2005). Numerical methods for multi-term fractional (arbitrary) orders differential equations. *Appl. Math. Comput.*, vol. 160, pp. 683–699.
[27] M.A. EL-Sayed, A.E.M. EL-Mesiry, H.A.A. EL-Saka (2004). Numerical solution for multi-term fractional (arbitrary) orders differential equations. *Comput. Appl. Math.*, vol. 23(1), pp. 33–54.
[28] A.M.A. El-Sayed, M.M. Saleh, E.A.A. Ziada (2010). Analytical and numerical solution of multi-term nonlinear differential equations of arbitrary orders. *J. Appl. Math. Comput.*, vol. 33, pp. 375–388.
[29] M. Javidi, N. Nyamoradi (2013). A numerical scheme for solving multi-term fractional differential equations. *Commun. Frac. Calc.*, vol. 4(1), pp. 38–49.
[30] K.D. Kucche, J.J. Trujillo (2017). Theory of system of nonlinear fractional differential equations. *Progr. Fract. Differ. Appl.*, vol. 3(1), pp. 7–18.
[31] W.K. Zahra, M.V. Daele (2017). Discrete spline methods for solving two point fractional Bagley–Torvik equation. *Appl. Math. Comput.*, vol. 296, pp. 42–55.
[32] S. Yuzbasi (2013). Numerical solution of the Bagley–Torvik equation by the Bessel collocation method. *Math. Meth. Appl. Sci.*, vol. 36, pp. 300–312.
[33] M.U. Rehman, R.A. Khan (2012). A numerical method for solving boundary value problems for fractional differential equations. *Appl. Math. Model.*, vol. 36, pp. 894–907.
[34] Q. Din, A.A. Elsadany, H. Khalil (2017). Neimark-Sacker Bifurcation and Chaos Control in a Fractional-Order Plant-Herbivore Model. *Discrete. Dynam. Nat. Soc.*, vol. 2017, Article ID 6312964, 15 pages.

[35] W.-C. Chen (2008). Nonlinear dynamics and chaos in a fractional-order financial system. *Chaos Soliton. Fract.*, vol. 36, pp. 1305–1314.

[36] A.A. Freihat, M. Zurigat, A.H. Handam (2015). The multi-step homotopy analysis method for modified epidemiological model for computer viruses. *Afr. Mat.*, vol. 26(3), pp. 585–595.

[37] F. Mazzia, F. Iavernaro (2003). Test set for initial value problem solvers, Tech. Rep 40. Department of Mathematics, University of Bari, Italy.

[38] H. Shintani (1982). Modified Rosenbrock methods for stiff systems. *Hiroshima Math. J.*, vol. 12, pp. 543–558.

[39] A. Elsaid (2012). Fractional differential transform method combined with the Adomian polynomials. *Appl. Math. Comput.*, vol. 218, pp. 6899–6911.

[40] E. Schafer (1975). A new approach to explain the 'high irradiance responses' of photomorphogenesis on the basis of phytochrome. *J. Math. Biology*, vol. 2, pp. 41–55.

6

Numerical Method for Simulation of Fractional Diffusion-Wave Equation

The following time fractional diffusion-wave equations are considered in this chapter:

$$\frac{\partial^{\alpha} U(x,t)}{\partial t^{\alpha}} + \frac{\partial U(x,t)}{\partial t} = \frac{\partial^2 U(x,t)}{\partial x^2} + q(x,t), \tag{6.1}$$

subject to the initial conditions

$$U(x,0) = f_0(x), \ \frac{\partial U(x,0)}{\partial t} = f_1(x), \ x \in [0,1], \tag{6.2}$$

and boundary conditions

$$U(x,0) = g_0(t), \ U(1,t) = g_1(t), \ t \in [0,1], \tag{6.3}$$

where x and t are the space and time variables, q is a given function in $L^2([0,1] \times [0,1]), f_0, f_1, g_0,$ and g_1 are given functions in $L^2[0,1]$, $\partial^{\alpha} U(x,t)/\partial t^{\alpha}$ is a Caputo time fractional derivative of order $\alpha(1 < \alpha < 2)$ of sufficiently smooth function $U(x,t)$, which is to be found out. If $\alpha = 1$, the model in Equations (6.1) through (6.3) turns to be traditional diffusion equation, whereas for $\alpha = 2$, it represents a traditional wave equation [1].

The fractional diffusion-wave equation has significant applications in signal analysis for transmission, reaction diffusion, propagation of electric signals, and the random walk of suspension flows [2, 3]. It is shown in [4–6] that fractional diffusion equations can provide an adequate and accurate description of the transport processes exhibiting anomalous diffusion. The nonlocal property of the fractional diffusion operator causes computational difficulties that have not been encountered in the context of second-order diffusion equations. The main goal of this chapter is develop an efficient numerical method using triangular functions to simulate fractional diffusion-wave equation given in Equations (6.1) to (6.3). The proposed numerical method is analyzed to verify whether the TF approximate solution can equal the exact solution in the limit of step size tends to zero.

6.1 The Proposed Numerical Method

The time fractional integration is performed on both sides of Equation (6.1):

$$U(x,t)-\left[U(x,0)+t\left.\frac{\partial U(x,t)}{\partial t}\right|_{t=0}\right]+{}_0J_t^{\alpha-1}U(x,t)-U(x,0)\frac{t^{\alpha-1}}{\Gamma(\alpha)}$$
$$={}_0J_t^{\alpha}\left(\frac{\partial^2 U(x,t)}{\partial x^2}\right)+{}_0J_t^{\alpha}q(x,t). \tag{6.4}$$

$$U(x,t)+{}_0J_t^{\alpha-1}U(x,t)-V(x,t)={}_0J_t^{\alpha}\left(\frac{\partial^2 U(x,t)}{\partial x^2}\right)+{}_0J_t^{\alpha}q(x,t), \tag{6.5}$$

where $V(x,t)=U(x,0)+t\left.\frac{\partial U(x,t)}{\partial t}\right|_{t=0}+U(x,0)\frac{t^{\alpha-1}}{\Gamma(\alpha)}$.

Let us expand $\partial^2 U(x,t)/\partial x^2$ into the triangular functions as given here:

$$\frac{\partial^2 U(x,t)}{\partial x^2}=T1_m^T(x)(F_1T1_m(t)+F_2T2_m(t))+T2_m^T(x)(F_3T1_m(t)+F_4T2_m(t)). \tag{6.6}$$

The second-order integration, with respect to x, is carried out on both sides of Equation (6.6):

$${}_0J_x^2\left(\frac{\partial^2 U(x,t)}{\partial x^2}\right)={}_0J_x^2(T1_m^T(x)(F_1T1_m(t)+F_2T2_m(t))+T2_m^T(x)(F_3T1_m(t)+F_4T2_m(t))), \tag{6.7}$$

$$U(x,t)-\left[U(0,t)+x\left.\frac{\partial U(x,t)}{\partial x}\right|_{x=0}\right]$$
$$=\left(T1_m^T(x)\left(P_1^2\right)^T+T2_m^T(x)\left(P_2^2\right)^T\right)(F_1T1_m(t)+F_2T2_m(t))$$
$$+\left(T1_m^T(x)\left(P_3^2\right)^T+T2_m^T(x)\left(P_4^2\right)^T\right)(F_3T1_m(t)+F_4T2_m(t)). \tag{6.8}$$

Putting $x=1$,

$$U(1,t)-\left[U(0,t)+\left.\frac{\partial U(x,t)}{\partial x}\right|_{x=0}\right]$$
$$=\left(T1_m^T(1)\left(P_1^2\right)^T+T2_m^T(1)\left(P_2^2\right)^T\right)(F_1T1_m(t)+F_2T2_m(t))$$
$$+\left(T1_m^T(1)\left(P_3^2\right)^T+T2_m^T(1)\left(P_4^2\right)^T\right)(F_3T1_m(t)+F_4T2_m(t)). \tag{6.9}$$

Equation (6.3) is employed in the above equation, and the result is provided in the following equation:

$$\left.\frac{\partial U(x,t)}{\partial x}\right|_{x=0} = g_1(t) - g_0(t) - \left(T1_m^T(1)\left(P_1^2\right)^T + T2_m^T(1)\left(P_2^2\right)^T\right)(F_1T1_m(t) + F_2T2_m(t)) - \left(T1_m^T(1)\left(P_3^2\right)^T + T2_m^T(1)\left(P_4^2\right)^T\right)(F_3T1_m(t) + F_4T2_m(t)). \quad (6.10)$$

The next equation can be obtained from Equations (6.8) and (6.10):

$$\begin{aligned} U(x,t) &= \left[U(0,t) + x\left.\frac{\partial U(x,t)}{\partial x}\right|_{x=0}\right] \\ &\quad + \left(T1_m^T(x)\left(P_1^2\right)^T + T2_m^T(x)\left(P_2^2\right)^T\right)(F_1T1_m(t) + F_2T2_m(t)) \\ &\quad + \left(T1_m^T(x)\left(P_3^2\right)^T + T2_m^T(x)\left(P_4^2\right)^T\right)(F_3T1_m(t) + F_4T2_m(t)) \\ &= g_0(t) + x(g_1(t) - g_0(t)) - xA + A_1(F_1T1_m(t) + F_2T2_m(t)) \\ &\quad + A_2(F_3T1_m(t) + F_4T2_m(t)) \\ &= \mathrm{H}(x,t) - [A_3A] + A_1(F_1T1_m(t) + F_2T2_m(t)) + A_2(F_3T1_m(t) + F_4T2_m(t)), \end{aligned} \quad (6.11)$$

where $\mathrm{H}(x,t) = g_0(t) + x(g_1(t) - g_0(t))$, $A_3 = T1_m^T(x)C + T2_m^T(x)D$, $C = \begin{bmatrix} c_0 & c_1 & \cdots & c_{m-1} \end{bmatrix}$, $D = \begin{bmatrix} d_0 & d_1 & \cdots & d_{m-1} \end{bmatrix}$, $c_i = ih_1$, $d_i = (i+1)h_1$, $A_1 = \left(T1_m^T(x)\left(P_1^2\right)^T + T2_m^T(x)\left(P_2^2\right)^T\right)$, $A_2 = \left(T1_m^T(x)\left(P_3^2\right)^T + T2_m^T(x)\left(P_4^2\right)^T\right)$, $A = V_0(F_1T1_m(t) + F_2T2_2(t)) + V_1(F_3T1_m(t) + F_4T2_m(t))$, $V_0 = \left(T1_m^T(1)\left(P_1^2\right)^T + T2_m^T(1)\left(P_2^2\right)^T\right)$, $V_1 = \left(T1_m^T(1)\left(P_3^2\right)^T + T2_m^T(1)\left(P_4^2\right)^T\right)$.

The two-variable function $H(x,t)$ is estimated in the TF domain as

$$H(x,t) = T1_m^T(x)\left(F_1^0T1_m(t) + F_2^0T2_m(t)\right) + T2_m^T(x)\left(F_3^0T1_m(t) + F_4^0T2_m(t)\right). \quad (6.12)$$

Equation (6.11) is rewritten as

$$\begin{aligned} U(x,t) &= T1_m^T(x)\left(F_1^0T1_m(t) + F_2^0T2_m(t)\right) + T2_m^T(x)\left(F_3^0T1_m(t) + F_4^0T2_m(t)\right) \\ &\quad - A_3V_0(F_1T1_m(t) + F_2T2_m(t)) - A_3V_1(F_3T1_m(t) + F_4T2_m(t)) \\ &\quad + A_1(F_1T1_m(t) + F_2T2_m(t)) + A_2(F_3T1_m(t) + F_4T2_m(t)). \end{aligned} \quad (6.13)$$

The above equation is expressed in a short form as given here:

$$U(x,t) = A_4 T1_m(t) + A_5 T2_m(t), \tag{6.14}$$

where $A_4 = T1_m^T(x)F_1^0 + T2_m^T(x)F_3^0 - A_3V_0F_1 - A_3V_1F_3 + A_1F_1 + A_2F_3$,

$$A_5 = T1_m^T(x)F_2^0 + T2_m^T(x)F_4^0 - A_3V_0F_2 - A_3V_1F_4 + A_1F_2 + A_2F_4.$$

Equation (6.6) is rewritten in the short form given here:

$$\begin{aligned}\frac{\partial^2 U(x,t)}{\partial x^2} &= T1_m^T(x)(F_1T1_m(t) + F_2T2_m(t)) + T2_m^T(x)(F_3T1_m(t) + F_4T2_m(t)),\\ &= A_6T1_m(t) + A_7T2_m(t),\end{aligned} \tag{6.15}$$

where $A_6 = T1_m^T(x)F_1 + T2_m^T(x)F_3$, $A_7 = T1_m^T(x)F_2 + T2_m^T(x)F_4$.

Likewise, the functions $q(x,t)$ and $V(x,t)$ are approximated in terms of triangular functions.

$$q(x,t) = A_8T1_m(t) + A_9T2_m(t),\ \ V(x,t) = A_{10}T1_m(t) + A_{11}T2_m(t), \tag{6.16}$$

where

$$\begin{aligned}A_8 &= T1_m^T(x)F_1^{10} + T2_m^T(x)F_3^{10}, \quad A_9 = T1_m^T(x)F_2^{10} + T2_m^T(x)F_4^{10},\\ A_{10} &= T1_m^T(x)F_1^{20} + T2_m^T(x)F_3^{20}, \quad A_{11} = T1_m^T(x)F_2^{20} + T2_m^T(x)F_4^{20}.\end{aligned}$$

The following equation can be obtained by employing Equations (6.14) to (6.16) in Equation (6.6):

$$\begin{aligned}A_4T1_m(t) + A_5T2_m(t) + {}_0J_t^{\alpha-1}(A_4T1_m(t) + A_5T2_m(t))\\ - (A_{10}T1_m(t) + A_{11}T2_m(t)) = A_{12},\end{aligned} \tag{6.17}$$

where $A_{12} = {}_0J_t^{\alpha}(A_6T1_m(t) + A_7T2_m(t)) + {}_0J_t^{\alpha}(A_8T1_m(t) + A_9T2_m(t))$.

Equation (1.97) is used in place of fractional order integrals in the above equation.

$$\begin{aligned}A_4T1_m(t) + A_5T2_m(t) + A_4\left(P_1^{\alpha-1}T1_m(t) + P_2^{\alpha-1}T2_m(t)\right)\\ + A_5\left(P_3^{\alpha-1}T1_m(t) + P_4^{\alpha-1}T2_m(t)\right) - A_{13} = A_{14},\end{aligned} \tag{6.18}$$

where $A_{13} = (A_{10}T1_m(t) + A_{11}T2_m(t))$, $A_{14} = A_6\left(P_1^{\alpha}T1_m(t) + P_2^{\alpha}T2_m(t)\right) + A_7\left(P_3^{\alpha}T1_m(t) + P_4^{\alpha}T2_m(t)\right) + A_{15}$,

$$A_{15} = A_8\left(P_1^\alpha T1_m(t) + P_2^\alpha T2_m(t)\right) + A_9\left(P_3^\alpha T1_m(t) + P_4^\alpha T2_m(t)\right).$$

By equating the coefficients of LHTF and RHTF vector, we get the following equations:

$$A_4 + A_4P_1^{\alpha-1} + A_5P_3^{\alpha-1} - A_{10} = A_6P_1^\alpha + A_7P_3^\alpha + A_8P_1^\alpha + A_9P_3^\alpha, \tag{6.19}$$

$$A_5 + A_4P_2^{\alpha-1} + A_5P_4^{\alpha-1} - A_{11} = A_6P_2^\alpha + A_7P_4^\alpha + A_8P_2^\alpha + A_9P_4^\alpha. \tag{6.20}$$

Equations (6.19) and (6.20) are expressed in terms of triangular functions $T1_m(x)$ and $T2_m(x)$.

$$\begin{aligned} &T1_m^T(x)\left[A_{16} + A_{16}P_1^{\alpha-1} + A_{18}P_3^{\alpha-1} - F_1^{20}\right] \\ &\quad + T2_m^T(x)\left[A_{17} + A_{17}P_1^{\alpha-1} + A_{19}P_3^{\alpha-1} - F_3^{20}\right] = A_{20}, \end{aligned} \tag{6.21}$$

$$\begin{aligned} &T1_m^T(x)\left[A_{18} + A_{16}P_2^{\alpha-1} + A_{18}P_4^{\alpha-1} - F_2^{20}\right] \\ &\quad + T2_m^T(x)\left[A_{19} + A_{17}P_2^{\alpha-1} + A_{19}P_4^{\alpha-1} - F_4^{20}\right] = A_{21}, \end{aligned} \tag{6.22}$$

where $A_{16} = F_1^0 - CV_0F_1 - CV_1F_3 + \left(P_1^2\right)^T F_1 + \left(P_3^2\right)^T F_3,$

$$A_{17} = F_3^0 - DV_0F_1 - DV_1F_3 + \left(P_2^2\right)^T F_1 + \left(P_4^2\right)^T F_3,$$

$$A_{18} = F_2^0 - CV_0F_2 - CV_1F_4 + \left(P_1^2\right)^T F_2 + \left(P_3^2\right)^T F_4,$$

$$A_{19} = F_4^0 - DV_0F_2 - DV_1F_4 + \left(P_2^2\right)^T F_2 + \left(P_4^2\right)^T F_4,$$

$$\begin{aligned} A_{20} = {} &T1_m^T(x)\left[F_1P_1^\alpha + F_2P_3^\alpha + F_1^{10}P_1^\alpha + F_2^{10}P_3^\alpha\right] \\ &+ T2_m^T(x)\left[F_3P_1^\alpha + F_4P_3^\alpha + F_3^{10}P_1^\alpha + F_4^{10}P_3^\alpha\right], \end{aligned}$$

$$\begin{aligned} A_{21} = {} &T1_m^T(x)\left[F_1P_2^\alpha + F_2P_4^\alpha + F_1^{10}P_2^\alpha + F_2^{10}P_4^\alpha\right] \\ &+ T2_m^T(x)\left[F_3P_2^\alpha + F_4P_{44}^{\alpha\alpha} + F_3^{10}P_2^\alpha + F_4^{10}P_4^\alpha\right]. \end{aligned}$$

By comparing the coefficients of $T1_m^T(x)$ and $T2_m^T(x)$, we get the following:

$$A_{16} + A_{16}P_1^{\alpha-1} + A_{18}P_3^{\alpha-1} - F_1^{20} = F_1P_1^\alpha + F_2P_3^\alpha + F_1^{10}P_1^\alpha + F_2^{10}P_3^\alpha, \tag{6.23}$$

$$A_{18} + A_{16}P_2^{\alpha-1} + A_{18}P_4^{\alpha-1} - F_2^{20} = F_1P_2^\alpha + F_2P_4^\alpha + F_1^{10}P_2^\alpha + F_2^{10}P_4^\alpha, \tag{6.24}$$

$$A_{17} + A_{17}P_1^{\alpha-1} + A_{19}P_3^{\alpha-1} - F_3^{20} = F_3P_1^\alpha + F_4P_3^\alpha + F_3^{10}P_1^\alpha + F_4^{10}P_3^\alpha, \tag{6.25}$$

$$A_{19} + A_{17}P_2^{\alpha-1} + A_{19}P_4^{\alpha-1} - F_4^{20} = F_3P_2^{\alpha} + F_4P_4^{\alpha} + F_3^{10}P_2^{\alpha} + F_4^{10}P_4^{\alpha}. \qquad (6.26)$$

The matrices F_1, F_2, F_3, and F_4 can be obtained by solving Equations (6.23) to (6.26). Utilizing the computed matrices in Equation (6.14) yields the TF approximate solution of fractional diffusion-wave equation.

6.2 Convergence Analysis

The given intervals $x \in [0,1]$ and $t \in [0,1]$ are split into an m number of subintervals of constant width h.

$$[0,h], [h,2h], [2h,3h], \ldots\ldots, [(m-1)h, mh]. \qquad (6.27)$$

Arbitrary subintervals $x \in [ih,(i+1)h)$ and $t \in [jh,(j+1)h)$, $i = 0,1,2,\ldots,$ $m-1$, $j = 0,1,2,\ldots,m-1$ are chosen.

Let $U(x,t)$ and $\tilde{U}(x,t)$ be the exact and TF approximate solution, respectively. The error between the exact and the approximate solution is defined in the following equation:

$$\varepsilon = \left\| U(x,t) - \tilde{U}(x,t) \right\|_{\infty}. \qquad (6.28)$$

The TF approximate for $U(x,t)$ in the selected subintervals is obtained by using Equation (1.78):

$$\begin{aligned}\tilde{U}(x,t) &= U(ih,jh)T1_i(x)T1_j(t) + U(ih,(j+1)h)T1_i(x)T2_j(t) \\ &\quad + U((i+1)h,jh)T2_i(x)T1_j(t) + U((i+1)h,(j+1)h)T2_i(x)T2_j(t).\end{aligned} \qquad (6.29)$$

Using Equations (1.65) and (1.66) in Equation (6.29), we get the following.

$$\begin{aligned}\tilde{U}(x,t) &= U(ih,jh)\left[1-\left(\frac{x-ih}{h}\right)\right]\left[1-\left(\frac{t-jh}{h}\right)\right] \\ &\quad + U(ih,(j+1)h)\left[1-\left(\frac{x-ih}{h}\right)\right]\left(\frac{t-jh}{h}\right) \\ &\quad + \mathrm{U}((i+1)h,jh)\left(\frac{x-ih}{h}\right)\left[1-\left(\frac{t-jh}{h}\right)\right] \\ &\quad + \mathrm{U}((i+1)h,(j+1)h)\left(\frac{x-ih}{h}\right)\left(\frac{t-jh}{h}\right).\end{aligned} \qquad (6.30)$$

The above equation can be rewritten in the following form:

$$\begin{aligned}\tilde{U}(x,t) &= U(ih,jh)+\frac{U((i+1)h,jh)-U(ih,jh)}{h}(x-ih)+\frac{U(ih,(j+1)h)-U(ih,jh)}{h}(t-jh)\\ &+\frac{U((i+1)h,(j+1)h)-U(ih,(j+1)h)-U((i+1)h,jh)+U(ih,jh)}{h^2}\\ &(x-ih)(t-jh),\end{aligned} \tag{6.31}$$

$$\begin{aligned}\tilde{U}(x,t) = U(ih,jh) &+ \left.\frac{\partial U(x,t)}{\partial x}\right|_{x=ih,t=jh}(x-ih)+\left.\frac{\partial U(x,t)}{\partial t}\right|_{x=ih,t=jh}(t-jh)\\ &+\left.\frac{\partial^2 U(x,t)}{\partial x\partial t}\right|_{x=ih,t=jh}(x-ih)(t-jh).\end{aligned} \tag{6.32}$$

The exact solution $U(x,t)$ is expanded by using Taylor series, and the first six terms of the resulting infinite series are considered.

$$\begin{aligned}U(x,t) = U(ih,jh) &+ \left.\frac{\partial U(x,t)}{\partial x}\right|_{x=ih,t=jh}(x-ih)+\left.\frac{\partial U(x,t)}{\partial t}\right|_{x=ih,t=jh}(t-jh)\\ &+\left.\frac{\partial^2 U(x,t)}{\partial x^2}\right|_{x=ih,t=jh}\frac{(x-ih)^2}{2}+\left.\frac{\partial^2 U(x,t)}{\partial t^2}\right|_{x=ih,t=jh}\frac{(t-jh)^2}{2}\\ &+\left.\frac{\partial^2 U(x,t)}{\partial x\partial t}\right|_{x=ih,t=jh}(x-ih)(t-jh).\end{aligned} \tag{6.33}$$

From Equations (6.28), (6.32), and (6.33),

$$\begin{aligned}\varepsilon &= \left\|\left.\frac{\partial^2 U(x,t)}{\partial x^2}\right|_{x=ih,t=jh}\frac{(x-ih)^2}{2}+\left.\frac{\partial^2 U(x,t)}{\partial t^2}\right|_{x=ih,t=jh}\frac{(t-jh)^2}{2}\right\|_\infty,\\ &\le \left\|\left.\frac{\partial^2 U(x,t)}{\partial x^2}\right|_{x=ih,t=jh}\right\|\left|\frac{(x-ih)^2}{2}\right|+\left\|\left.\frac{\partial^2 U(x,t)}{\partial t^2}\right|_{x=ih,t=jh}\right\|\left|\frac{(t-jh)^2}{2}\right|.\end{aligned} \tag{6.34}$$

The following are assumed:

$$M_1 = \left\| \frac{\partial^2 U(x,t)}{\partial x^2} \bigg|_{x=ih,t=jh} \right\|, \; M_2 = \left\| \frac{\partial^2 U(x,t)}{\partial t^2} \bigg|_{x=ih,t=jh} \right\|. \tag{6.35}$$

From Equations (6.34) and (6.35),

$$\varepsilon \le \left| M_1 \frac{(x-ih)^2}{2} + M_2 \frac{(t-jh)^2}{2} \right|. \tag{6.36}$$

Because $x \in [ih, (i+1)h)$ and $t \in [jh, (j+1)h)$, the terms $(x-ih)$ and $(t-jh)$ are always less than h.

Then Equation (6.36) becomes

$$\varepsilon \le \left| M_1 \frac{h^2}{2} + M_2 \frac{h^2}{2} \right|. \tag{6.37}$$

As the step size decreases to zero,

$$\lim_{h \to 0} \varepsilon = 0. \tag{6.38}$$

The error between the exact and TF solution becomes negligible when sufficiently large number of subintervals are selected.

References

[1] W. Deng, C. Li, Q. Guo (2007). Analysis of fractional differential equations with multi-orders. *Fractals.*, vol. 15, pp. 173–182.

[2] M. Aslefallah, E. Shivanian (2015). Nonlinear fractional integro-differential reaction–Diffusion equation via radial basis functions. *Eur. Phys. J. Plus*, vol. 47, pp. 1–9.

[3] M.M. Meerschaert, C. Tadjeran (2006). Finite difference approximations for two-sided space-fractional partial differential equations. *Appl. Numer. Math.*, vol. 56, pp. 80–90.

[4] R. Metler, J. Klafter (2004). The restaurant at the end of random walk: Recent developments in the description of anomalous transport by fractional dynamics. *J. Phys. A*, vol. 37, pp. 161–208.

[5] G.M. Zaslavsky (2002). Chaos, fractional kinetics, and anomalous transport. *Phys. Rep.*, vol. 371, pp. 461–580.

[6] Z. Wang, S. Vong (2014). Compact difference schemes for the modified anomalous fractional sub-diffusion equation and the fractional diffusion-wave equation. *J. Comput. Phys.*, vol. 277, pp. 1–15.

7

Identification of Fractional Order Linear and Nonlinear Systems from Experimental or Simulated Data

As was seen with integer calculus, fractional calculus (FC) has two benefits: first, FC can mathematically represent long memory processes; this enables FC to explain hereditary properties of real materials; second, FC can describe real processes such as thermal diffusion in a wall [1, 2], semi-infinite lossy (RC) transmission line [3], the rotor skin effect of induction machine [4], and viscoelastic systems [5], which demonstrate fractional behavior in a brief and accurate way. Many physical processes initially modeled by integer calculus have been later described more efficiently by FC. Specifically, the romantic relationship between male and female partners [6], peoples' emotions such as happiness [7], computer viruses spread in a network of computers [8], rheological behavior of glucose and insulin in type 2 diabetic patients [9], interaction between a plant and an insect [10], vibration of large membranes [11], evolution of diclofenac in a small number of healthy adults [12], human calcaneal fat pad [13], electrical impedance of wine samples [14], and so on have been fruitfully modeled by FC. Because the fractional order derivative and fractional order integral are lacking clear physical meaning (although some attempts made to give physical interpretation [15]), it is problematic to model real systems using FC based on mechanism analysis. One alternative is to estimate the parameters (fractional differential orders, coefficients of fractional order operators, and initial conditions) of a fractional order model, which can exactly reproduce physical system's behavior, from measured input and output data of the real process under consideration. Fractional order system (FOS) identification is more complex than the integer order system identification as FOS identification problem involves an extra task of estimating order of differential operators; hence, the existing classical (integer-order) system identification methods cannot be directly extended to identify FOS from measured data.

Fractional order system identification can be performed in the frequency domain and the time domain. Dzielinski et al. [16] used an identification method based on least squares and total least squares, respectively, to identify fractional order models from frequency response data obtained from a laboratory setup of ultracapacitors. However, they chose some of the model parameters manually to reduce the complexity of the identification problem and to avoid the dependency of results on initial conditions. Valerio et al. [17] formulated a frequency domain system identification method based on the Levy identification method to estimate parameters of fractional commensurate transfer function, and applied the method to an engineering problem of generating electricity from sea waves. Valerio and Tejado [18] extended Levy's identification method to noncommensurable fractional order models. In [19], order distributions based frequency domain system identification method was developed to approximate the parameters of commensurate fractional order system.

A new identification technique based on equation error was proposed by Djouambi et al. [20] to provide recursive parameter estimates of the fractional order models. In [20], a fractional order ARX that was obtained by discretization of the continuous time fractional order differential equations was used as the identification model and the parameters of the fractional order ARX model were estimated by extending the classical algorithms by extending recursive least squares algorithm and recursive instrumental variable algorithm. Their method estimates only model parameters, not the fractional differential orders that have to be supplied by the user. Poinot and Trigeassou [21] developed an identification method based on an output-error technique to estimate the parameters including fractional differential orders of the fractional order system and applied their method to model the dynamics of a real heat transfer system. Maachou et al. [22] extended the Volterra series to fractional derivatives and used the generalized orthogonal functions as Volterra kernels to identify a model, which captures system uncertainties, of nonlinear thermal diffusion phenomena for large temperature variations involving variable thermal parameters such as thermal diffusivity and thermal conductivity, and showed that the identified model is always more accurate than the finite element model. Stochastic optimization algorithm, that is, particle swarm optimization (PSO), was used in [23] to identify fractional order models of a circulating fluidized bed boiler, and a new variant of PSO called quantum parallel PSO was proposed in [24] to estimate parameters of fractional order chaotic systems. Du et al. [25] introduced composite differential evolution algorithm for the identification of fractional order chaotic system with not only unknown parameters and fractional differential orders but also unknown initial values and fractional order model structure, and applied it to a set of fractional-order chaotic systems such as Lorenz, Lu, Chen, Rossler, Arneodo, and Volta chaotic systems. Many other works addressing time domain system identification problems were reported in [26–31].

Orthogonal functions such as block pulse functions (BPFs), Haar wavelets (HW), and Laguerre orthogonal functions also have applications in fractional order system identification. In [32], as the classical Laguerre orthogonal functions fail to capture the nonexponential aperiodic multimode behavior of the fractional differential systems, the fractional Laguerre orthogonal functions were synthesized by applying the Gram-Schmidt orthogonalization procedure for approximating fractional systems and for fractional commensurate system identification. In [33, 34], free FOS identification method based on HW and BPFs, respectively, were developed. In this chapter, a new arbitrary order (note that the order can be integer and noninteger) identification method based on triangular orthogonal functions is proposed to estimate parameters including fractional differential orders, coefficients of fractional operators and initial conditions of linear and nonlinear fractional order systems. To the best of our knowledge, it is the first time that TFs have been used in parameter identification of arbitrary (integer as well as noninteger order) linear and nonlinear systems from simulated input-output data.

7.1 Fractional Order System (FOS) Identification using TFs

7.1.1 Linear FOS Identification

Consider a fractional order model described by the fractional order transfer function (FOTF)

$$G(s) = \frac{Y(s)}{U(s)} = \frac{b_m s^{\beta_m} + b_{m-1} s^{\beta_{m-1}} + \ldots\ldots + b_0 s^{\beta_0}}{a_n s^{\alpha_n} + a_{n-1} s^{\alpha_{n-1}} + \ldots\ldots + a_0 s^{\alpha_0}}, \tag{7.1}$$

where $\beta_0 < \beta_1 < \ldots < \beta_m$, $\alpha_0 < \alpha_1 < \ldots < \alpha_n$, and $\alpha_n > \beta_m$.

The goal of this chapter is to identify the set of model parameters including the fractional differential orders θ ($\theta = [a_0, \ldots, a_n, b_0, \ldots, b_m, \alpha_0, \ldots, \alpha_n, \beta_0, \ldots, \beta_m]$) according to the measured input and output data, that is, the identified fractional order model gives the same response as the original fractional order plant for a particular input excitation signal.

The output and input of the linear fractional order model in Equation (7.1) can be expressed via TFs as

$$\tilde{y}(t) = y(t) \approx C^T T1_m(t) + D^T T2_m(t), \tag{7.2}$$

$$u(t) \approx C_0^T T1_m(t) + D_0^T T2_m(t). \tag{7.3}$$

In the same fashion, the Riemann-Liouville fractional order integral of the output and input of Equation (7.1) can be estimated by using Equation (1.97):

$${}_0J_t^\alpha \tilde{y}(t) = (C^T P_1^\alpha + D^T P_3^\alpha) T1_m(t) + (C^T P_2^\alpha + D^T P_4^\alpha) T2_m(t), \tag{7.4}$$

$${}_0J_t^\alpha u(t) \approx (C_0^T P_1^\alpha + D_0^T P_3^\alpha) T1_m(t) + (C_0^T P_2^\alpha + D_0^T P_4^\alpha) T2_m(t). \tag{7.5}$$

Recall the FOTF,

$$\frac{Y(s)}{U(s)} = \frac{b_m s^{\beta_m} + b_{m-1} s^{\beta_{m-1}} + \ldots\ldots + b_0 s^{\beta_0}}{a_n s^{\alpha_n} + a_{n-1} s^{\alpha_{n-1}} + \ldots\ldots + a_0 s^{\alpha_0}}. \tag{7.6}$$

Rewriting Equation (7.6),

$$\frac{Y(s)}{U(s)} = \frac{b_m s^{\beta_m - \alpha_n} + b_{m-1} s^{\beta_{m-1} - \alpha_n} + \ldots\ldots + b_0 s^{\beta_0 - \alpha_n}}{a_n + a_{n-1} s^{\alpha_{n-1} - \alpha_n} + \ldots\ldots + a_0 s^{\alpha_0 - \alpha_n}}. \tag{7.7}$$

Rearranging Equation (7.7) gives

$$\begin{aligned} a_n Y(s) + a_{n-1} \frac{Y(s)}{s^{\alpha_n - \alpha_{n-1}}} + \ldots\ldots + a_0 \frac{Y(s)}{s^{\alpha_n - \alpha_0}} = b_m \frac{U(s)}{s^{\alpha_n - \beta_m}} + b_{m-1} \frac{U(s)}{s^{\alpha_n - \beta_{m-1}}} \\ + \ldots\ldots + b_0 \frac{U(s)}{s^{\alpha_n - \beta_0}}. \end{aligned} \tag{7.8}$$

As per the Laplace transform of the R-L fractional order integral,

$$\begin{aligned} & a_n L(\tilde{y}(t)) + a_{n-1} L\left({}_0 J_t^{\alpha_n - \alpha_{n-1}} \tilde{y}(t)\right) + \cdots\cdots + a_0 L\left({}_0 J_t^{\alpha_n - \alpha_0} \tilde{y}(t)\right) \\ & = b_m L\left({}_0 J_t^{\alpha_n - \beta_m} u(t)\right) + b_{m-1} L\left({}_0 J_t^{\alpha_n - \beta_{m-1}} u(t)\right) + \cdots\cdots + b_0 L\left({}_0 J_t^{\alpha_n - \beta_0} u(t)\right). \end{aligned} \tag{7.9}$$

Applying the inverse Laplace transform to Equation (7.9),

$$\begin{aligned} & a_n \tilde{y}(t) + a_{n-1} {}_0 J_t^{\alpha_n - \alpha_{n-1}} \tilde{y}(t) + \cdots\cdots + a_0 {}_0 J_t^{\alpha_n - \alpha_0} \tilde{y}(t) \\ & \quad = b_m {}_0 J_t^{\alpha_n - \beta_m} u(t) + b_{m-1} {}_0 J_t^{\alpha_n - \beta_{m-1}} u(t) + \cdots\cdots + b_0 {}_0 J_t^{\alpha_n - \beta_0} u(t). \end{aligned} \tag{7.10}$$

Using Equations (7.2) to (7.5),

$$\begin{aligned} & a_n\left(C^T T1_m(t) + D^T T2_m(t)\right) + a_{n-1}\left(\left(C^T P_1^{\gamma_1} + D^T P_3^{\gamma_1}\right) T1_m(t) + \left(C^T P_2^{\gamma_1} + D^T P_4^{\gamma_1}\right) T2_m(t)\right) + \cdots \\ & \cdots + a_0\left(\left(C^T P_1^{\gamma_2} + D^T P_3^{\gamma_2}\right) T1_m(t) + \left(C^T P_2^{\gamma_2} + D^T P_4^{\gamma_2}\right) T2_m(t)\right) = b_m\left(C_0^T P_1^{\gamma_3} + D_0^T P_3^{\gamma_3}\right) T1_m(t) + \\ & b_m\left(C_0^T P_2^{\gamma_3} + D_0^T P_4^{\gamma_3}\right) T2_m(t) + b_{m-1}\left(C_0^T P_1^{\gamma_4} + D_0^T P_3^{\gamma_4}\right) T1_m(t) + b_{m-1}\left(C_0^T P_2^{\gamma_4} + D_0^T P_4^{\gamma_4}\right) T2_m(t) + \cdots \\ & \cdots + b_0\left(C_0^T P_1^{\gamma_5} + D_0^T P_3^{\gamma_5}\right) T1_m(t) + \left(C_0^T P_2^{\gamma_5} + D_0^T P_4^{\gamma_5}\right) T2_m(t), \end{aligned} \tag{7.11}$$

where $\gamma_1 = \alpha_n - \alpha_{n-1}$, $\gamma_2 = \alpha_n - \alpha_0$, $\gamma_3 = \alpha_n - \beta_m$, $\gamma_4 = \alpha_n - \beta_{m-1}$, $\gamma_5 = \alpha_n - \beta_0$.

Equating the coefficients of $T1_m(t)$,

$$a_n C^T + a_{n-1}\left(C^T P_1^{\gamma_1} + D^T P_3^{\gamma_1}\right) + \ldots + a_0\left(C^T P_1^{\gamma_2} + D^T P_3^{\gamma_2}\right)$$
$$= b_m\left(C_0^T P_1^{\gamma_3} + D_0^T P_3^{\gamma_3}\right) + b_{m-1}\left(C_0^T P_1^{\gamma_4} + D_0^T P_3^{\gamma_4}\right) + \ldots + b_0\left(C_0^T P_1^{\gamma_5} + D_0^T P_3^{\gamma_5}\right). \tag{7.12}$$

Equating the coefficients of $T2_m(t)$,

$$a_n D^T + a_{n-1}\left(C^T P_2^{\gamma_1} + D^T P_4^{\gamma_1}\right) + \ldots + a_0\left(C^T P_2^{\gamma_2} + D^T P_4^{\gamma_2}\right)$$
$$= b_m\left(C_0^T P_2^{\gamma_3} + D_0^T P_4^{\gamma_3}\right) + b_{m-1}\left(C_0^T P_2^{\gamma_4} + D_0^T P_4^{\gamma_4}\right) + \ldots + b_0\left(C_0^T P_2^{\gamma_5} + D_0^T P_4^{\gamma_5}\right). \tag{7.13}$$

Rearranging Equations (7.12) and (7.13) in the following form,

$$\underbrace{\begin{bmatrix} C^T & D^T \end{bmatrix}}_{X} \underbrace{\begin{bmatrix} A_1 & A_2 \\ B_1 & B_2 \end{bmatrix}}_{A} = \underbrace{\begin{bmatrix} Z_1 & Z_2 \end{bmatrix}}_{Z}, \tag{7.14}$$

where $A_1 = a_n I + a_{n-1} P_1^{\gamma_1} + \ldots + a_0 P_1^{\gamma_2}$, $B_1 = a_{n-1} P_3^{\gamma_1} + \ldots + a_0 P_3^{\gamma_2}$, $A_2 = a_{n-1} P_2^{\gamma_1} + \ldots + a_0 P_2^{\gamma_2}$,

$B_2 = a_n I + a_{n-1} P_4^{\gamma_1} + \ldots + a_0 P_4^{\gamma_2}$, I is an identity matrix of size $m \times m$,

$Z_1 = b_m\left(C_0^T P_1^{\gamma_3} + D_0^T P_3^{\gamma_3}\right) + b_{m-1}\left(C_0^T P_1^{\gamma_4} + D_0^T P_3^{\gamma_4}\right) + \ldots + b_0\left(C_0^T P_1^{\gamma_5} + D_0^T P_3^{\gamma_5}\right)$,

$Z_2 = b_m\left(C_0^T P_2^{\gamma_3} + D_0^T P_4^{\gamma_3}\right) + b_{m-1}\left(C_0^T P_2^{\gamma_4} + D_0^T P_4^{\gamma_4}\right) + \ldots + b_0\left(C_0^T P_2^{\gamma_5} + D_0^T P_4^{\gamma_5}\right)$.

Rewriting Equation (7.14),

$$X = ZA^{-1}. \tag{7.15}$$

From Equations (7.2) and (7.14), the approximation for the output of the fractional order model in Equation (7.1) can be obtained.

The linear FOS identification problem is mathematically defined by

$$J = \min_{\hat{\theta} \in \Lambda} \left\|y(t) - \tilde{y}(t)\right\|_2, \tag{7.16}$$

where $y(t)$ and $\tilde{y}(t)$ are the measured output and the output of the fractional order model in Equation (7.1), respectively, $\hat{\theta}$ is the estimation obtained by the proposed TFs-based linear FOS identification method for the set of true model parameters θ, is the admitted search range of model parameters.

7.1.2 Nonlinear FOS Identification

Consider the SISO nonlinear FOS system described by

$$ {}_0^C D_t^{\alpha} y(t) + \sum_{k=1}^{r} a_k {}_0^C D_t^{\beta_k} y(t) + bF(t, y(t)) = \sum_{j=1}^{q} c_j {}_0^C D_t^{\gamma_j} u(t) + du(t), \tag{7.17} $$

where a_k, b, c_k, d are arbitrary constants, $y(t)$ and $u(t)$ are the output and input signals of the nonlinear FOS, $F(y(t))$ is a nonlinear function, $\alpha > \beta_1 > \ldots > \beta_r$, $\alpha > \gamma_1 > \ldots > \gamma_q$, where α, β_k, γ_k are positive nonintegers.

The set of model parameters including the fractional differential orders and the initial conditions to be identified is

$$ \begin{aligned} \theta = [a_1, \ldots, a_r, b, c_1, \ldots, c_q, d, \alpha, \beta_1, \ldots, \beta_r, \gamma_1, \ldots, \gamma_q, y(0), y^{(1)}(0), \\ y^{(2)}(0), \ldots, y^{(\lfloor \alpha \rfloor)}(0)]. \end{aligned} \tag{7.18} $$

Equation (7.17) can be written as

$$ \begin{aligned} & y(t) + \sum_{k=1}^{r} a_{k0} J_t^{\alpha-\beta_k} y(t) + b_0 J_t^{\alpha} F(t, y(t)) \\ & = \sum_{j=1}^{q} c_{j0} J_t^{\alpha-\gamma_j} u(t) + d_0 J_t^{\alpha} u(t) + A_1(t) + \sum_{k=1}^{r} a_{k0} J_t^{\alpha-\beta_k} A_2(t) - \sum_{j=1}^{q} c_{j0} J_t^{\alpha-\gamma_j} A_3(t), \end{aligned} \tag{7.19} $$

where

$$ A_1(t) = \sum_{f=0}^{\lfloor \alpha \rfloor} \frac{t^f}{\Gamma(f+1)} y^{(f)}(0), $$

$$ A_2(t) = \sum_{g=0}^{\lfloor \beta_k \rfloor} y^{(g)}(0) \frac{t^g}{\Gamma(g+1)}, \quad A_3(t) = \sum_{h=0}^{\lfloor \gamma_j \rfloor} \frac{t^h}{\Gamma(h+1)} u^{(h)}(0). $$

In the TF domain, the input and output signals, the nonlinear function, A_1, A_2 and A_3 are estimated as

$$ \tilde{y}(t) = y(t) \approx C^T T1_m(t) + D^T T2_m(t), \tag{7.20} $$

$$ u(t) \approx C_0^T T1_m(t) + D_0^T T2_m(t), \tag{7.21} $$

$$F(t, y(t)) \approx C_n^T T1_m(t) + D_n^T T2_m(t), \tag{7.22}$$

$$A_1(t) \approx C_1^T T1_m(t) + D_1^T T2_m(t), \tag{7.23}$$

$$A_2(t) \approx C_2^T T1_m(t) + D_2^T T2_m(t), \tag{7.24}$$

$$A_3(t) \approx C_3^T T1_m(t) + D_3^T T2_m(t). \tag{7.25}$$

Equation (7.19) is modified using Equations (1.97) and Equations (7.20) to (7.25):

$$\begin{aligned}
&C^T T1_m(t) + D^T T2_m(t) + \sum_{k=1}^{r} a_k((C^T P_1^{\psi_1} + D^T P_3^{\psi_1}) T1_m(t) + (C^T P_2^{\psi_1} + D^T P_4^{\psi_1}) T2_m(t)) + \\
&b((C_n^T P_1^{\psi_2} + D_n^T P_3^{\psi_2}) T1_m(t) + (C_n^T P_2^{\psi_2} + D_n^T P_4^{\psi_2}) T2_m(t)) = C_1^T T1_m(t) + D_1^T T2_m(t) + \\
&\sum_{j=1}^{q} c_j((C_0^T P_1^{\psi_3} + D_0^T P_3^{\psi_3}) T1_m(t) + (C_0^T P_2^{\psi_3} + D_0^T P_4^{\psi_3}) T2_m(t)) + d(C_0^T P_1^{\psi_2} + D_0^T P_3^{\psi_2}) T1_m(t) + \\
&d(C_0^T P_2^{\psi_2} + D_0^T P_4^{\psi_2}) T2_m(t) + \sum_{k=1}^{r} a_k((C_2^T P_1^{\psi_4} + D_2^T P_3^{\psi_4}) T1_m(t) + (C_2^T P_2^{\psi_4} + D_2^T P_4^{\psi_4}) T2_m(t)) - \\
&\sum_{j=1}^{q} c_j((C_3^T P_1^{\psi_5} + D_3^T P_3^{\psi_5}) T1_m(t) + (C_3^T P_2^{\psi_5} + D_3^T P_4^{\psi_5}) T2_m(t)),
\end{aligned} \tag{7.26}$$

where $\psi_1 = \alpha - \beta_k, \psi_2 = \alpha, \psi_3 = \alpha - \gamma_j, \psi_4 = \alpha - \beta_k, \psi_5 = \alpha - \gamma_j$. Comparing the coefficients of $T1_m(t)$ and $T2_m(t)$,

$$\begin{aligned}
&C^T + \sum_{k=1}^{r} a_k (C^T P_1^{\psi_1} + D^T P_3^{\psi_1}) + b(C_n^T P_1^{\psi_2} + D_n^T P_3^{\psi_2}) \\
&= C_1^T + \sum_{j=1}^{q} c_j (C_0^T P_1^{\psi_3} + D_0^T P_3^{\psi_3}) + dC_0^T P_1^{\psi_2} + \\
&dD_0^T P_3^{\psi_2} + \sum_{k=1}^{r} a_k (C_2^T P_1^{\psi_4} + D_2^T P_3^{\psi_4}) - \sum_{j=1}^{q} c_j (C_3^T P_1^{\psi_5} + D_3^T P_3^{\psi_5}),
\end{aligned} \tag{7.27}$$

$$\begin{aligned}
&D^T + \sum_{k=1}^{r} a_k (C^T P_2^{\psi_1} + D^T P_4^{\psi_1}) + b(C_n^T P_2^{\psi_2} + D_n^T P_4^{\psi_2}) \\
&= D_1^T + \sum_{j=1}^{q} c_j (C_0^T P_2^{\psi_3} + D_0^T P_4^{\psi_3}) + dC_0^T P_2^{\psi_2} + \\
&dD_0^T P_4^{\psi_2} + \sum_{k=1}^{r} a_k (C_2^T P_2^{\psi_4} + D_2^T P_4^{\psi_4}) - \sum_{j=1}^{q} c_j (C_3^T P_2^{\psi_5} + D_3^T P_4^{\psi_5}).
\end{aligned} \tag{7.28}$$

This system of nonlinear algebraic equations can be solved for the coefficients vectors C^T and D^T of the approximation of the output signal $y(t)$ of the nonlinear FOS in Equation (7.17).

The nonlinear FOS identification problem can be defined by

$$J = \min_{\hat{\theta} \in \Omega} \|y(t) - \tilde{y}(t)\|_2, \tag{7.29}$$

where $\hat{\theta}$ is the estimation for the set of true model parameters (θ) including fractional differential orders and the initial conditions, Ω is the considered search range of optimization variables.

The objective function for the linear FOS identification problem in Equation (7.16) and for the nonlinear FOS identification problem in Equation (7.29) can be solved by using any traditional or nontraditional optimization methods. In this chapter, the MATLAB® built-in *particleswarm* function for particle swarm optimization method (PSO) is used to minimize the error between the measured output and the output of the respective fractional order model.

7.2 Simulation Examples

To demonstrate the effectiveness of the proposed TFs-based fractional order system identification method, five identification problems are given in this section. In every problem, the proposed method is tested for noise-free output data (case 1) and noisy output data (case 2), which is obtained by adding Gaussian white noise with zero mean and signal-to-noise ratio of 100 to the generated output data.

Case study 7.1: Identification of Linear Single Input Single Output (SISO) FOS

Consider the following fractional order system (FOS):

$$G(s) = \frac{1}{a_2 s^{\beta_2} + a_1 s^{\beta_1} + a_0}. \tag{7.30}$$

The true values of the model parameters are $a_0 = 1$, $a_1 = 0.5$, $a_2 = 0.8$, $\beta_1 = 0.88$, and $\beta_2 = 2.23$. The time interval considered is $t \in [0, 10]$. The step size h (number of subintervals, $(t_f - t_0)/h$) is taken as 0.1. The FOS in Equation (7.30) is simulated over the given time interval for a unit step input signal, and the corresponding output is recorded and treated as simulated output data. The lower bound (*lb*) and upper bound (*ub*) on the optimization variables (i.e., model parameters) are selected as $lb = [0,1 \quad 0.1 \quad 0.1 \quad 0.1 \; 0.1]$ and $ub = [1.5 \quad 1 \quad 1 \quad 1 \quad 3]$. The swarm size and maximum number of iterations are chosen as 30 and 50, respectively. For the values of remaining parameters of PSO, the default ones specified in the *particleswarm* function are used. The simulations are performed 10 times independently in case 1 (noise-free simulated output data (SNR=∞)) and case 2 (noisy simulated output data (SNR=100)), and the obtained results are recorded. The swarm size and the maximum number of iterations are increased to 150 and 100, respectively, to estimate the model parameters (θ) in case 2. The statistical analysis of results of 10 individual runs in each case is carried out as given in Tables 7.1 to 7.3. In case 1, PSO produced feasible solution in each run and the obtained optimal solution in every run is close to the neighborhood of the global optimum. In Table 7.4, the estimations that give the smallest (or best) J are compared with the results obtained via PSO and Haar wavelets, respectively, in [27] and [33]. The identification results obtained via the proposed method are superior to those in [27] and [33]. Even in case 2 (Table 7.4), the optimal values producing the best J are acceptable. The step response, the frequency response, and the sinusoidal response of the identified models shown in Figures 7.1 to 7.3, respectively, match the corresponding responses of the true model.

Case study 7.2: Identification of Linear SISO Integer Order System (IOS)

Consider the integer order system

TABLE 7.1

Statistical analysis for simulated output data without noise

Problem	Mean J	Standard deviation (J)	Best J	Mean CT (s)
CS 7.1	3.361163110e-06	1.11145101659e-06	1.02233354e-06	4.9288979
CS 7.2	4.085447587e-07	7.803827595e-08	2.84154589e-07	5.420573
CS 7.3	1.824419580e-10	0	1.82441958e-10	1.6239221
CS 7.4	2.566367743e-06	7.9456516222e-06	1.84274948e-07	47.6412485

CS: case study, CT: CPU time

TABLE 7.2

Statistical analysis for simulated output data with noise

Problem	Mean J	Standard deviation (J)	Best J	Mean CT (s)
CS 7.1	0.00447204233	0.00264406822	1.46241361e-04	44.8798759
CS 7.2	0.00500512958	0.00188340221	7.32184472e-04	4.4249965
CS 7.3	7.65006980e-06	5.19523833e-06	2.8088279e-06	7.5776005
CS 7.4	4.24139783e-05	2.11993255e-05	7.4724047e-06	45.5453173

TABLE 7.3

Identification results of case study 7.1

Parameter	True value (θ)	Identified value (SNR=∞) Mean $\hat{\theta}$	Identified value (SNR=∞) Std. $\hat{\theta}$	Identified value (SNR=100) Mean $\hat{\theta}$	Identified value (SNR=100) Std. $\hat{\theta}$
a_0	1	0.999999280	3.00989147e-07	1.000301439	0.000967583
a_1	0.5	0.499998121	7.37666541e-07	0.500632043	0.002233931
a_2	0.8	0.800001612	7.75566022e-07	0.799138691	0.002162041
β_1	0.88	0.879995210	1.91138739e-06	0.882203144	0.006230875
β_2	2.23	2.229997687	7.40176925e-07	2.230859201	0.002861208

Std.: standard deviation

TABLE 7.4

Comparison of identification results of case study 7.1

Parameter	True value	Identified vale (SNR=∞) PM	Identified vale (SNR=∞) Ref. [27]	Identified vale (SNR=∞) Ref. [33]	Identified vale by PM (SNR=100)
a_0	1	0.999999	0.999941	0.999942	1.000028
a_1	0.5	0.499999	0.499879	0.499879	0.500038
a_2	0.8	0.800000	0.800132	0.800132	0.799927
β_1	0.88	0.879999	0.879624	0.879625	0.880185
β_2	2.23	2.229999	2.229834	2.229835	2.230067

PM: proposed method

$$G(s) = \frac{1}{a_2 s^{\beta_2} + a_1 s^{\beta_1} + a_0}. \tag{7.31}$$

The true values of model parameters are $a_0 = 2$, $a_1 = 3$, $a_2 = 1$, $\beta_1 = 1$, and $\beta_2 = 2$. The step response data of the true model is collected on the time interval $t \in [0, 10]$ using the sample time (step size h) of 0.1. The lower and upper bounds on the optimization variables are considered as $lb = [0.1 \quad 0.1 \quad 0.1 \quad 0.1 \quad 0.1]$ and $ub = [2.5 \quad 3.5 \quad 1.5 \quad 1.5 \quad 2.5]$. The swarm size and maximum number of iterations are chosen as 20 and 50, respectively. Similar to case study 7.1, 10 independent runs are conducted in each case and statistical analysis is performed. In Tables 7.1, 7.2, and 7.5, it can be seen that reasonable approximations are obtained. In [34], the identification problem in Equation (7.31) was addressed with block pulse functions (BPFs) on the same time interval but the number of subintervals considered was 150. Their method took 1300 seconds to give estimations

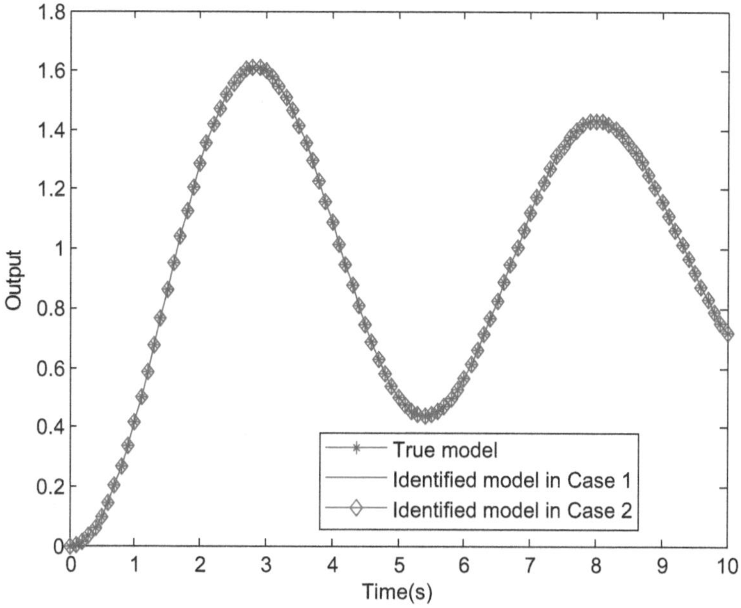

FIGURE 7.1
Step responses of the true model and the identified model in case 1 and 2 (case study 7.1)

and the objective function (mean squared error) considered was minimized to 0.00707. As shown in Table 7.6, the identification results with the best J obtained in cases 1 and 2 via the proposed method are better than the results obtained in [34]. Because the block pulse functions are piecewise constant and the integration involved in the determination of coefficients, the system identification method proposed in [34] requires larger number of subintervals hence more CPU usage. By contrast, the triangular orthogonal functions are piecewise linear polynomials and the computation of coefficients of LHTF and RHTF vector does not need integration, the proposed method identifies the model parameters with better accuracy at lower computational time. The step and frequency response of the identified models are compared with that of true model in Figures 7.4 and 7.5. To further test the identified models, a sinusoidal input is given to the identified models as well as the true model, and the respective responses are shown in Figure 7.6.

Case study 7.3: Identification of Linear Multi-Input Single Output IOS

It is required to estimate the parameters of the model of a two-input one-output system described by

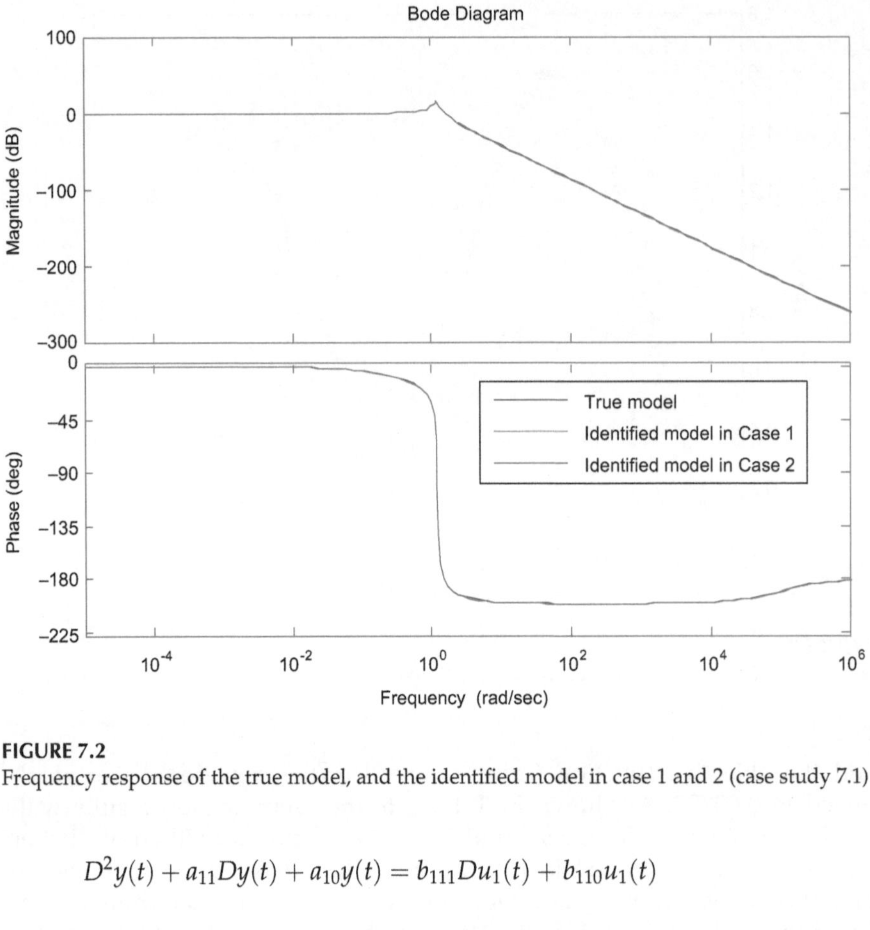

FIGURE 7.2
Frequency response of the true model, and the identified model in case 1 and 2 (case study 7.1)

$$D^2y(t) + a_{11}Dy(t) + a_{10}y(t) = b_{111}Du_1(t) + b_{110}u_1(t)$$

$$+b_{121}Du_2(t) + b_{120}u_2(t),\ t\in[0,1], \qquad (7.32)$$

where $D = d/dt$.

With the input functions $u_1(t) = t$ and $u_2(t) = t^3$, the output data are generated with zero initial conditions to estimate the parameters whose true values are $b_{111} = 1$, $b_{110} = 0.5$, $b_{121} = 0$, $b_{120} = 1$, $a_{11} = 1.5$, $a_{10} = 0.5$.

In this case study, the integer differential orders are assumed to be known and the model parameters and the initial conditions have to be estimated.

The lower and upper bounds considered are $lb = [0.1\ \ 0.1\ \ -0.5\ \ 0.1\ \ 0.1\ \ 0.1\ \ -0.5\ \ -0.5]$ and $ub = [1.5\ \ 1\ \ 0.5\ \ 1.5$ $210.50.5]$. The swarm size and maximum number of iterations are selected as 50 and 100, respectively, for case 1 and 250 and 100, respectively, for case 2. In [35], the model parameters and the initial conditions were estimated by

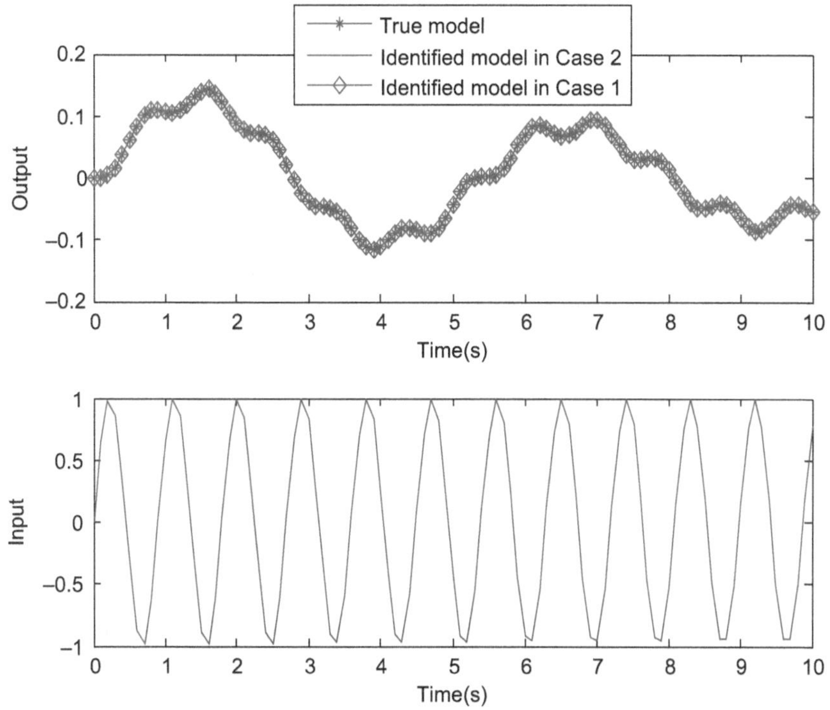

FIGURE 7.3
Sinusoidal response of the true model and the identified model in case 1 and 2 (case study 7.1)

TABLE 7.5
Identification results of case study 7.2

Parameter	True value	Identified value (SNR=∞)		Identified value (SNR=100)	
		Mean $\hat{\theta}$	Std. $\hat{\theta}$	Mean $\hat{\theta}$	Std. $\hat{\theta}$
a_0	2	1.999998840	2.69627629e-07	1.99126923057	0.0104319188279
a_1	3	2.999990194	2.15846355e-06	2.90549757403	0.0876360245248
a_2	1	1.000009887	2.38351564e-06	1.08979935706	0.0854964686759
β_1	1	0.999997749	5.15574426e-07	0.98054245125	0.0201357357368
β_2	2	1.999994157	1.26849376e-06	1.94893691313	0.0514933424953

Legendre polynomials, Tchebycheff polynomials of the first kind (Tcheby I), Tchebycheff polynomials of the second kind (Tcheby II), Gegenbauer polynomials, sine-cosine (S-C) function and Poission moment functionals (PMF) using the step size of 0.125. The estimations offered by those methods in [35] are tabulated in Table 7.7. The same step size used in [35] is employed here. Similar

TABLE 7.6

Comparison of identification results of case study 7.2

Parameter	True value	Identified value (SNR=∞) Proposed method	BPFs based method in Ref. [34]	Identified value by PM (SNR=100)
a_0	2	1.99999927414142	1.9667	1.99839560544489
a_1	3	2.99999310399643	2.9697	2.98540235893405
a_2	1	1.00000640094727	1.0020	1.01633339188226
β_1	1	0.99999851640331	0.9650	0.99668737368073
β_2	2	1.99999590399738	1.9641	1.99058353933181

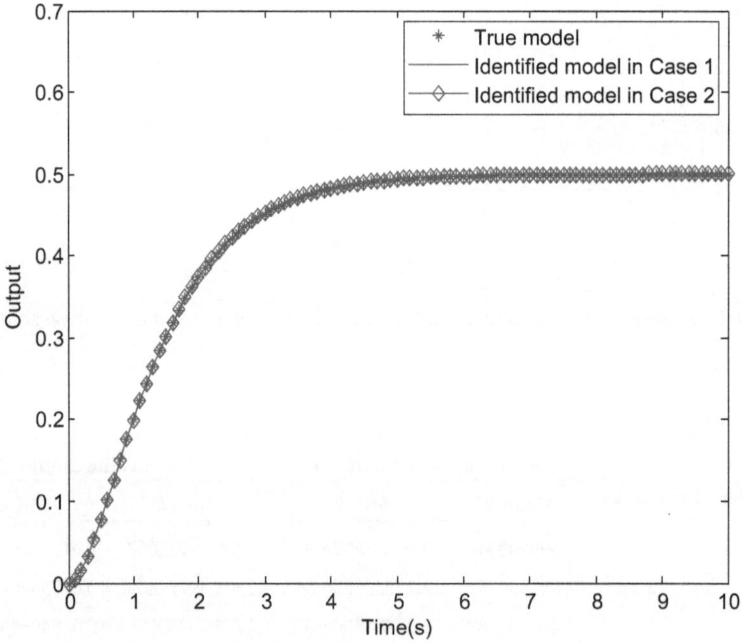

FIGURE 7.4
Step responses of the true model and the identified model in case 1 and 2 (case study 7.2)

to previous problems, the simulations are performed 10 times separately in each case and the statistical analysis of results (Tables 7.1 and 7.2) reveals that PSO does well in locating the optimal solution. In case 1, as presented in Table 7.8, the optimal values of the model parameters and the initial conditions are close to the true values and are better than those in Table 7.7. In comparison to the

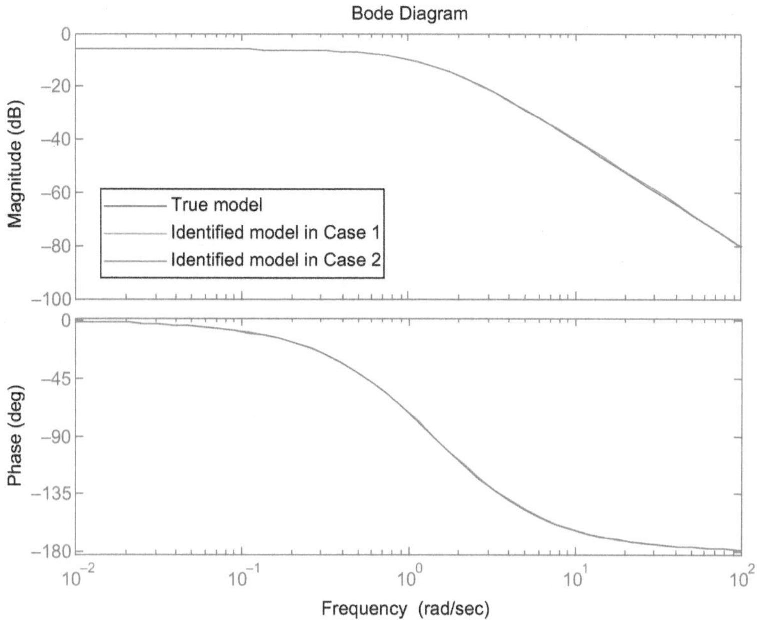

FIGURE 7.5
Frequency response of the true model and the identified model in case 1 and 2 (case study 7.2)

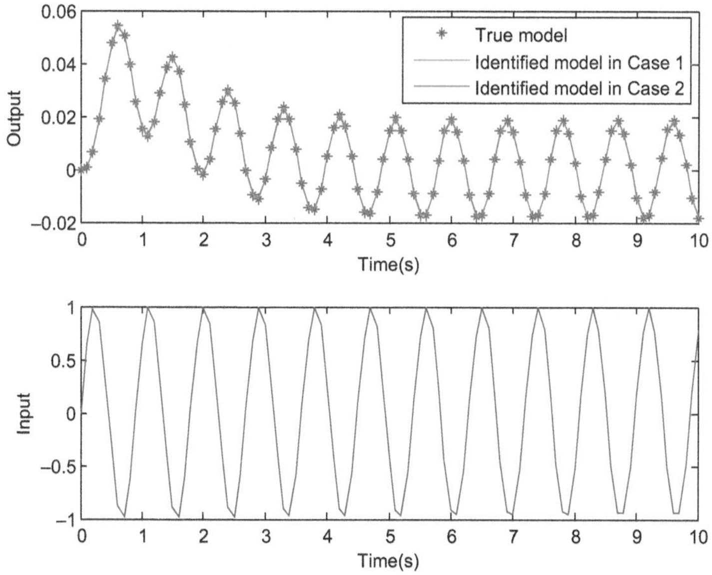

FIGURE 7.6
Sinusoidal response of the true model and the identified model in case 1 and 2 (case study 7.2)

TABLE 7.7

Identification results of case study 7.3 obtained via various methods

Parameter	True value	Legendre	Tcheby I	Tcheby II	Gegenbauer	S-C	PMF
a_{10}	0.5	0.500817	0.500906	0.500722	0.5008179	0.4664833	0.4959574
a_{11}	1.5	1.498943	1.498849	1.499065	1.4989434	1.4488530	1.4983069
b_{110}	0.5	0.498943	0.456864	0.499056	0.4989434	0.4664833	0.4970313
b_{111}	1	1	1.020197	1.000375	1.0000006	0.9659020	0.9982612
b_{120}	1	0.999536	0.999496	0.999589	0.9995365	0.9329667	1.0000062
b_{121}	0	0.000332	0.001029	0.000293	0.0003320	0	–0.59e-06
y(0)	0	0	0	–.3e-07	0	0.0020706	0
y′(0)	0	–0.2e-07	0.20e-04	0.21e-06	–0.2e-07	–0.000373	0

TABLE 7.8

Identification results of case study 7.3 obtained by the proposed method

Parameter	True value	Identified value (SNR=∞)		Identified value (SNR=100)	
		Mean $\hat{\theta}$	Std. $\hat{\theta}$	Mean $\hat{\theta}$	Std. $\hat{\theta}$
a_{10}	0.5	0.50000000005	0	0.49712174624	0.00636299419
a_{11}	1.5	1.5000000005	0	1.49386549231	0.00684639383
b_{110}	0.5	0.50000000005	0	0.49464177834	0.00417865713
b_{111}	1	1.0000000005	0	0.99992266060	0.00048468763
b_{120}	1	1.000000000005	2.34055564e-16	0.99946105126	0.00323518810
b_{121}	0	–2.6047238e-26	3.66153967e-26	–4.920951e-26	4.1145012e-26
y(0)	0	–3.9677108e-26	4.43173117e-26	1.1874081e-26	7.8700376e-26
y′(0)	0	–4.412055075e-26	4.36825349e-26	3.3799007e-26	7.5836219e-26

optimal solution found in case 1, the estimations obtained according to the output data with noise are less accurate but considerable. Figure 7.7 presents the validation of the identified models.

Case study 7.4: Identification of Nonlinear SISO FOS

Consider the nonlinear FOS described by the following model:

$$a\,{}_0^C D_t^{\alpha} x(t) + b\,{}_0^C D_t^{\alpha_2} x(t) + c\,{}_0^C D_t^{\alpha_1} x(t) + e\big(x(t)\big)^3 = u(t),\ \ t\in[0,1] \tag{7.33}$$

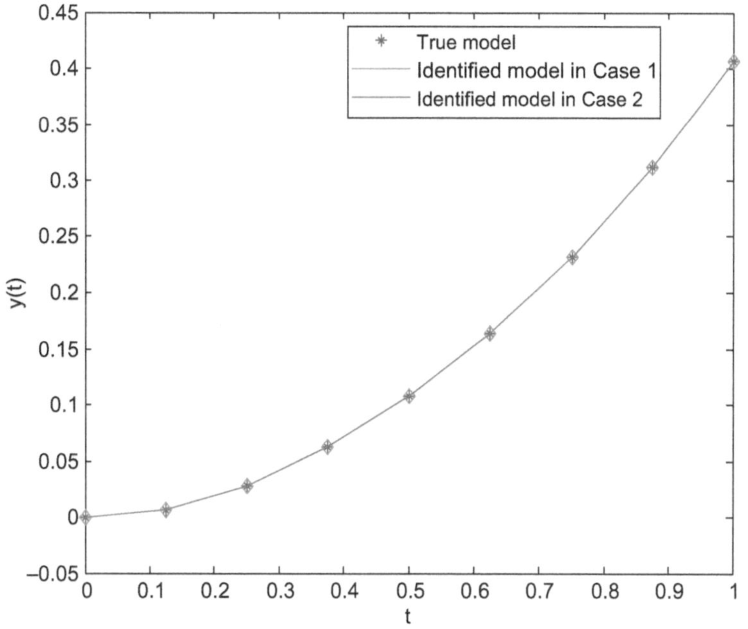

FIGURE 7.7
Comparison of true output and estimated output (case study 7.3)

The input function is

$$u(t) = \left((2t)/\Gamma(2)\right) + \left(4/\Gamma(3.9238)\right)t^{2.9238} + \left(1/\Gamma(3.9980)\right)t^{2.9980} + \left(t^3/3\right)^3. \quad (7.34)$$

The true values of the model parameters and the initial conditions are $a = 1$, $b = 2$, $c = 0.5$, $e = 1$, $\alpha_1 = 0.00196$, $\alpha_2 = 0.07621$, $\alpha = 2$, $x(0) = 0$, $x'(0) = 0$. The fractional differential orders and the model parameters are required to be estimated. The output data are generated using the zero initial conditions.

The lower and upper bounds considered are $lb = [0.1 \quad 0.1 \quad 0.1 \quad 0.1 10^{-4} 10^{-4} 0.1]$ and $ub = [1.5 \quad 2.5 \quad 1 \quad 1.5 \quad 0.01 \quad 0.1 \quad 2.5]$. The swarm size and the maximum number of iterations are taken as 50 and 50, respectively. The step size of 0.1 is used. As given in Tables 7.1, 7.2, and 7.9, the performance of PSO is almost similar in both cases and the optimal solutions found in both cases are more or less similar and acceptable. The accuracy of the estimations can be improved by reducing the step size. The outputs of the identified models match the output of the true model as shown in Figure 7.8.

TABLE 7.9
Identification results of case study 7.4

Parameter	True value	Identified value (SNR=∞)		Identified value (SNR=100)	
		Mean $\hat{\theta}$	Std. $\hat{\theta}$	Mean $\hat{\theta}$	Std. $\hat{\theta}$
α_1	0.00196	0.0019450441	1.21454423e-08	0.00195502711	2.6981118e-05
α_2	0.07621	0.0756878187	7.59424086e-05	0.07436286978	0.00054844549
α	2	1.9998254248	0.000468904204	1.99706380480	0.00178506782
a	1	1.0003682318	0.000952473033	1.00593572126	0.00360081511
b	2	1.9993920984	0.003729553236	1.99989092861	0.01811417846
c	0.5	0.4998574441	0.003051195760	0.46243263008	0.02545073567
e	1	0.9921399688	0.001132749484	0.98749337244	0.01515788747

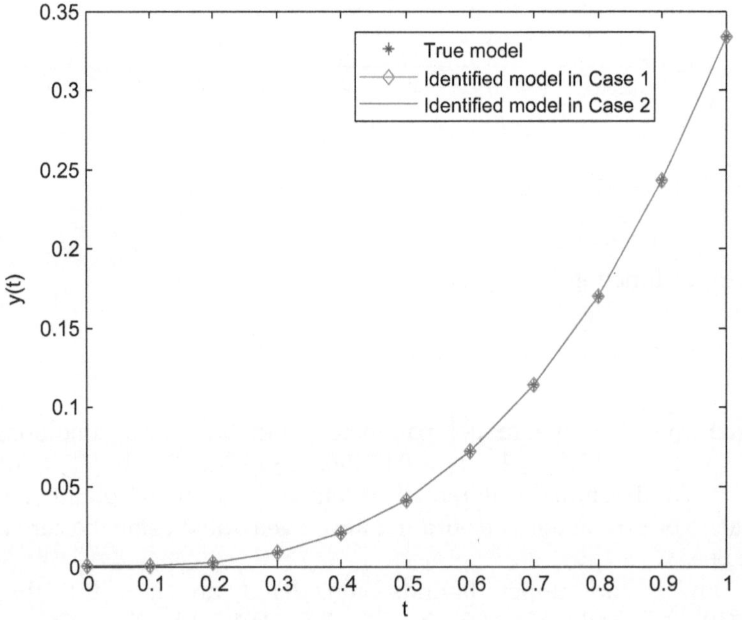

FIGURE 7.8
Comparison of true output and estimated output (case study 7.4)

Case study 7.5: Verification of applicability of proposed identification method for sinusoidal signal, square wave signal, Sawtooth wave signal, step signal, pseudo random binary signal

Consider the linear FOS described by

TABLE 7.10

Statistical analysis for noise-free simulated output data (case study 7.5)

Input type	Mean J	Standard deviation (J)	Best J	Mean CT
Step	7.80555132e-08	7.63274019e-08	7.86952254e-09	9.7140878
PRBS	1.56696448e-08	7.40978964e-09	7.30177798e-09	7.1297108
Sinusoidal	4.43132635e-08	1.78701298e-08	1.97452356e-08	10.2198509
Square	2.06192187e-08	1.74147162e-08	7.27657120e-10	7.3738363
Sawtooth	1.59210433e-08	7.64984749e-09	4.99289024e-09	7.21872379

TABLE 7.11

Statistical analysis for simulated output data with noise (case study 7.5)

Input type	Mean J	Standard deviation (J)	Best J	Mean CT
Step	7.57407780e-05	2.01295433e-06	7.17837069e-05	10.4025511
PRBS	1.64223674e-05	3.93352274e-07	1.58956358e-05	7.936823
Sinusoidal	1.68114334e-05	7.81583165e-06	1.19070296e-05	11.3233347
Square	1.79132502e-05	7.01000439e-07	1.71211969e-05	9.6582151
Sawtooth	1.04252495e-05	4.30269756e-07	9.758813736e-06	9.5195279

$$G(s) = \frac{1}{a_0 s^{\alpha} + a_1}. \qquad 7.35$$

The true values of the model parameters are $a_0 = 1$, $a_1 = 1$, and $\alpha = 0.7$.

The time interval and the step size considered are $t \in [0, 10]$ and $h = 0.1$, respectively. The swarm size and maximum number of iterations are 50 and 100, respectively. In case studies 7.1 and 7.2, the model parameters are estimated as per the step response of the true model. In case studies 7.3 and 7.4, the model is solved using the given true values of model parameters and initial conditions to get the identification data. Here the proposed method is tested to know whether it works for various types of input excitation signal. The input excitation signals considered are step signal, pseudo random binary signal, sinusoidal wave signal, square wave signal, and sawtooth wave. The fractional order system in Equation (7.35) is excited with one input signal at a time and the corresponding response is recorded. For each pair of input-output data, the respective noisy data is created using Gaussian white noise (mean is zero and SNR is 100). Tables 7.10 to 7.12 and Figures 7.9 to 7.13 demonstrate the applicability of the proposed method to any kind of input excitation signal.

TABLE 7.12

Identification results of case study 7.5

Input type	θ	True value	Identified value (SNR=∞)		Identified value (SNR=100)	
			Mean $\hat{\theta}$	Std. $\hat{\theta}$	Mean $\hat{\theta}$	Std. $\hat{\theta}$
Step	a_0	1	0.99999999	2.815457e-08	0.99999728	1.0489272e-05
	a_1	1	0.99999996	5.799533e-08	0.99999882	9.7438416e-06
	α	0.7	0.69999995	9.894450e-08	0.69999664	1.5871707e-05
PRBS	a_0	1	1.00000000	1.296592e-08	0.99999964	2.8754828e-06
	a_1	1	0.99999999	1.302996e-08	1.00000077	2.3274777e-06
	α	0.7	0.69999999	4.9334381e-09	0.70000050	9.3056241e-07
Sinusoidal	a_0	1	1.00000012	5.9720978e-08	1.00002127	2.75071137e-05
	a_1	1	0.99999969	1.7153998e-07	0.99995474	5.81363495e-05
	α	0.7	0.6999999528	2.16159761e-08	0.69999302	7.39074320e-06
Square	a_0	1	1.00000001	1.79369736e-08	1.00000191	4.32837845e-06
	a_1	1	0.99999998	2.55880412e-08	0.99999711	7.01784010e-06
	α	0.7	0.6999999944	7.41244038e-09	0.69999932	1.62060728e-06
Sawtooth	a_0	1	1.00000000	2.32351211e-08	1.00000090	5.03397339e-06
	a_1	1	0.99999999	2.86191808e-08	1.00000014	3.63760823e-06
	α	0.7	0.69999999	9.26990376e-09	0.69999971	1.95008688e-06

7.3 MATLAB Codes for Simulation Examples

The source codes developed in MATLAB for case studies 7.1 to 7.5 are provided in this section.

Program 7.1

```
%%%%%%%%%%%%%%%%%% MATLAB code for case study 7.1 %%%%%%%%%%%%%
clc
clear all
warning('off')
global noise
aa=[0;1];SS=[30 150]; % swarm size a
MItr=[50 100];   % maximum iterations
for i=1:1:2
noise=aa(i);ss=SS(i);mitr=MItr(i);fun = @Objective_Fun;
nvars = 5;lb = [0.1,0.1,0.1,0.1,0.1];ub = [1.5,1,1,1,3];
tic
options = optimoptions('particleswarm','SwarmSize',ss,'MaxIter',mitr);
options.HybridFcn = @fmincon;
[x,fval,exitflag] = particleswarm(fun,nvars,lb,ub,options);
toc
```

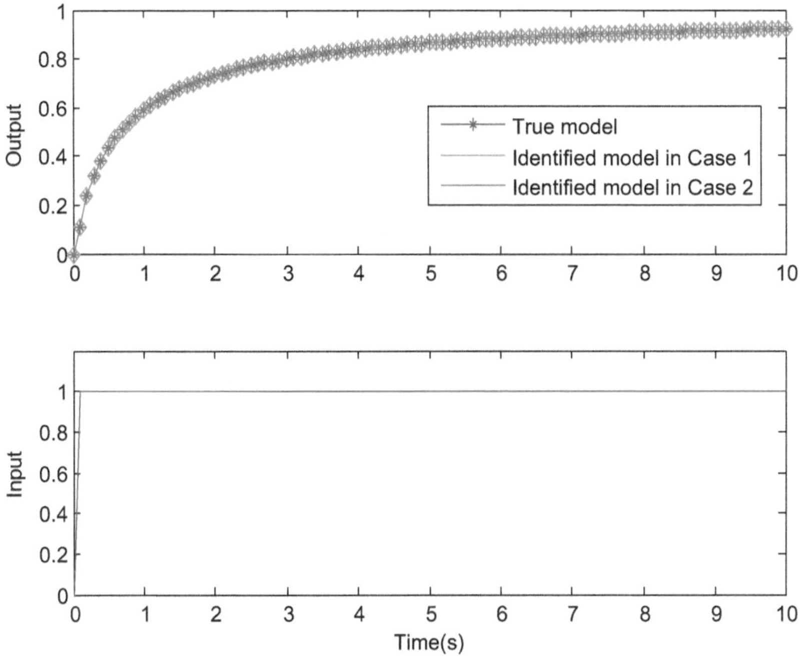

FIGURE 7.9
Identified system validation for case study 7.5 (for step input signal)

```
if i==1
x1=x;
else
x2=x;
end
end
[x1' x2']  % estimated mdel parameters in case 1 and 2
%%%%%%%%%%%%%%%%%%%%%%%%%%%% Objective_Fun.m %%%%%%%%%%%%%%%%%%%
function f=Objective_Fun(x)
global noise
t0=0;T=10;m=100;h=(T-t0)/m;t=[t0:h:T];
Ytrue=[0,0.00291500231356065,0.0135676134995406,0.0331165050895373,
0.0619850706325505,0.100213976692808,0.147568197000851,0.20359188301576-
5,
0.267644731341395,0.338930496663512,0.416521888459663,0.499383740776359,
0.586395325734925,0.676372185458934,0.768087595047261,0.860293623062204,
0.951741672441163,1.04120233765689,1.12748438965642,1.20945269097567,
1.28604484494870,1.35628639227781,1.41930438350723,1.47433917578808,
1.52075432569294,1.55804447587919,1.58584116137955,1.60391649055909,
1.61218468572039,1.61070149840495,1.59966154410837,1.57939362991279,
1.55035417599245,1.51311885765186,1.46837261813836,1.41689822359901,
1.35956354993821,1.29730780674035,1.23112691565761,1.16205826959176,
1.09116510452911,1.01952071798832,0.948192766723057,0.878227871656251,
```

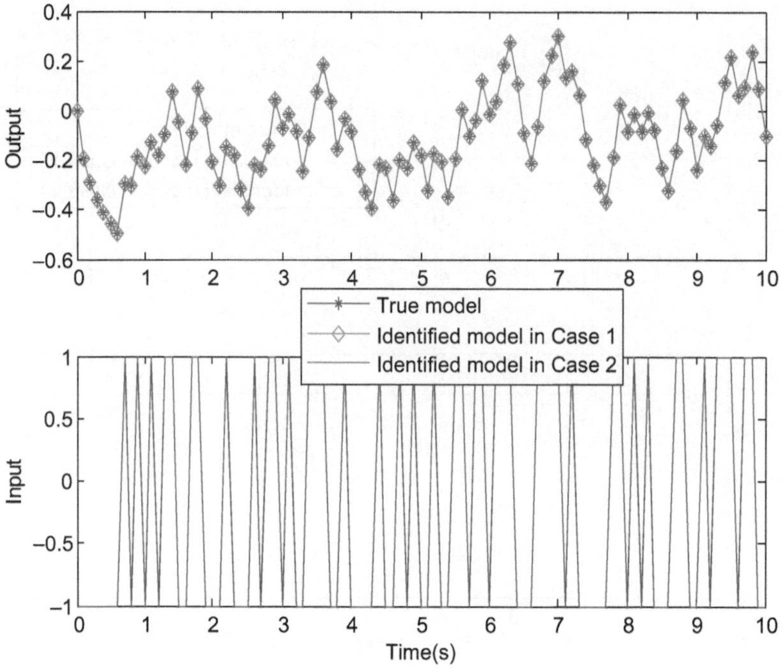

FIGURE 7.10
Identified system validation for case study 7.5 (for PRBS input signal)

```
0.810636750120878,0.746380084506120,0.686355322559306,0.631384588117759,
0.582203862213351,0.539453573619986,0.503670715326107,0.475282579471007,
0.454602178346725,0.441825393518016,0.437029869328718,0.440175641423995,
0.451107465793665,0.469558789592045,0.495157281948823,0.527431821479683,
0.565820817515976,0.609681724464415,0.658301593403544,0.710908492217915,
0.766683615401829,0.824773897254266,0.884304937591445,0.944394047359154,
1.00416322260647,1.06275185913327,1.11932902665696,1.17310513040643,
1.22334279949342,1.26936685501912,1.31057322642273,1.34643670181137,
1.37651741665758,1.40046600500490,1.41802735790235,1.42904295486164,
1.43345175539913,1.43128965885857,1.42268756142715,1.40786805924432,
1.38714086547711,1.36089702695533,1.32960204216811,1.29378799690833,
1.25404484644407,1.21101098361209,1.16536324057386,1.11780647804494,
1.06906291954987,1.01986138964288,0.970926614110423,0.922968736932795,
0.876673203361797,0.832691150945108,0.791630440859485,0.754047450647952,
0.720439736590037];
if noise==0 % noise-free simulated output data
  Ytrue=Ytrue;
else        % simulated output data with noise
  Ytrue=awgn(Ytrue,100,'measured');
endg
a0=x(1);a1=x(2);a2=x(3);beta1=x(4);beta2=x(5);alpha1=beta2-beta1;
alpha2=beta2;I=eye(m,m);
```

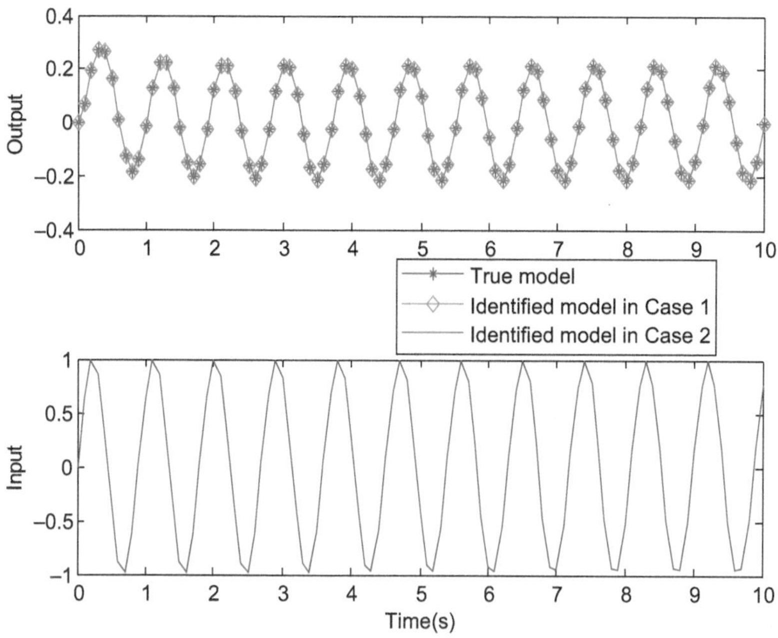

FIGURE 7.11
Identified system validation for case study 7.5 (for sinusoidal input signal)

```
[P1alph1,P2alph1]=TOF1(t0,T,m,alpha1);[P3alph1,P4alph1]=TOF12(t0,T,m,
alpha1);
[P1alph2,P2alph2]=TOF1(t0,T,m,alpha2);[P3alph2,P4alph2]=TOF12(t0,T,m,
alpha2);
C00=[1 ones(1,lenth(t)-1)]; % unit step input
C01=C00(1:end-1);D01=C00(2:end);
A1=a2*I+a1*P1alph1+a0*P1alph2;B1=a1*P3alph1+a0*P3alph2;
A2=a1*P2alph1+a0*P2alph2;B2=a2*I+a1*P4alph1+a0*P4alph2;
Z1=(C01*P1alph2+D01*P3alph2);Z2=(C01*P2alph2+D01*P4alph2);A=[A1     A2;
B1 B2];
Z=[Z1 Z2];X=Z/A;x1=X(1:m);x2=X(m+1:end);Y=[x1 x2(end)];e=abs(Ytrue-Y);
f=norm(e,2);
end
%%%%%%%%%%%%%%%%%%%%%%%%%%%%%%%%%%%%%%%%%%%%%%%%%%%%%%%%%%%%%%%%%%%%%%%%%%%
```

Program 7.2

```
%%%%%%%%%%%%%%% MATLAB code for case study 7.2 %%%%%%%%%%%%%%%%%%%%%%%
clc
clear all
warning('off')
global noise
aa=[0;1];
for i=1:1:2
noise=aa(i);fun = @Objective_Fun;
```

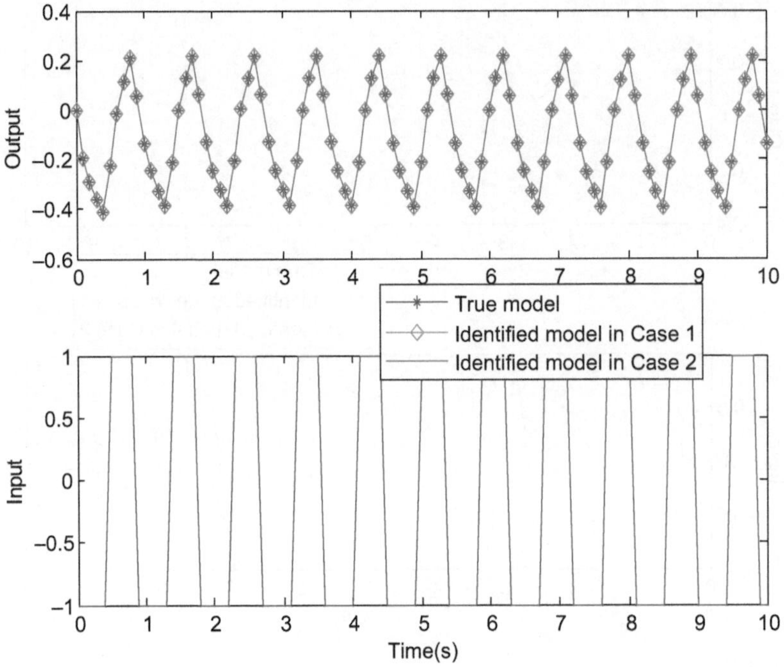

FIGURE 7.12
Identified system validation for case study 7.5 (for square wave input signal)

```
nvars = 5;lb = [0.1,0.1,0.1,0.1,0.1];ub = [2.5,3.5,1.5,1.5,2.5];
tic
options = optimoptions('particleswarm','SwarmSize',20,'MaxIter',50);
options.HybridFcn = @fmincon;
[x,fval,exitflag] = particleswarm(fun,nvars,lb,ub,options);
toc
if i==1
x1=x;
else
x2=x;
end
end
[x1' x2']  % estimated mdel parameters in case 1 and 2
%%%%%%%%%%%%%%%%%%%%%%%%%%%%%%% Objective_Fun.m %%%%%%%%%%%%%%%%%%%%
function f=Objective_Fun(x)
global noise
t0=0;T=10;m=100;h=(T-t0)/m;t=[t0:h:T];
%%%% simulated data
Ytrue=
[0,0.00433526011560694,0.0161381937251495,0.0332616672560513,0.05402-
47863900817,0.0771207684131202,0.101542290920112,0.126521065533607,
0.151478984342430,0.175988675986952,0.199741707793529,0.222522996509490,
```

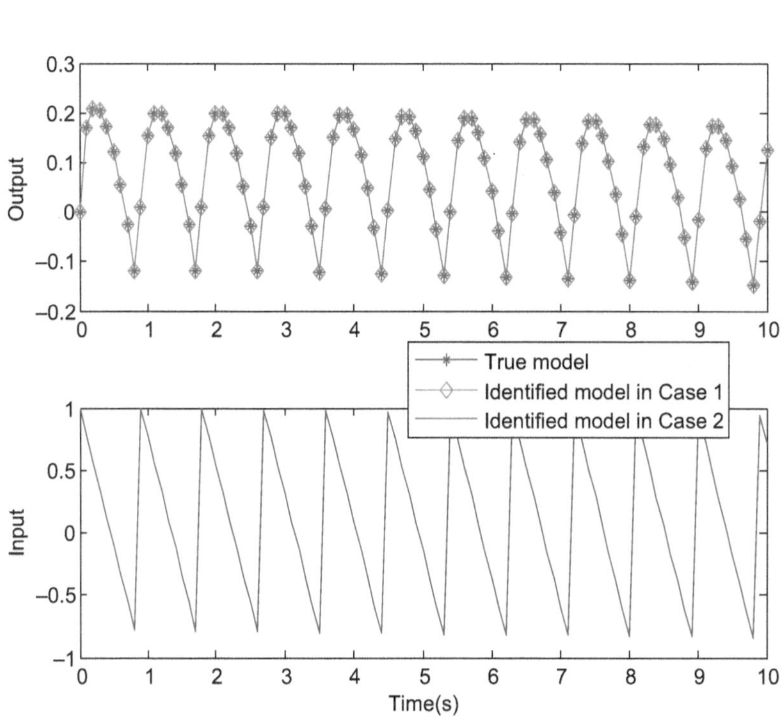

FIGURE 7.13
Identified system validation for case study 7.5 (for sawtooth wave input signal)

```
0.244190256429227,0.264657530998237,0.283882031297882,0.301853649473956,
0.318586633162486,0.334113003177772,0.348477375157096,0.361732909769133,
0.373938168156608,0.385154691677574,0.395445159513707,0.404872005782398,
0.413496400609251,0.421377518158438,0.428572029671851,0.435133771779952,
0.441113550245253,0.446559047313593,0.451514807329817,0.456022280509512,
0.460119908980450,0.463843242606686,0.467225074839853,0.470295591032490,
0.473082523399888,0.475611308213918,0.477905241922325,0.479985633765729,
0.481871953156818,0.483581970628431,0.485131891579190,0.486536482370863,
0.487809188580239,0.488962245395606,0.490006780286824,0.490952908178244,
0.491809819424294,0.492585860934246,0.493288610821359,0.493924946966372,
0.494501109889273,0.495022760319701,0.495495031846494,0.495922579013491,
0.496309621211958,0.496659982701948,0.496977129075432,0.497264200454505,
0.497524041698190,0.497759229872162,0.497972099216977,0.498164763832494,
0.498339138279252,0.498496956281257,0.498639787699832,0.498769053933802,
0.498886041888414,0.498991916642893,0.499087732935635,0.499174445575308,
0.499252918876779,0.499323935211782,0.499388202756558,0.499446362510674,
0.499498994655144,0.499546624311370,0.499589726756828,0.499628732148432,
0.499664029799754,0.499695972053925,0.499724877790171,0.499751035598673,
0.499774706654700,0.499796127320634,0.499815511501425,0.499833052776785,
0.499848926331204,0.499863290700996,0.499876289355503,0.499888052128388,
0.499898696513041,0.499908328834999,0.499917045313042,0.499924933019714,
0.499932070750309,0.499938529809649,0.499944374723903,0.499949663884900,
0.499954450133180];
```

```
if noise==0 % noise-free simulated output data
  Ytrue=Ytrue;
else        % simulated output data with noise
  Ytrue=awgn(Ytrue,100,'measured');
end
a0=x(1);a1=x(2);a2=x(3);beta1=x(4);beta2=x(5);
alpha1=beta2-beta1;alpha2=beta2;I=eye(m,m);
[P1alph1,P2alph1]=TOF1(t0,T,m,alpha1);
[P3alph1,P4alph1]=TOF12(t0,T,m,alpha1);
[P1alph2,P2alph2]=TOF1(t0,T,m,alpha2);
[P3alph2,P4alph2]=TOF12(t0,T,m,alpha2);
C00=[1 ones(1,length(t)-1)]; % unit step input
C01=C00(1:end-1);D01=C00(2:end);A1=a2*I+a1*P1alph1+a0*P1alph2;
B1=a1*P3alph1+a0*P3alph2;A2=a1*P2alph1+a0*P2alph2;
B2=a2*I+a1*P4alph1+a0*P4alph2;Z1=(C01*P1alph2+D01*P3alph2);
Z2=(C01*P2alph2+D01*P4alph2);A=[A1 A2;B1 B2];Z=[Z1 Z2];X=Z/A;x1=X(1:m);
x2=X(m+1:end);Y=[x1 x2(end)];e=abs(Ytrue-Y);f=norm(e,2);
end
%%%%%%%%%%%%%%%%%%%%%%%%%%%%%%%%%%%%%%%%%%%%%%%%%%%%%%%%%%%%
```

Program 7.3

```
%%%%%%%%%%%%%%%%% MATLAB code for case study 7.3 %%%%%%%%%%%%
clc
clear all
warning('off')
global noise
aa=[0;1];SS=[50 100]; % swarm size a
MItr=[250 100];   % maximum iterations
for i=1:1:2
noise=aa(i);ss=SS(i);mitr=MItr(i);fun = @Objective_Fun;
nvars = 8;lb = [0.1,0.1,-0.5,0.1,0.1,0.1,-0.5,-0.5];
ub = [1.5,1,0.5,1.5,2,1,0.5,0.5];
tic
options = optimoptions('particleswarm','SwarmSize',ss,'MaxIter',mitr);
options.HybridFcn = @fmincon;
[x,fval,exitflag] = particleswarm(fun,nvars,lb,ub,options);
toc
if i==1
x1=x;
else
x2=x;
end
end
[x1' x2']  % estimated mdel parameters in case 1 and 2
%%%%%%%%%%%%%%%%%%%%%%%% Objective_Fun.m %%%%%%%%%%%%%%%%%%%
function f=Objective_Fun2(x)
global noise
t0=0;T=1;m=8;h=(T-t0)/m;t=[t0:h:T];
Ytrue=[0,0.00728764120095125,0.0284917257073214,0.0621928809352597,

0.107460030647818,0.163940640833033,0.231931385724944,0.312432082366-
219,
  0.407185373763093];
```

```
if noise==0 % noise-free simulated output data
  Ytrue=Ytrue;
else      % simulated output data with noise
  Ytrue=awgn(Ytrue,100,'measured');
end
b111=x(1);b110=x(2);b121=x(3);b120=x(4);a11=x(5);a10=x(6);
beta1t=1;beta2t=2;y10=x(7);y20=x(8);alpha1t=beta2t-beta1t;
alpha2t=beta2t;
I=eye(m,m);
[P1alph1t,P2alph1t]=TOF1(t0,T,m,alpha1t);
[P3alph1t,P4alph1t]=TOF12(t0,T,m,alpha1t);
[P1alph2t,P2alph2t]=TOF1(t0,T,m,alpha2t);
[P3alph2t,P4alph2t]=TOF12(t0,T,m,alpha2t);
C00=(b111+b110*t+2*b121*(t.^2)+b120*(t.^3));   % unit step input
C01t=C00(1:end-1);D01t=C00(2:end);C00t=(y10+y20*t)+a11*y10*t;
C02t=C00t(1:end-1);D02t=C00t(2:end);A1t=I+a11*P1alph1t+a10*P1alph2t;
B1t=a11*P3alph1t+a10*P3alph2t;A2t=a11*P2alph1t+a10*P2alph2t;
B2t=I+a11*P4alph1t+a10*P4alph2t;Z1t=C02t+(C01t*P1alph2t
+D01t*P3alph2t);
Z2t=D02t+(C01t*P2alph2t+D01t*P4alph2t);At=[A1t A2t;B1t B2t];
Zt=[Z1t Z2t];Xt=Zt/At;x1t=Xt(1:m);x2t=Xt(m+1:end);
Y=[x1t x2t(end)];e=abs(Ytrue-Y);f=norm(e,2);
end
%%%%%%%%%%%%%%%%%%%%%%%%%%%%%%%%%%%%%%%%%%%%%%%%%%%%%%%%%%%%%%%
```

Program 7.4

```
%%%%%%%%%%%%%% MATLAB code for case study 7.4 %%%%%%%%%%%%%%%%%%%%%
clc
clear all
warning('off')
global noise
aa=[0;1];
for i=1:1:2
noise=aa(i);fun = @Objective_Fun;nvars = 7;
lb      =       [0.1,0.1,0.1,0.1,0.0001,0.0001      0.1];ub      =
[1.5,2.5,1,1.5,0.01,0.1,2.5];
tic
options = optimoptions('particleswarm','SwarmSize',50,'MaxIter',50);
options.HybridFcn = @fmincon;
[x,fval,exitflag] = particleswarm(fun,nvars,lb,ub,options);
toc
if i==1
x1=x;
else
x2=x;
end
end
[x1' x2']  % estimated mdel parameters in case 1 and 2
%%%%%%%%%%%%%%%%%%%%%%%%%%%% Objective_Fun.m %%%%%%%%%%%%%%%%%%
function [Y,Ytrue,f]=Objective_Fun(x)
global noise
t0=0;T=1;m=10;h=(T-t0)/m;t=[t0:h:T];
```

```
alpha1t=x(1);alpha2t=x(2);alphat=x(3);Aa=x(4);Bb=x(5);Cc=x(6);
Ee=x(7);
alpha1=alphat;alpha2=alphat-alpha2t;alpha3=alphat-alpha1t;
[P1alph1,P2alph1]=TOF1(t0,T,m,alpha1);[P3alph1,P4alph1]=TOF12(t0,T,m,
alpha1);
[P1alph2,P2alph2]=TOF1(t0,T,m,alpha2);[P3alph2,P4alph2]=TOF12(t0,T,m,
alpha2);
[P1alph3,P2alph3]=TOF1(t0,T,m,alpha3);[P3alph3,P4alph3]=TOF12(t0,T,m,
alpha3);
b=zeros(1,11);Y=fsolve(@Problem_Fun,b);
function f=Problem_Fun(x)
for i=1:1:length(t)
  C11(i)=x(i);
end
C1=C11(1:end-1);D1=C11(2:end);
C00=[(2*1)/gamma(4-2)]*(t.^(3-2))+[(2*2)/(gamma(4-0.07621))]*(t.^(3-
0.07621))+[(2*0.5)/(gamma(4-0.00196))]*(t.^(3-0.00196))+1*(((1/3)*(t.
^3)).^3);
C01=C00(1:end-1);D01=C00(2:end);
 e1=[(1/Aa)*(C01*P1alph1+D01*P3alph1)-(Bb/Aa)*(C1*P1alph2+D1*P3alph2)-
(Cc/Aa)*((C1)*P1alph3+(D1)*P3alph3)-(Ee/Aa)*((C1.^3)*P1alph1+(D1.^3)
*P3alph1)-C1];
f1=[(1/Aa)*(C01*P2alph1+D01*P4alph1)-(Bb/Aa)*(C1*P2alph2+D1*P4alph2)-
(Cc/Aa)*((C1)*P2alph3+(D1)*P4alph3)-(Ee/Aa)*((C1.^3)*P2alph1+(D1.^3)
*P4alph1)-D1];
f=[e1;f1];
end
Ytrue=(1/3)*(t.^3);
if noise==0 % noise-free simulated output data
  Ytrue=Ytrue;
else        % simulated output data with noise
  Ytrue=awgn(Ytrue,100,'measured');
end
e=abs(Ytrue-Y);f=norm(e,2);
end
%%%%%%%%%%%%%%%%%%%%%%%%%%%%%%%%%%%%%%%%%%%%%%%%%%%%%%%%%%%%%
```

Program 7.5

```
%%%%%%%%%%%%%% MATLAB code for case study 7.5 %%%%%%%%%%%%%%%%%%%%%
clc
clear all
warning('off')
global noise signal
sgl=[1,2,3,4,5];aa=[0;1];X=zeros(5,2);   fun = @Objective_Fun;
nvars = 3;lb = [0.1,0.1,0.1];ub = [1.5,1.5,1];
for j=1:1:5
signal=sgl(j);
for i=1:1:2
noise=aa(i);
tic
options = optimoptions('particleswarm','SwarmSize',50,'MaxIter',50);
options.HybridFcn = @fmincon;
[x,fval,exitflag] = particleswarm(fun,nvars,lb,ub,options);
```

```
toc
X(j,i)=x;   % storing estimated parameters' values
end
end
%%%%%%%%%%%%%%%%%%%%%%%%%%%% Objective_Fun.m %%%%%%%%%%%%%%%%%%%
function f=Objective_Fun(x)
global noise signal
t0=0;T=10;m=100;h=(T-t0)/m;t=[t0:h:T];
%%%% simulated data
y0=[0,0.114393332145893,0.240927367856667,0.321622053648463,
   0.383008843377024,0.432504819209789,0.473753585295820,
   0.508894310018167,0.539315375057147,0.565978507656557,
   0.589581271026989,0.610647339780858,0.629580524388900,
   0.646698979874801,0.662257851704272,0.676464815962907,
   0.689491064901357,0.701479269079818,0.712549472391638,
   0.722803537602257,0.732328553008436,0.741199480122732,
   0.749481237434875,0.757230358841309,0.764496326937631,
   0.771322654743325,0.777747770639981,0.783805747834868,
   0.789526909866515,0.794938336451519,0.800064288589943,
   0.804926567789166,0.809544821175675,0.813936801888386,
   0.818118592304491,0.822104796208121,0.825908704877456,
   0.829542441165030,0.833017084927206,0.836342782580646,
   0.839528843096289,0.842583822361973,0.845515597533801,
   0.848331432742282,0.851038037308269,0.853641617450211,
   0.856147922318383,0.858562285071361,0.860889659608016,
   0.863134653482373,0.865301557457326,0.867394372090801,
   0.869416831696901,0.871372425979080,0.873264419595331,
   0.875095869882244,0.876869642936411,0.878588428228391,
   0.880254751902592,0.881870988898698,0.883439374014540,
   0.884962012016380,0.886440886890868,0.887877870322174,
   0.889274729468766,0.890633134106193,0.891954663195472,
   0.893240810929356,0.894492992305194,0.895712548265919,
   0.896900750447942,0.898058805570917,0.899187859499362,
   0.900289001005283,0.901363265256617,0.902411637054109,
   0.903435053838403,0.904434408484839,0.905410551904191,
   0.906364295464576,0.907296413247802,0.908207644154874,
   0.909098693870238,0.909970236697271,0.910822917273837,
   0.911657352177107,0.912474131425994,0.913273819888715,
   0.914056958601855,0.914824066008277,0.915575639118601,
   0.916312154602281,0.917034069813405,0.917741823754788,
   0.918435837985501,0.919116517475265,0.919784251409503,
   0.920439413948028,0.921082364940599,0.921713450602256,
   0.922333004150880];
y1=[0,-0.194468664648018,-0.288129912251252,-0.356742013374338,
   -0.410988885924266,-0.455642983357287,-0.493356851820328,
   -0.297006823378897,-0.301009469376687,-0.188834427227769,
   -0.225344147496995,-0.131888325685471,-0.180636417874422,
   -0.0957757123385291,0.0779353254978906,-0.0465276885851954,
   -0.221328307952495,-0.0912253210359857,0.0911209741444051,
   -0.0310579110952231,-0.205678569950675,-0.304984942651532,
   -0.147846621057940,-0.179363330488785,-0.316378251265945,
   -0.392863305496993,-0.220123307151213,-0.240243839937645,
   -0.139780779728380,0.0436381089822938,-0.0743022575651655,
```

```
  -0.0156984768432475,-0.0867016215476560,-0.246521603670781,
  -0.109122864296087,0.0774363169167683,0.186718543057399,
  0.0381853439947826,-0.151524350978130,-0.0320692640317578,
  -0.0866658197647949,-0.239633768464212,-0.327907881585889,
  -0.393050147165189,-0.215899355947624,-0.234148221705428,
  -0.361733921646210,-0.202418044305141,-0.228791079300964,
  -0.131834266636753,-0.179394720031141,-0.323092958421625,
  -0.173860108798555,-0.207025398528652,-0.343659594540057,
  -0.190235006501131,0.00816380470862979,-0.102209085042027,
  -0.0389764059602771,0.121967228904928,-0.0112539201201742,
  0.0362527306771858,0.185624839326676,0.272300048590687,
  0.108259165465043,-0.0928233532197897,-0.210863483352473,
  -0.0678300155141846,0.118395156903334,0.225558975243647,
  0.303161158418847,0.135007122376006,0.160164251791012,
  0.0645084105610621,-0.114488181824130,-0.221308320374442,
  -0.300533519311322,-0.363305250873927,-0.186106375083311,
  0.0235321423348750,-0.0810695834702669,-0.0148791223621604,
  -0.0813431093070413,-0.00961113579173901,-0.0750474463047607,
  -0.232267780372384,-0.322486642927166,-0.159807946116458,
  0.0411162342941191,-0.0690945639432100,-0.235467012538271,
  -0.100048430981619,-0.143005850069715,-0.0583447974854215,
  0.113364842286461,0.215140219750992,0.0619869092745788,
  0.0978743188813311,0.239236097295478,0.0910970460325439,
  -0.102006396398582];
y2=[0,0.0736942078704400,0.194244342118096,0.275423270969206,
  0.266612787582055,0.164296327333898,0.0112819218290297,
  -0.124400271388436,-0.181977285001341,-0.136790676386897,
  -0.0120639576234989,0.131907763006576,0.226038293979243,
  0.224887495149910,0.127991893813540,-0.0199479174544796,
  -0.150114171484338,-0.201956371896499,-0.151683714175586,
  -0.0234659853797123,0.121924346402248,0.215686561099993,
  0.213343270458820,0.115653427210885,-0.0317491648264246,
  -0.159821987440170,-0.208588858770186,-0.155350474471930,
  -0.0253626608390617,0.120039745208172,0.212287812442750,
  0.207825970657360,0.108595488017178,-0.0388797578316546,
  -0.165375262708910,-0.211524214414750,-0.155739477678889,
  -0.0243671139069673,0.120703911250502,0.211148474146972,
  0.204339126902989,0.103393967066696,-0.0442902896356689,
  -0.169330120221863,-0.212987706877833,-0.154796574586857,
  -0.0221875885787968,0.122411942217383,0.210938559690708,
  0.201703741297179,0.0990004595309002,-0.0489160570746181,
  -0.172523683453494,-0.213730897321636,-0.153198274790289,
  -0.0194342571678679,0.124612691243886,0.211159615722288,
  0.199468348650069,0.0950052088884597,-0.0531299836787049,
  -0.175296404410835,-0.214065469715317,-0.151230469604739,
  -0.0163695352289663,0.127065355639889,0.211590776713867,
  0.197430308420796,0.0912219585337022,-0.0571038443137366,
  -0.177807356046571,-0.214139250321109,-0.149030752640996,
  -0.0131213884898090,0.129651229726961,0.212122228401539,
  0.195488131751485,0.0875567140914444,-0.0609251454611711,
  -0.180138538651051,-0.214029475608488,-0.146671974816952,
  -0.00975832246933245,0.132306284717968,0.212694490768670,
  0.193586712091307,0.0839582614802878,-0.0646418979369711,
```

```
  -0.182335450886390,-0.213779608054698,-0.144195698465441,
  -0.00631982246643568,0.134993435336177,0.213273118776824,
  0.191694218214944,0.0803970372884386,-0.0682819605023768,
  -0.184424818038526,-0.213415601532442,-0.141627136369488,
  -0.00283008863269999];
y3=[0,-0.194468664648018,-0.288129912251252,-0.356742013374338,
  -0.410988885924266,-0.226856319065502,-0.0115021161069947,
  0.117450619626243,0.211940145955813,0.0572125110246456,
  -0.135533363497361,-0.246511800486969,-0.326362537564910,
  -0.388277772430592,-0.209590616168464,0.00177599733933551,
  0.127737957976338,0.219941262588858,0.0634388640409340,
  -0.130702262153078,-0.242789146365608,-0.323528204277499,
  -0.386160553610655,-0.208055816588864,0.00283541419737916,
  0.128407800332046,0.220290856285157,0.0635246049976452,
  -0.130834215125935,-0.243100807145019,-0.323988160494759,
  -0.386742716707227,-0.208738436391704,0.00207053239445326,
  0.127575922231114,0.219404823147855,0.0625952424928437,
  -0.131797763773523,-0.244090808275792,-0.324998065419202,
  -0.387766976017586,-0.209772345794849,0.00103096057693980,
  0.126534066562760,0.218363543317748,0.0615569553124488,
  -0.132831020176222,-0.245117320078095,-0.326016396918482,
  -0.388775930307133,-0.210770931203604,4.35591877093466e-05,
  0.125558512399021,0.217400368774653,0.0606065801427475,
  -0.133768273209113,-0.246041211693585,-0.326926759648438,
  -0.389672658399337,-0.211653971876441,-0.000825786672801809,
  0.124702829935395,0.216558285198745,0.0597780028180121,
  -0.134583460727875,-0.246843145903739,-0.327715593830831,
  -0.390448559773818,-0.212417119126675,-0.00157636779283950,
  0.123964619514049,0.215832244234573,0.0590639256734377,
  -0.135285782859916,-0.247533923937558,-0.328395039867914,
  -0.391116886309741,-0.213074538367237,-0.00222309106653784,
  0.123328382257987,0.215206284855118,0.0584480382008010,
  -0.135891801918158,-0.248130275339519,-0.328981921425801,
  -0.391694492714712,-0.213643061049393,-0.00278271808388811,
  0.122777466305455,0.214663898888063,0.0579140047043309,
  -0.136417656867759,-0.248648122063204,-0.329491926642355,
  -0.392196819552661,-0.214137869068875,-0.00327016330537928,
  0.122297231364307,0.214190725170183,0.0574477465661285,
  -0.136877141708006];
y4=[0,0.168979894673137,0.208958448147661,0.205907713767019,
  0.174813735377039,0.123098440966434,0.0552519945592022,
  -0.0257016325214670,-0.117586170350040,0.0100238281984152,
  0.153898551798797,0.199095878576244,0.199507552188911,
  0.170653577178029,0.120396356968334,0.0535021453862761,
  -0.0268343501264949,-0.118328614720849,0.00951540013143400,
  0.153515128763588,0.198760767447632,0.199166578284259,
  0.170268498270530,0.119940346166241,0.0529566379508329,
  -0.0274818916651946,-0.119086305611570,0.00864270094481850,
  0.152524963377319,0.197652446350074,0.197940709079463,
  0.168926632578669,0.118484713857952,0.0513899464445246,
  -0.0291566092879222,-0.120865805631511,0.00676178584043435,
  0.150546058391893,0.195578985214722,0.195776097239870,
  0.166674220043366,0.116147775495518,0.0489716680274101,
```

```
  -0.0316531406080463,-0.123437607435657,0.00411758774673784,
  0.147832228489708,0.192798178303801,0.192930859614977,
  0.163766991520534,0.113180892057447,0.0459473649084165,
  -0.0347327255381895,-0.126570430098127,0.000933481318584853,
  0.144598705868966,0.189517024379015,0.189603780251414,
  0.160395617146002,0.109766781152012,0.0424920073836743,
  -0.0382279050805956,-0.130104069234305,-0.0026373141628832 2,
  0.140992000986366,0.185875603476063,0.185928785757925,
  0.156688143027663,0.106027875275197,0.0387226737621681,
  -0.0420267041521665,-0.133931411156199,-0.00649231411011089,
  0.137110191884626,0.181967799862487,0.181995769682423,
  0.152730665497814,0.102046657724978,0.0347184094437349,
  -0.0460533488492489,-0.137979795455065,-0.0105618216618972,
  0.133020154123942,0.177857802696526,0.177866362684262,
  0.148582377967028,0.0978799995960585,0.0305338721480460,
  -0.0502552915659989,-0.142198686756143,-0.0147972208826012,
  0.128768672176583,0.173590648408879,0.173583932268205,
  0.144285054060692,0.0935681518354825,0.0262078577102568,
  -0.0545951274457373,-0.146552010292318,-0.0191637092712703,
  0.124389331204253];
if signal==1
Ytrue=y0;         % output data for step signal
elseif signal==2
Ytrue=y1;         % output data for PRBS signal
elseif signal==3
Ytrue=y2;         % output data for sinusoidal signal
elseif signal==4
Ytrue=y3;        % output data for square signal
elseif signal==5
Ytrue=y4;        % output data for Sawtooth signal
end
if noise==0
Ytrue=Ytrue;    % noise-free simulated output data
else
Ytrue=awgn(Ytrue,100,'measured');  % simulated output data with noise
end
a0=x(1);a1=x(2);alpha=x(3);I=eye(m,m);[P1alph1,P2alph1]=TOF1(t0,T,m,
alpha);
[P3alph1,P4alph1]=TOF12(t0,T,m,alpha);u0=[0 ones(1,length(t)-1)]';
u1=[-1;-1;-1;-1;-1;-1;-1;1;-1;1;-1;1;-1;1;1;-1;-1;1;1;-1;-1;-1;1;-1;-
1;-1;
       1;-1;1;1;-1;1;-1;-1;1;1;1;-1;-1;1;-1;-1;-1;-1;1;-1;-1;1;-1;1;-
1;-1;1;
        -1;-1;1;1;-1;1;1;-1;1;1;1;-1;-1;-1;1;1;1;1;-1;1;-1;-1;-1;-1;-
1;1;1;-1;
  1;-1;1;-1;-1;-1;1;1;-1;-1;1;-1;1;1;1;-1;1;1;-1;-1];
u2=[0;0.644217687237691;0.985449729988460;0.863209366648874;
  0.334988150155905;-0.350783227689620;-0.871575772413589;
  -0.982452612624333;-0.631266637872321;0.0168139004843497;
  0.656986598718789;0.988168233877001;0.854598908088280;
  0.319098362349352;-0.366479129251928;-0.879695759971670;
  -0.979177729151317;-0.618137112237032;0.0336230472211367;
  0.669569762196602;0.990607355694870;0.845746831142933;
```

```
  0.303118356745701;-0.382071417184009;-0.887567033581506;
  -0.975626005468158;-0.604832822406284;0.0504226878068148;
  0.681963620068136;0.992766405835908;0.836655638536056;
  0.287052651327728;-0.397555683121436;-0.895187367819682;
  -0.971798445743862;-0.591357529865124;0.0672080725254749;
  0.694164668252245;0.994644773877838;0.827327900595377;
  0.270905788307869;-0.412927549240543;-0.902554608210187;
  -0.967696132133806;-0.577715044445729;0.0839744556917468;
  0.706169457180334;0.996241928754864;0.817766254526438;
  0.254682332844028;-0.428182669496151;-0.909666671833528;
  -0.963320224473761;-0.563909223250257;0.100717096992508;
  0.717974592771644;0.997557418907805;0.807973403666985;
  0.238386871748885;-0.443316730850327;-0.916521547915634;
  -0.958671959951993;-0.549943969560348;0.117431262827103;
  0.729576737392858;0.998590872411771;0.797952116722631;
  0.222024012193082;-0.458325454491772;-0.923117298396360;
  -0.953752652759472;-0.535823231733515;0.134112227645657;
  0.740972610801724;0.999341997081311;0.787705226984118;
  0.205598380402601;-0.473204597045565;-0.929452058477419;
  -0.948563693718306;-0.521551002086912;0.150755275285162;
  0.752158991074479;0.999810580553025;0.777235631526222;
  0.189114620350892;-0.487949951772918;-0.935524037149618;
  -0.943106549888530;-0.507131315768526;0.167355700302807;
  0.763132715516779;0.999996490345607;0.766546290388704;
  0.172577392445818;-0.502557349760487;-0.941331517699235;
  -0.937382764153321;-0.492568249616379;0.183908809306343;
  0.773890681557889];
u3=[-1;-1;-1;-1;-1;1;1;1;1;-1;-1;-1;-1;-1;1;1;1;1;-1;-1;-1;-1;-
1;1;1;1;1;
          -1;-1;-1;-1;-1;1;1;1;1;-1;-1;-1;-1;-1;1;1;1;1;-1;-1;-1;-1;-
1;1;1;1;1;
          -1;-1;-1;-1;-1;1;1;1;1;-1;-1;-1;-1;-1;1;1;1;1;-1;-1;-1;-1;-
1;1;1;1;1;
  -1;-1;-1;-1;-1;1;1;1;1;-1;-1;-1;-1;-1;1;1;1;1;-1;-1];
u4=[1;0.777183079671347;0.554366159342693;0.331549239014039;
  0.108732318685386;-0.114084601643268;-0.336901521971921;
  -0.559718442300575;-0.782535362629228;0.994647717042119;
  0.771830796713465;0.549013876384811;0.326196956056158;
  0.103380035727505;-0.119436884601149;-0.342253804929802;
  -0.565070725258456;-0.787887645587110;0.989295434084237;
  0.766478513755583;0.543661593426930;0.320844673098276;
  0.0980277527696227;-0.124789167559031;-0.347606087887685;
  -0.570423008216338;-0.793239928544991;0.983943151126355;
  0.761126230797702;0.538309310469048;0.315492390140395;
  0.0926754698117414;-0.130141450516913;-0.352958370845566;
  -0.575775291174220;-0.798592211502872;0.978590868168475;
  0.755773947839820;0.532957027511166;0.310140107182512;
  0.0873231868538600;-0.135493733474794;-0.358310653803448;
  -0.581127574132101;-0.803944494460755;0.973238585210593;
  0.750421664881937;0.527604744553285;0.304787824224631;
  0.0819709038959768;-0.140846016432675;-0.363662936761330;
  -0.586479857089982;-0.809296777418638;0.967886302252710;
  0.745069381924058;0.522252461595404;0.299435541266750;
```

```
  0.0766186209380955;-0.146198299390557;-0.369015219719209;
  -0.591832140047865;-0.814649060376517;0.962534019294829;
  0.739717098966175;0.516900178637522;0.294083258308868;
  0.0712663379802159;-0.151550582348440;-0.374367502677092;
  -0.597184423005745;-0.820001343334400;0.957181736336949;
  0.734364816008291;0.511547895679641;0.288730975350987;
  0.0659140550223327;-0.156902865306321;-0.379719785634975;
  -0.602536705963626;-0.825353626292280;0.951829453379066;
  0.729012533050412;0.506195612721758;0.283378692393104;
  0.0605617720644531;-0.162255148264201;-0.385072068592859;
  -0.607888988921509;-0.830705909250163;0.946477170421186;
  0.723660250092532;0.500843329763875;0.278026409435224;
  0.0552094891065700;-0.167607431222084;-0.390424351550738;
  -0.613241271879392;-0.836058192208046;0.941124887463303;
  0.718307967134649];
if signal==1
u=u0;         % step signal
elseif signal==2
u=u1;         % for PRBS signal
elseif signal==3
u=u2;         % sinusoidal signal
elseif signal==4
u=u3;        % square signal
elseif signal==5
u=u4;        % Sawtooth signal
end
C00=u; % input signal
C01=C00(1:end-1);D01=C00(2:end);A1=a0*I+a1*P1alph1;B1=a1*P3alph1;
A2=a1*P2alph1;B2=a0*I+a1*P4alph1;Z1=(C01*P1alph1+D01*P3alph1);
Z2=(C01*P2alph1+D01*P4alph1);A=[A1 A2;B1 B2];
Z=[Z1 Z2];X=Z/A;x1=X(1:m);x2=X(m+1:end);Y=[x1 x2(end)];
e=abs(Ytrue-Y);f=norm(e,2);
end
```

7.4 Summary of Chapter Deliverables

The following is the summary of the key findings of this chapter.

- The proposed TFs-based fractional order (integer as well as noninteger) system identification method is a free identification method, that is, the method produces estimations for model parameters including arbitrary differential orders and initial conditions.
- Linear commensurate and incommensurate nonlinear arbitrary order systems can be identified by the proposed method.
- The method works well for any kind of input excitation signal.
- Multi-input multi-output (MIMO) fractional order as well as integer-order systems can be identified by using the proposed method.

References

[1] A. Benchellal, T. Poinot, J.C. Trigeassou (2007). Modelling and Identification of Diffusive Systems using Fractional Models. In J. Sabatier, O.P. Agrawal, J. A.T. Machado (eds) *Advances in Fractional Calculus*. Dordrecht: Springer.

[2] A. Benchellal, T. Poinot, J.-C. Trigeassou (2006). Approximation and identification of diffusive interfaces by fractional models. *Signal Process.*, vol. 86(10), pp. 2712–2727.

[3] J.C. Wang (1987). Realizations of generalized Warburg impedance with RC ladder networks and transmission lines. *J. Electrochem. Soc.*, vol. 134(8), pp. 1915–1920.

[4] A. Jalloul, K. Jelassi, P. Melchior, J.-C. Trigeassou (2011). Fractional identification of rotor skin effect in induction machines. *Int. J. Comput. Sci.*, vol. 8(1), pp. 57–67.

[5] R.L. Bagley, R.A. Calico (1991). Fractional order state equations for the control of viscoelasticallydamped structures. *J. Guid. Cont. Dyn.*, vol. 14(2), pp. 304–311.

[6] W.M. Ahmad, R. El-Khazali (2007). Fractional-order dynamical models of love. *Chaos Soliton Fract*, vol. 33, pp. 1367–1375.

[7] L. Song, S. Xu, J. Yang (2010). Dynamical models of happiness with fractional order. *Commun. Nonlinear Sci*, vol. 15(3), pp. 616–627.

[8] A.A. Freihat, M. Zurigat, A.H. Handam (2015). The multi-step homotopy analysis method for modified epidemiological model for computer viruses. *Afrika Mat.*, vol. 26, pp. 585–597.

[9] Y. Cho, I. Kim, D. Sheen (2015). A fractional-order model for MINMOD Millennium. *Math. Biosci.*, vol. 262, pp. 36–45.

[10] Q. Din, A.A. Elsadany, H. Khalil (2017). Neimark-Sacker bifurcation and chaos control in a fractional-order Plant-Herbivore model. *Discrete. Dyn. Nat. Soc.*, vol. 2017, Article ID 6312964, 15.

[11] H.M. Srivastava, D. Kumar, J. Singh (2017). An efficient analytical technique for fractional model of vibration equation. *Appl. Math. Model.*, vol. 45, pp. 192–204.

[12] J.K. Popovic, M.T. Atanackovic, A.S. Pilipovic, M.R. Rapaic, S. Pilipovic, T.M. Atanackovic (2010). A new approach to the compartmental analysis in pharmacokinetics: Fractional time evolution of diclofenac. *J. Pharmacokinet. Phar.*, vol. 37(2), pp. 119–134.

[13] A.D. Freed, K. Diethelm (2006). Fractional calculus in biomechanics: A 3D viscoelastic model using regularized fractional derivative kernels with application to the human calcaneal fat pad. *Biomech. Model. Mechan.*, vol. 5, pp. 203–215.

[14] A.M. Lopes, J.A.T. Machado, E. Ramalho (2017). On the fractional-order modeling of wine. *Eur. Food. Res. Technol.*, vol. 243(6), pp. 921–929.

[15] N. Heymans, I. Podlubny (2005). Physical interpretation of initial conditions for fractional differential equations with Riemann-Liouville fractional derivatives. *Rheol. Acta.*, vol. 45(5), pp. 765–771.

[16] A. Dzielinski, D. Sierociuk, G. Sarwas, I. Petras, I. Podlubny, T. Skovranek (2011). Identification of the fractional-order systems: A frequency domain approach. *Acta. Montan. Slovaca.*, vol. 16(1), pp. 26–33.

[17] D. Valerio, M. Ortigueira, J. Sa Da Costa (2008). Identifying a transfer function from a frequency response. *J. Comput. Nonlinear Dyn.*, vol. 3(2), pp. 021207-1–021207-7. DOI: 10.1115/1.2833906.

[18] D. Valério, I. Tejado (2015). Identifying a non-commensurable fractional transfer function from a frequency response. *Signal Process.*, vol. 107, pp. 254–264.

[19] P. Nazarian, M. Haeri, M.S. Tavazoei (2010). Identifiability of fractional order systems using input output frequency contents. *ISA T.*, vol. 49(2), pp. 207–214.
[20] A. Djouambi, A. Voda, A. Charef (2012). Recursive prediction error identification of fractional order models. *Commun. Nonlinear. Sci.*, vol. 17(6), pp. 2517–2524.
[21] T. Poinot, J.-C. Trigeassou (2004). Identification of fractional systems using an output-error technique. *Nonlinear Dyn.*, vol. 38, pp. 133–154.
[22] A. Maachou, R. Malti, P. Melchior, J.-L. Battaglia, A. Oustaloup (2014). Nonlinear thermal system identification using fractional Volterra series. *Contr. Eng. Pract.*, vol. 29, pp. 50–60.
[23] D. Wang, X. Wang, P. Han (2010). Identification of thermal process using fractional order transfer function based on intelligent optimization. In: *Proceedings of 2010 IEEE/ASME International Conference on Mechatronic and Embedded Systems and Applications*, Qingdao, pp. 498–503.
[24] Y. Huang, F. Guo, Y. Li, Y. Liu (2015). Particle estimation of fractional-order chaotic systems by using quantum parallel particle swarm optimization algorithm. *PLoS ONE*, vol. 10(1), pp. e0114910. DOI: 10.1371/journal.pone.0114910.
[25] W. Du, Q. Miao, L. Tong, Y. Tang (2017). Identification of fractional order system with unknown initial values and structure. *Phys. Lett. A*, vol. 381(23), pp. 1943–1949.
[26] M. Aoun, R. Malti, O. Cois, A. Oustaloup (2002). System identification using fractional Hammerstein models. *IFAC Proc. Vol.*, vol. 35(1), pp. 265–269.
[27] D. Maiti, M. Chakraborty, A. Konar (2008). A novel approach for complete identification of dynamic fractional order systems using stochastic optimization algorithms and fractional calculus. In: *Proceedings of IEEE* international conference on Electrical *and Computer* Engineering *(ICECE 2008)*, Dhaka, Bangladesh, pp. 867–872.
[28] S. Victor, R. Malti, P. Melchior, A. Oustaloup (2011). Instrumental variable identification of hybrid fractional Box-Jenkins models. *IFAC Proc. Vol.*, vol. 44 (1), pp. 4314–4319.
[29] D.-Y. Liu, T.-M. Laleg-Kirati, O. Gibaru, W. Perruquetti (2013). Identification of fractional order systems using modulating functions method. In: *Proceedings of 2013 American Control Conference (ACC)*, Washington, DC, United States, pp. 1679–1684.
[30] O. Cois, A. Oustaloup, T. Poinot, J.-L. Battaglia (2001). Fractional state variable filter for system identification by fractional model. In: *Proceedings of 2001 European Control Conference (ECC)*, Porto, pp. 2481–2486.
[31] D.V. Ivanov, A.V. Ivanov (2017). Identification fractional linear dynamic systems with fractional errors-in-variables. *J. Phys. Conf. Ser.*, vol. 803(1), pp. pages 012057–012057012058.
[32] M. Aoun, R. Malti, F. Levron, A. Oustaloup (2007). Synthesis of fractional Laguerre basis for system approximation. *Automatica*, vol. 43, pp. 1640–1647.
[33] Y. Li, X. Meng, B. Zheng, Y. Ding (2015). Parameter identification of fractional order linear system based on Haar wavelet operational matrix. *ISA T.*, vol. 59, pp. 79–84.
[34] Y. Tang, H. Liu, W. Wang, Q. Lian, X. Guan (2015). Parameter identification of fractional order systems using block pulse functions. *Signal Process.*, vol. 107, pp. 272–281.
[35] K.B. Datta, B.M. Mohan (1995). *Orthogonal Functions in Systems and Csontrol.* Singapore: World Scientific Publishing Co. Pte Ltd.

8

Design of Fractional Order Controllers using Triangular Strip Operational Matrices

All controller design techniques involve explicit or implicit use of information concerning the transient behavior of a process. This information can be a system of coupled differential and/or partial differential equations. The real processes themselves change with time due to feedstock variations, catalyst deactivation, changes in market conditions of product, and so on. It is rare that these changes are reflected in their corresponding mathematical models. Consequently, the information used for a controller design is never perfect. Such a controller, which is designed by an approximate description of the process, is likely to be unsuccessful when it is applied to an actual process. Therefore, the ultimate objective of any controller design or tuning method is to develop a robust controller that shows satisfactory performance in the real operating environment where the plant shows different dynamic behavior than that of its model [1]. Six decades ago, Bode [2] encountered this robustness problem while designing a feedback amplifier. The design objective was to make the feedback amplifier insensitive to variations in the amplifier gain. He solved the problem through the proposition of an open loop transfer function $(w_{gc}/s)^{\gamma}$, where γ is in the range $(1, 2)$ and w_{gc} is the gain crossover frequency, called Bode's ideal transfer function.

In fact, Bode's ideal transfer function is a fractional integrator of order γ. Thus, the use of fractional calculus concepts in automatic control was begun with Bode. Oustaloup [3] attempted to bridge the gap between the fractional calculus and the automatic control. The author developed three generations of fractional order controller called CRONE controller. Podlubny [4] proposed a fractional $PI^{\lambda}D^{\mu}$ controller and also proved that this controller structure is more suitable to regulate a fractional system in comparison to classical PID controller. In addition to controller gains (K_C, K_I, K_d), the fractional $PI^{\lambda}D^{\mu}$ controller has two more unknowns; the order of integral, λ, and the order of derivative, μ, which allow the controller to be eligible to meet industrial process control needs [5]. However, the ability of satisfying conflicting performance requirements is ensured at the cost of complexity in the selection of controller parameters. A good review of tuning and implementation methods for $PI^{\lambda}D^{\mu}$ as well as for other forms of fractional order controller such as CRONE, TID, and fractional order lead-lag compensator can be found in [6]. The literature on the application of fractional calculus in automatic control is not limited and ever increasing [7–11].

To tackle plant uncertainty issues, many controller design methods are developed. The convenient one among these methods is designing a robust fractional $PI^{\lambda}D^{\mu}$ controller with a loop shaping approach using the specified frequency domain characteristics such as gain crossover frequency, phase margin, and flat phase constraint [12]. Employing nongradient optimization techniques, for example, particle swarm optimization for the design of an H_{∞}-optimal fractional $PI^{\lambda}D^{\mu}$ controller, is a proven and promising approach [13, 14]. In [15], the authors developed a robust design method to tune the third-generation CRONE controller for the nominal process model of a hydraulic system and also implemented it into a Programmable Logic Controller in order to apply it to control the real hydraulic system. Bouafoura and Braiek [16] introduced a new application of orthogonal functions for tuning fractional PID controller to make sure that the fractional PID control system follows the dynamics of the specified reference system.

This chapter addresses the robustness issue by proposing a simple tuning technique aimed to produce a robust fractional PID controller exhibiting iso-damping property during the reparameterization of a plant. The required robustness property is achieved by letting the fractional PID control system imitate the dynamics of a reference system having Bode's ideal transfer function in its forward path. Thus, the goal of accomplishing a flat phase angle around the gain crossover frequency is defined as an H_{∞}-optimal control problem (nonlinear programming) in which the objective is to shape the fractional open loop as an ideal Bode's control loop. The most time taken performing this type of method is the recurrent simulation of fractional PID controller system until the optimization problem is solved. To avoid such time-consuming simulations, fractional differential systems are transformed into algebraic ones by the use of triangular strip operational matrices. The H_{∞}-optimal control problem is then changed to an ∞-norm minimization of a parameter varying (i.e., unknowns, $K_C, K_I, K_d, \lambda, \mu$) square matrix. Global optimization techniques, the Luus-Jaakola direct search multipass method (LJ), and the particle swarm optimization (PSO) are employed to find the optimum values of fractional PID controller parameters, which make sure that the closed loop system is insensitive to variations in the plant gain. For the sake of simplicity, the robust fractional order PID controller is designed using triangular strip operational matrices and a reference control system. However, the developed technique can be used for designing different forms of fractional order PID controller.

8.1 Triangular Strip Operational Matrices–Based Fractional Order Controller Design Method

To design a robust controller, a simple design method is developed in this work. The idea is to allow the fractional order PID control system shown in Figure 8.1 to mimic the dynamics of a reference system (Figure 8.2) with Bode's ideal transfer function in its forward path.

The ideal Bode's control loop is

$$L(s) = \left(\frac{w_{gc}}{s}\right)^{\gamma}, \tag{8.1}$$

where w_{gc} is the gain crossover frequency, γ is a noninteger number that lies in the range $(1, 2)$.

The magnitude of ideal loop is

$$|L(jw)| = 20\gamma \log\left(\frac{w_{gc}}{w}\right) \tag{8.2}$$

and the phase angle is

$$\angle L(jw) = -\frac{\gamma\Pi}{2}. \tag{8.3}$$

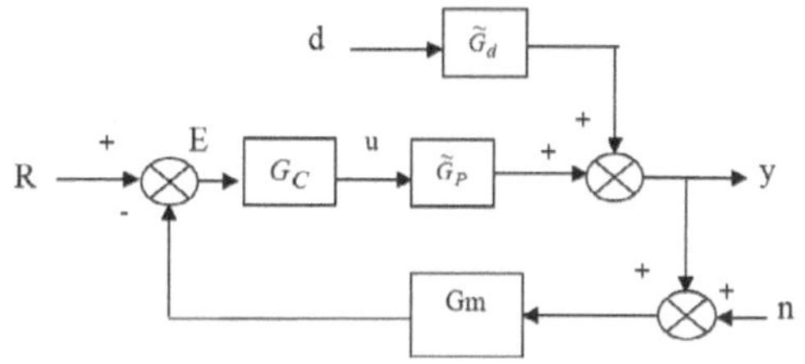

FIGURE 8.1
Feedback control system

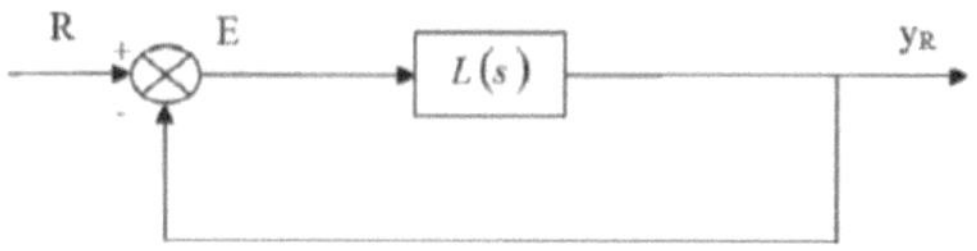

FIGURE 8.2
Unity feedback reference control system

The magnitude and the phase Bode plot of ideal open loop are shown in Figure 8.3.

The magnitude curve is a straight line with slope of $-20\gamma\ dB/dec$ whereas the phase Bode plot remains constant in the region around the gain crossover frequency, that is,

$$\left.\frac{d\angle L(s)}{ds}\right|_{s=jw_{gc}} = 0 \ or\ \angle\left.\frac{dL(s)}{ds}\right|_{s=jw_{gc}} = \angle L(jw). \tag{8.4}$$

This characteristic of flat phase angle is called iso-damping. The phase margin of ideal control loop is constant ($\Pi - 0.5\gamma\Pi$) and is independent of loop gain. Therefore, when the loop gain varies, the magnitude curve moves up or down but the phase angle curve stays unchanged indicating that the reference control system demonstrates robust stability and robust performance in the event of gain uncertainty. As a result of its uncompromising behavior in all circumstances, the open loop transfer function has been called Bode's ideal transfer function or ideal Bode's control loop. In a

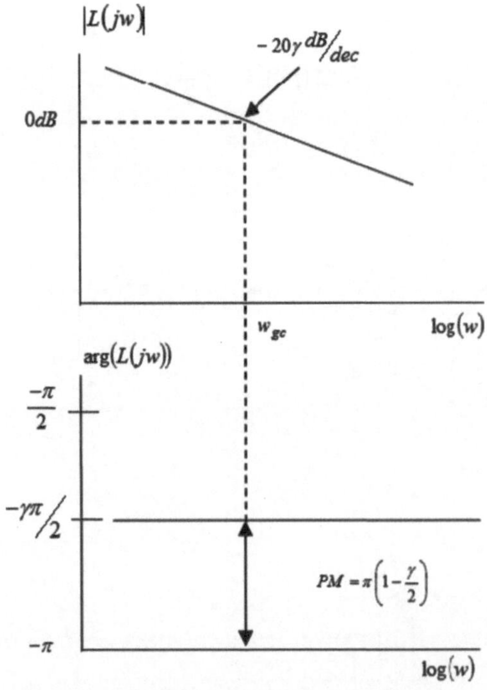

FIGURE 8.3
Frequency response of Bode's ideal transfer function

time domain, the effect of the constant phase margin can be recognized from the constant overshoots of step responses to different values of gain. The step responses of the closed loop reference control system to various values of γ are computed and plotted in Figure 8.4. Regardless of the value of γ (in the range $(1, 2)$), the ideal Bode's control loop always owns iso-damping property. But to what extent it holds acceptable relative stability depends on the selection of γ. It can be seen in Figure 8.4 and in Table 8.1 that the large values of γ cause the reference control system to oscillate hence the large settling time and the large overshoot whereas the impact of small values (near 1) is reflected in somewhat large rise time. The reference system has a good transient response if γ is in the range $(1, 1.5]$.

The goal of attaining robustness via the imitation of the dynamics of the reference control system is defined mathematically as an H_∞-optimal control problem:

$$J = \min_{K_C, K_I, K_d, \lambda, \mu} \|Y_R(s) - Y(s)\|_\infty. \tag{8.5}$$

where Y_R is the output of the reference system, Y is the output of plant model $\tilde{G}_P$.

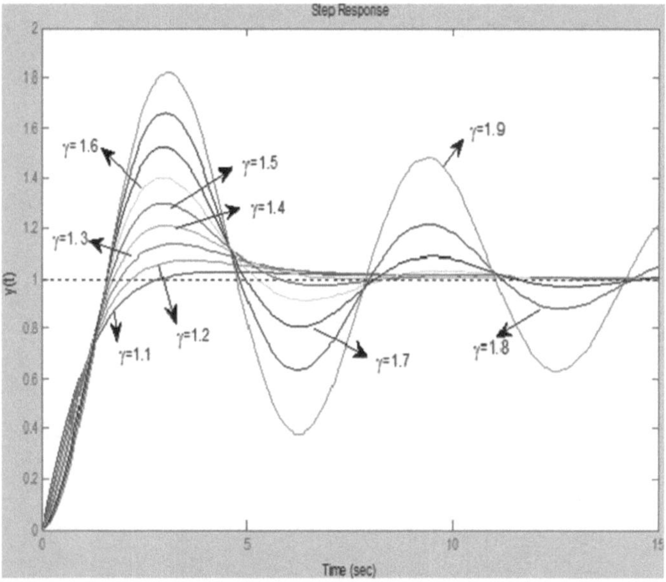

FIGURE 8.4
Unit step responses of reference control system to different values of γ

TABLE 8.1

Time and frequency domain characteristics of reference system

γ	t_r	t_s	OS	Pu	t_p	PM
1.1	1.7800	6.2789	2.7784	1.0278	4.4059	81.0114
1.2	1.5323	8.6737	8.4177	1.0742	3.5618	72.0228
1.3	1.3752	8.1551	13.5252	1.1353	3.1984	63.0342
1.4	1.2691	6.1150	21.0276	1.2103	3.0266	54.0456
1.5	1.1945	8.3413	28.9440	1.2994	2.9444	45.0570
1.6	1.1417	10.9548	40.2955	1.4030	2.8833	36.0685
1.7	1.0986	13.6940	52.2684	1.5227	3.0206	28.0799
1.8	1.0825	22.7302	65.8005	1.6580	2.9028	18.0914
1.9	1.0589	48.3479	81.5959	1.8160	3.1416	8.1030

t_r: rise time, t_s: settling time, OS: overshoot, Pu: peak value, t_p: peak time, PM: phase margin

For the sake of simplicity, it is assumed that the measuring device transfer function has negligible dynamics $G_m = 1$ and noise is zero, $n = 0$. The process considered here is a fractional order process. The fractional order plant model $\tilde{G}_P$ has the following general form:

$$\tilde{G}_P(s) = \frac{Y(s)}{U(s)} = \frac{b_m s^{\beta_m} + b_{m-1} s^{\beta_{m-1}} + \cdots\cdots + b_0 s^{\beta_0}}{a_n s^{\alpha_n} + a_{n-1} s^{\alpha_{n-1}} + \cdots\cdots + a_0 s^{\alpha_0}} \tag{8.6}$$

and the controller chosen is a fractional order $PI^{\lambda}D^{\mu}$ controller of the form

$$G_C = \frac{U(s)}{E(s)} = K_C + \frac{K_I}{s^{\lambda}} + K_d s^{\mu}. \tag{8.7}$$

The relationship between the reference signal R and the output of the fractional plant model $\tilde{G}_P(s)$ is

$$G_{CL}(s) = \frac{Y(s)}{R(s)} = \frac{G_C(s)\tilde{G}_P(s)}{1 + G_C(s)\tilde{G}_P(s)} \Rightarrow Y(s) = \frac{G_C(s)\tilde{G}_P(s)}{1 + G_C(s)\tilde{G}_P(s)} R(s), \tag{8.8}$$

$$\begin{aligned} Y(s) &= \frac{\left(K_C + \frac{K_I}{s^{\lambda}} + K_d s^{\mu}\right)\left(\frac{b_m s^{\beta_m} + b_{m-1} s^{\beta_{m-1}} + \cdots\cdots + b_0 s^{\beta_0}}{a_n s^{\alpha_n} + a_{n-1} s^{\alpha_{n-1}} + \cdots\cdots + a_0 s^{\alpha_0}}\right)}{1 + \left(K_C + \frac{K_I}{s^{\lambda}} + K_d s^{\mu}\right)\left(\frac{b_m s^{\beta_m} + b_{m-1} s^{\beta_{m-1}} + \cdots\cdots + b_0 s^{\beta_0}}{a_n s^{\alpha_n} + a_{n-1} s^{\alpha_{n-1}} + \cdots\cdots + a_0 s^{\alpha_0}}\right)} R(s), \\ &= \frac{A}{A + B} R(s), \end{aligned} \tag{8.9}$$

where $A = \left(K_C s^{\lambda} + K_I + K_d s^{\lambda+\mu}\right)\left(b_m s^{\beta_m} + b_{m-1} s^{\beta_{m-1}} + \cdots\cdots + b_0 s^{\beta_0}\right)$,

$$B = (a_n s^{\alpha_n} + a_{n-1} s^{\alpha_{n-1}} + \cdots\cdots + a_0 s^{\alpha_0}).$$

Applying the inverse Laplace transform and replacing all fractional order differential operators in the resulting equation with their respective triangular strip operational matrices (derived in Chapter 1) leads to

$$Y_N = \frac{A_N}{B_N + A_N} R_N, \tag{8.10}$$

where $A_N = \left(K_C B_N^{\lambda} + K_I I + K_d B_N^{\lambda+\mu}\right)\left(b_m B_N^{\beta_m} + \cdots\cdots + b_0 B_N^{\beta_0}\right)$,

$$B_N = \left(a_n B_N^{\alpha_n+\lambda} + \cdots\cdots + a_0 B_N^{\alpha_0+\lambda}\right).$$

In the reference control system, the reference signal R and the output Y_R are related as

$$G_{CLR}(s) = \frac{Y_R(s)}{R(s)} = \frac{L(s)}{1+L(s)} \Rightarrow Y_R(s) = \frac{(w_{gc})^{\gamma}}{s^{\gamma} + (w_{gc})^{\gamma}} R(s). \tag{8.11}$$

Similarly to Equation (8.10), $Y_R(s)$ is estimated with the triangular strip operational matrices:

$$Y_{RN} = \frac{(w_{gc})^{\gamma} I}{B_N^{\gamma} + I(w_{gc})^{\gamma}} R_N. \tag{8.12}$$

The shapes of the reference signal R or the setpoint changes can be steps, pulses or ramps depending on the user's choice. Replacing $Y(s)$ and $Y_R(s)$ with their respective estimations Y_N and Y_{RN}, respectively, in Equation (8.5) leads to

$$\begin{aligned} \tilde{J} &= \min_{K_C, K_I, K_d, \lambda, \mu} \|Y_{RN} - Y_N\|_{\infty}, \\ &= \min_{K_C, K_I, K_d, \lambda, \mu} \left\| \underbrace{\frac{(w_{gc})^{\gamma} I}{B_N^{\gamma} + I(w_{gc})^{\gamma}} R_N - \frac{A_N}{B_N + A_N} R_N}_{C} \right\|_{\infty}, \end{aligned} \tag{8.13}$$

where I is the identity matrix.

The constraints that must be satisfied during the optimization are

$$l_1 \leq K_C \leq u_1, l_2 \leq K_I \leq u_2, l_3 \leq K_d \leq u_3, l_4 \leq \lambda \leq u_4, l_5 \leq \mu \leq u_5. \quad (8.14)$$

The triangular strip operational matrices greatly simplified the objective function of the original H_∞-optimal control problem by the transformation of time domain developments into algebraic equations. The goal of tracking the dynamics of robust reference control system is now changed to the minimization of ∞-norm of the parameter varying square matrix C, which contains information about the deviation of Y from Y_R. The control design method devised here can be extended to integer order processes and can be used to track the dynamics of any type of reference system.

8.2 Constrained Nonlinear Optimization

Clearly, the H_∞- optimal control problem in Equation (8.5) or (8.13) is a constrained nonlinear programming problem (NLP) that has to be solved to find out the values of controller parameters including λ and μ. Global optimization techniques including the Luus-Jaakola direct search multipass optimization method and nature-inspired particle swarm optimization method are used here [17, 18]. Although these techniques follow different strategies to locate the global optimum, they have a common attribute of ease of implementation.

8.2.1 Luus-Jaakola (LJ) Multipass Optimization Method

The initial guess for a p-dimensional vector can be made on the center of a search space as $0.5(a+b)$, where a and b are the lower and the upper bound, respectively, on p-dimensional optimization variables. To search the entire physical region, the initial region size vector is chosen as $0.5(a+b)$. This method first picks R points randomly in the whole search space and then finds the best one among them, which gives the minimum objective function value. Once the first best point is found (i.e., local minimum in the first iteration), the search region is reduced by 5 percent. In the reduced search space around this local minimum, again R random points are taken and similarly the best one which offers a better (smaller) objective function value compared to that given by the previous best point, is identified (the second iteration). As the optimization process progresses (one iteration after another), the boundaries of the search space keep on compressing. The search space eventually becomes the close neighborhood of the global optimum. Unlike the traditional optimization algorithms, LJ is insensitive to the choice of initial guess and does not need the gradient of objective function. To reduce the number of function evaluations, hence, CPU time, the random points that fall outside the physical search area, are rejected.

The pseudo code for finding the global optimum of Equation (8.5) or (8.13) is given here.

Pseudo Code for LJ Optimization Method

Choose the number of iterations to be used in each pass N_T, number of passes N_P, initial region size vector r^1, region reduction factor η, number of random points R, and tolerance φ.

%Algorithm begins%

$$K^{(0)} = 0.5(a + b).$$

$$K^* = K^{(0)}.$$

$$J_{old} = 10^{50}.$$

for $q = 1 : 1 : N_P$
for $j = 1 : 1 : N_T$
for $i = 1 : 1 : R$

Generate random points in p-dimensional space around the best value K^* (found from previous iteration) using the following relation.

$$K^j = K^* + D^* r^j,$$

where D^* is a diagonal matrix containing diagonal elements in the range $[-0.5, 0.5]$.

Check the feasibility of each element of K^j with respect to the bounds given in Equation (8.14). If any element violates the constraints, discard that set. Otherwise, compute the value of objective function J using K^j.

if $J < J_{old}$

$$J_{old} = J.$$

$$K_S^* = K^j.$$

end
end

$$r^{j+1} = r^j \eta.$$

$$K^* = K_S^*.$$

end
FOR $k = 1 : 1 : p$

Determination of region size for each parameter from its size of variation.

$$r_k^{q+1} = \left|K^{*q} - K^{*1}\right|,$$

where K^{*q} is the best value of ith parameter obtained at the end of qth pass, K^{*1} is the best value of ith parameter employed at the beginning of qth pass.

To prevent search space from collapse,

if $r_k^{q+1} < \varphi$

$$r_k^{q+1} = \varphi.$$

end

end

$$r^1 = \left[r_1^{q+1}, \ldots\ldots, r_p^{q+1}\right]^T.$$

if $q > 5$

Check whether the improvement of objective function value J for three successive passes is less than φ.

if 'yes'

$$\varphi = 0.5 * \varphi.$$

end

end

if $\varphi < 10^{-10}$

Algorithm gets terminated.

end

end %Algorithm stops%

8.2.2 Particle Swarm Optimization Method

The particle swarm optimization (PSO) is also a stochastic global optimization method but derived from the simulation of social behavior of a flock of birds or fish shoaling. PSO is initialized with a set of candidate solutions or a population of particles whose initial positions and velocities are arbitrarily chosen in the physical search region. These initial locations of particles are their respective personal best positions (local minima). At each initial position, the objective function is evaluated and the position which provides the lowest objective function value is discovered, this is termed as the swarm's best known position. Each particle starts flying from its initial position to a new position computed from the velocity and position updating equations. The new position of each particle is determined in such a way that the particle is directed towards the swarm's previous best known position by the influence of its previous local best known position. Once better positions are found, the particles' local best known positions are updated and also the swarm's previous best known position is updated. By doing so, the swarm eventually moves to the vicinity of global optimum. Like Luus-Jaakola, PSO is a gradient-free method and its performance is unaltered by the selection of initial candidate solutions.

Pseudo Code for PSO

Select the size of population (S), number of iterations ($Maxitr$), cognition factor (c_1), social factor (c_2), inertial weight (w), and the lower and upper bounds on n_d optimization variables a and b.

%PSO begins%

for $i = 1 : 1 : S$

$$pos(i, n_d) = a + (b - a)rand(1, n_d).$$

$$vel(i, n_d) = a + (b - a)rand(1, n_d).$$

end

$$Pbest\_Pos = pos.$$

for $i = 1 : 1 : S$

$$best(i) = J(pos(i, :)).$$

end

$$[gbest, I] = \min(pbest).$$

$$gbest\_pos = pos(I, :).$$

for $w = 1 : 1 : Maxitr$

for $i = 1 : 1 : S$

$v(i,d) = w * vel(,d) + c_1 rand()(pbest\_pos(i,d) - pos(i,d)) + c_2 rand()(gbest\_pos(1,d) - pos(i,d))$.

IF $v(i,d) > b(d) - pos(i,d)$ or $v(i,d) < a(d) - pos(i,d)$

$$vel(i,d) = 0.$$

end

$$pos(i,d) = pos(i,d) + vel(i,d).$$

end
if $J(pos(i,:)) < pbest(i)$

$$pbest(i) = J(pos(i,:)).$$

$$pbest\_pos(i,:) = pos(i,:).$$

end
if $J(pos(i,:)) < gbest$

$$gbest = J(pos(i,:)).$$

$$gbest\_pos = pos(i,:).$$

end
end
end %PSO stops%

8.3 Simulation Examples

In this section, we implement the proposed method on heating furnace model, automatic voltage regulator (AVR) system, and some integer and fractional order models.

8.3.1 Design of Robust Fractional $PI^{\lambda}D^{\mu}$ Controller for a Heating Furnace System

We consider the following heating furnace fractional model:

$$G(s) = \frac{k_p}{\tau_1 s^{1.31} + \tau_2 s^{0.97} + \tau_3}, \tag{8.15}$$

where $k_p = 1$, $\tau_1 = 14994$, $\tau_2 = 6009.5$, $\tau_3 = 1.69$. The parameters of the Bode's ideal system are selected as $\gamma = 4/3$, $w_{gc} = 1 rad/\sec$. So that the ideal loop owns phase margin of $\pi/3$,

$$L(s) = \left(\frac{1}{s}\right)^{4/3}. \tag{8.16}$$

The parameters of LJ are $N_P = 50$, $N_T = 80$, $R = 40$, $\varphi = 10^{-6}$, $\eta = 0.95$, $a = [0, 0, 0, 0.001, 0.001]$, $b = [10^4, 10^4, 10^5, 1, 1]$ and the parameters of PSO are $S = 50$, $Maxitr = 100$, $c_1 = 1.49$, $c_2 = 1.49$, $w = 1.1$, $a = [0, 0, 0, 0.001, 0.001]$, $b = [10^4, 10^4, 10^5, 1, 1]$.

The original H_∞-optimization problem (Equation (8.5)) of making the fractional open loop behave like Bode's ideal control loop in the event of gain uncertainties is solved by using LJ and PSO (the respective MATLAB® codes are provided in Programs 8.1 and 8.2). By employing each optimization technique, 10 independent runs are carried out to verify whether the techniques can reproduce feasible solutions. The statistical measures of results of all runs are computed and tabulated in Table 8.2. Both LJ and PSO attained partial success in repetitively finding the feasible solutions but PSO is computationally more effective than LJ. The overall best feasible solution found via each method is not very accurate. However, by altering the values of parameters of PSO or LJ that is increasing the size of swarm and the number of iterations, changing c_1 and/or c_2, and so on, or taking more random points, iterations, and so on in case of LJ, one can get better solutions but at the cost of higher CPU time (CT). So we follow the proposed design method to construct an approximation for the original optimization problem. After introducing the triangular strip operational matrices (TSOM) in Equation (8.5), the argument of H_∞-norm in Equation (8.5), that is, the fractional order transfer function became a square matrix that can be minimized with less effort. For generating the triangular strip operational matrices (Program 1.3), we consider the step size of 0.5. The MATLAB codes for LJ and PSO optimization of Equation (8.13) are provided in Programs 8.3 and 8.4. Similar to the original problem, 10 independent runs are conducted and Table 8.2 presents the statistical analysis. It is evidently shown that the use of triangular strip operational matrices is beneficial to get more accurate feasible solution at much lower computational time (CT). Here also, PSO dominates LJ in locating the global optimum. It should be noted that the same initial data used for PSO and LJ for solving the original optimization problem is taken into account for minimizing the TSOM-based H_∞-optimal control problem in Equation (8.13). The optimum values of $PI^\lambda D^\mu$ controller parameters

TABLE 8.2

Comparison of optimal solutions of J and $\tilde{J}$ for heating furnace system

	Original H_∞-optimization				TSOM-based H_∞-optimization			
Method	**Mean J**	**Std.**	**Best J**	**CT(s)**	**Mean J**	**Std.**	**Best J**	**CT(s)**
PSO	0.0029	0.0026	5.75e-04	10.045	5.46e-05	3.06e-05	2.72e-07	0.5747
LJ	0.00339	0.0028	4.01e-04	2.95e02	1.80e-06	2.70e-05	5.37e-06	33.376

TABLE 8.3

Optimal values of controller parameters for heating furnace system

Method	K_C	K_I	K_d	λ	μ
PSO	0.02208	8668.622	12338.418	0.30245	0.007
LJ	0.30628	8.320e+03	1.168e+04	0.2859	0.01215

(Table 8.3) obtained by minimizing Equation (8.13) are only considered for closed loop simulations.

The designed fractional $PI^{\lambda}D^{\mu}$ controllers are

$$FOPID1 = 0.02208 + \frac{8669.622}{s^{0.30245}} + 12337.418s^{0.007}, \text{ using PSO.} \tag{8.17}$$

$$FOPID2 = 0.30628 + \frac{9320}{s^{0.2859}} + 11680s^{0.01215}, \text{ using LJ.} \tag{8.18}$$

As seen in Figure 8.5, the Bode phase plot of fractional open loop with each controller remains constant in the region ($[0.294, 10.3]$) around the gain crossover frequency that makes sure that the $PI^{\lambda}D^{\mu}$ controllers earned desired iso-damping property hence the heating furnace fractional control system can tolerate variations in k_p. Both the reference system and the closed loop fractional system are simulated for a unit step change in the setpoint, and the resulting responses are compared in Figure 8.6. The $PI^{\lambda}D^{\mu}$ control systems perfectly imitated the dynamics of robust reference system; therefore, when k_p experiences changes, the overshoots of step responses of the heating furnace closed loop systems will remain almost constant.

To examine whether the designed fractional $PI^{\lambda}D^{\mu}$ controllers perform well in the face of unexpected gain variations, the process gain (k_p) is varied in the range $[0.5, 1.5]$ using the step size of 0.1. Though the gain variations influenced the Bode magnitude plots (Figures 8.7 and 8.8), they could not hamper the flat phase angle characteristic of fractional open loops. For further investigation, robustness measures are calculated and tabulated in Table 8.4. As the conditions for robust stability and robust performance are satisfied, the designed (FOPID1 and FOPID2) controllers can maintain stability and exhibit satisfactory performance even in the presence of gain uncertainties. As we observe in Figure 8.9 and Table 8.4, both controllers can suppress the impact of unanticipated disturbances. A step change is made in the setpoint at time $t = 0$ and in the disturbance

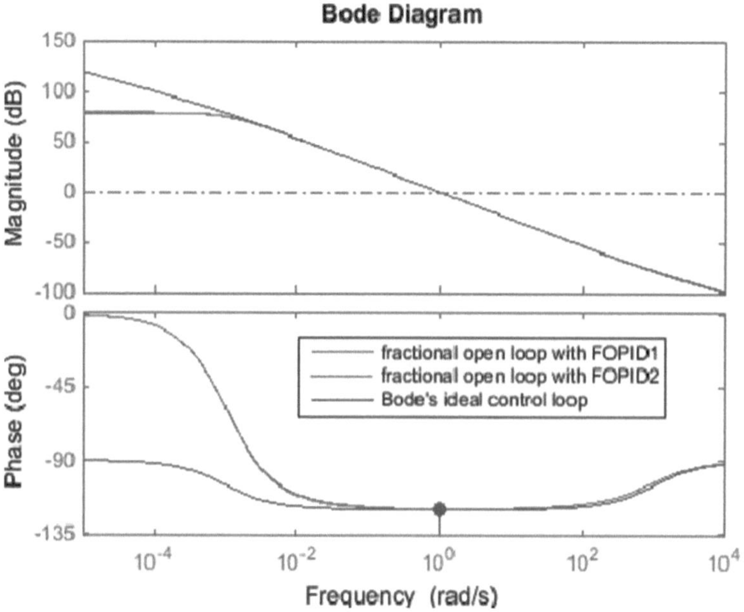

FIGURE 8.5
Comparison of frequency response of ideal control loop and fractional open loop

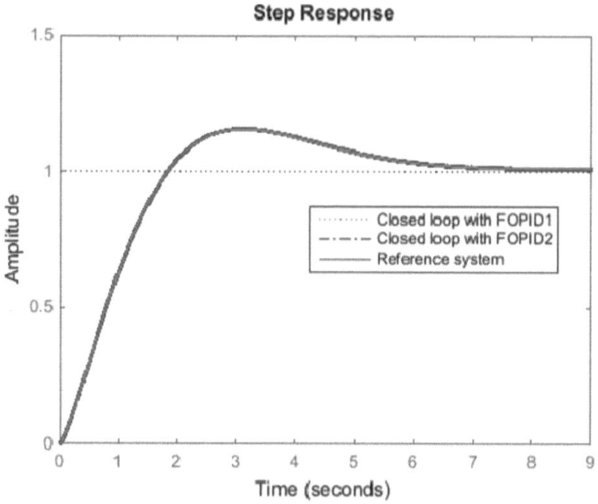

FIGURE 8.6
Step response of reference system and fractional $PI^{\lambda}D^{\mu}$ control systems

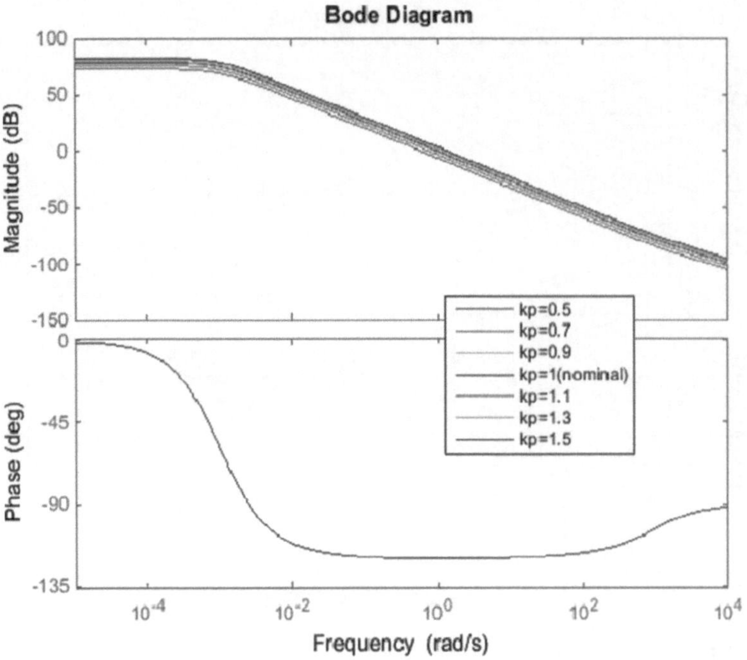

FIGURE 8.7
Frequency response of fractional open loop using FOPID1 to different plant gain

input at $t = 25$ seconds. Figure 8.10 highlights that the heating furnace control system can simultaneously track the changes in the command signal, and suppress the effect of the changes in the disturbances. Because the controllers possess iso-damping property, the overshoots of step responses are almost at the same level.

For the same heating furnace system, the following fractional $PI^{\lambda}D^{\mu}$ controllers were designed using various tuning methodologies in [5, 16, 19, and 20],

$$FOPID3 = 736.8054 - \frac{0.5885}{s^{0.6}} - 818.420s^{0.35} \quad \text{(Reference[5])} \tag{8.19}$$

$$FOPID4 = 7714.9739 + \frac{107.0099}{s^{0.6}} + 287.7011s^{0.35} \quad \text{(Reference[16])} \tag{8.20}$$

$$FOPID5 = 1924.7 - \frac{111.8922}{s^{0.325}} - 653.2185^{0.325} \quad \text{(Reference[19])} \tag{8.21}$$

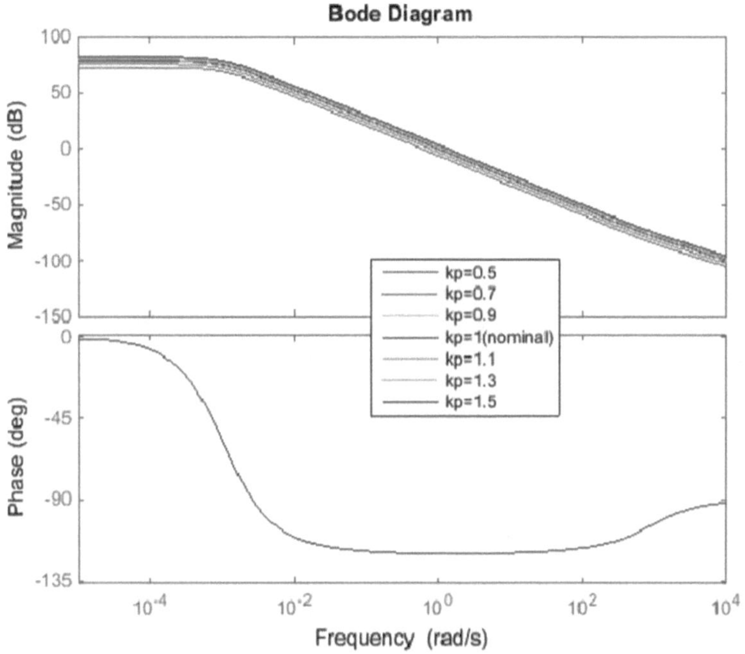

FIGURE 8.8
Frequency response of fractional open loop using FOPID2 to different plant gain

TABLE 8.4
Robustness analysis of FOPID1 and FOPID2

	FOPID1					
k_p	**0.9**	**0.7**	**0.5**	**1.1**	**1.3**	**1.5**
RS	0.11523	0.34569	0.57616	0.11523	0.34569	0.57616
RP	0.46983	0.70030	0.46983	0.46983	0.70030	0.93076
Ms	1.1535	1.1515	1.1545	1.1539	1.1503	1.1503
	FOPID2					
k_p	0.9	0.7	0.5	1.1	1.3	1.5
RS	0.11525	0.34577	0.57629	0.11525	0.34577	0.57629
RP	0.46852	0.69904	0.92955	0.46852	0.69904	0.92955
Ms	1.1511	1.1547	1.1541	1.1523	1.1547	1.1527

RS-Robust stability, RP-Robust performance, Ms-Maximum sensitivity peak

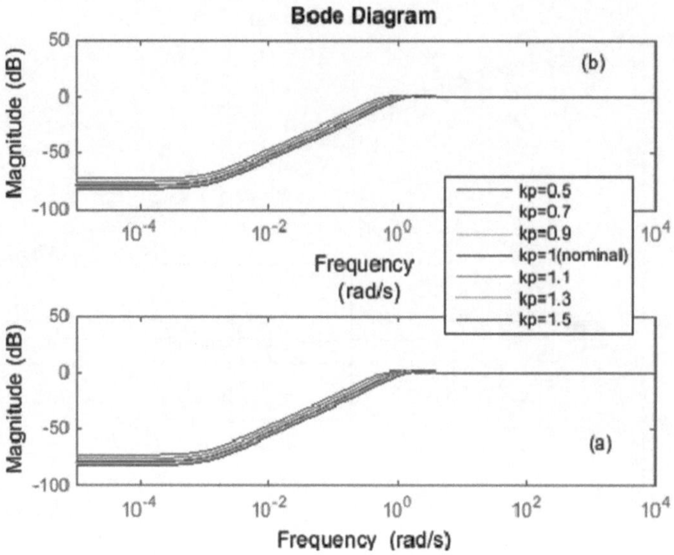

FIGURE 8.9
Sensitivity analysis with FOPID1 (Figure (a)) and FOPID2 (Figure (b))

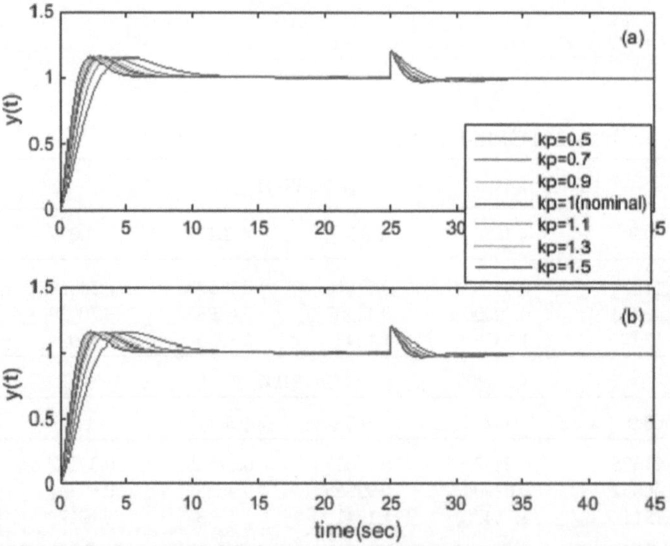

FIGURE 8.10
Iso-damped closed loop response of heating furnace (FOPID1-Figure (a) and FOPID2-Figure (b))

$$FOPID6 = 100\left(10 + \frac{1}{s^{0.5}} + s^{0.31}\right) \quad \text{(Reference[20])} \tag{8.22}$$

The fractional controllers given above are also tested against uncertainties in k_p. Figures 8.11 and 8.12 show the transient response of respective closed loop systems to variations in k_p. Here the setpoint is changed at $t = 0$ and the disturbances enter the heating furnace system at $t = 250$ seconds. To quantify the performance of each controller, the step response characteristics (for setpoint tracking) such as rise time (t_r), overshoot (OS), settling time (t_s), peak time (t_p), and the performance index; integral of squared error (ISE) are computed and tabulated in Tables 8.5 (nominal case) and 8.6 (uncertain case) and Tables 8.7 and 8.8, respectively. Comparing with the performance of FOPID3, FOPID4, FOPID5, and FOPID6, the designed fractional controllers; FOPID1 and FOPID2 are quickly responding to changes in the reference input R and suppressing the disturbances as soon as they enter the heating furnace system. As k_p increases from its nominal value, the proposed design method not only guarantees better time domain performance but also

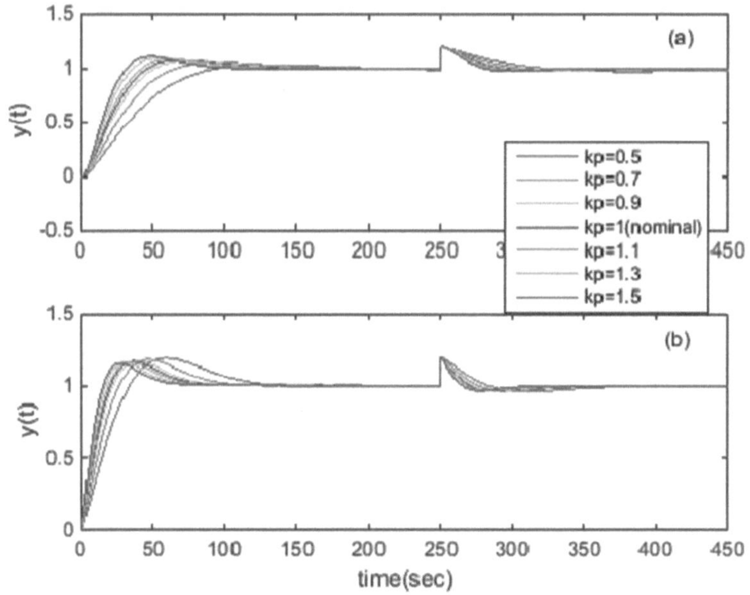

FIGURE 8.11
Closed loop response of heating furnace (FOPID3-Figure (a) and FOPID4-Figure (b))

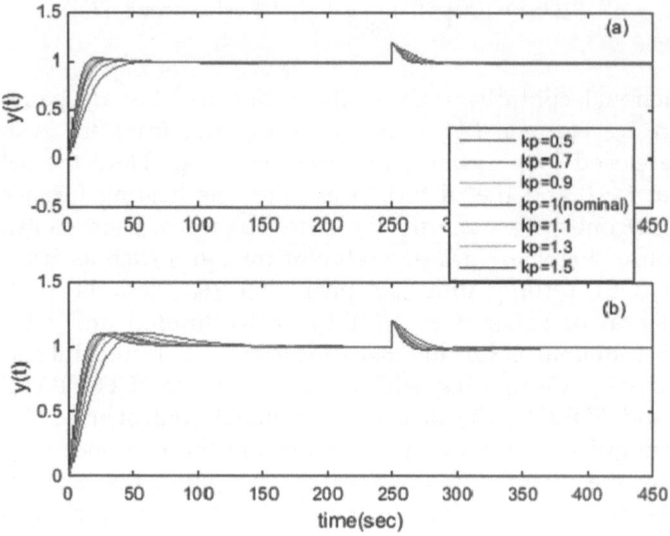

FIGURE 8.12
Closed loop response of heating furnace (FOPID5-Figure (a) and FOPID6-Figure (b))

makes the heating furnace system faster while keeping the overshoot almost at constant level and when k_p drops below its nominal value, the designed controllers still provide better transient response over the fractional controllers designed in [16, 5, 19, and 20]. It is noticed from Tables 8.9 through 8.11 that even under uncertainties in other parameters; τ_1, τ_2, τ_3 (simultaneously changed), FOPID1 and FOPID2 are still

TABLE 8.5
Step response characteristics of various controllers in nominal case

Controller	t_r (0.1 to 0.9)	OS (%)	t_s ($\pm$2%)	t_p(s)
FOPID1	1.3348	15.8644	6.7116	3.1200
FOPID2	1.3343	15.7884	6.6865	3.1100
FOPID3	30.5251	8.0009	128.5561	68.4700
FOPID4	15.3678	18.1374	81.1817	36.7200
FOPID5	14.7950	3.3733	48.2821	33.7200
FOPID6	14.7953	10.8445	80.8073	34.5300

TABLE 8.6

Step response characteristics of various controllers in uncertain case

k_p	Controller	t_r	OS	t_s	t_p
0.9 (–10%)	FOPID1	1.4440	15.8544	8.2579	3.3700
	FOPID2	1.4439	15.7673	8.2316	3.3700
	FOPID3	33.6079	8.4223	138.6678	76.1900
	FOPID4	16.5708	18.5075	86.4632	38.5700
	FOPID5	16.3205	3.1057	50.5587	36.8100
	FOPID6	16.0766	10.9339	88.3485	38.5100
0.7 (–30%)	FOPID1	1.7436	15.8187	8.7476	4.0700
	FOPID2	1.74363	15.7095	8.7189	4.0600
	FOPID3	42.2901	8.1673	168.6056	95.2400
	FOPID4	18.8086	18.3736	100.5336	48.2600
	FOPID5	20.6583	2.5037	58.3226	48.0500
	FOPID6	18.5872	11.1451	104.9243	45.6600
0.5 (–50%)	FOPID1	2.2440	15.7554	11.2319	5.2300
	FOPID2	2.2448	15.6148	11.2015	5.2300
	FOPID3	58.5637	5.7257	218.1683	128.000
	FOPID4	25.0877	18.4735	123.1496	58.7800
	FOPID5	28.4164	1.7860	42.0995	66.0200
	FOPID6	25.4817	11.4130	133.4486	58.3000
1.1 (+10%)	FOPID1	1.2428	15.8761	6.2529	2.900
	FOPID2	1.2423	15.8058	6.2289	2.900
	FOPID3	28.9822	8.5540	118.1819	63.9300
	FOPID4	14.3508	16.8002	76.6878	34.300
	FOPID5	13.5431	3.6228	44.4051	30.3800
	FOPID6	13.7228	10.7637	75.2742	32.0300
1.3 (+50%)	FOPID1	1.0967	15.8898	5.5234	2.5600
	FOPID2	1.0961	15.8329	5.5013	2.5600

(Continued)

TABLE 8.6 (Cont.)

k_p	Controller	t_r	OS	t_s	t_p
	FOPID3	24.0277	10.5978	103.2071	55.3400
	FOPID4	12.7195	16.2062	68.4080	30.4200
	FOPID5	11.6087	4.0765	38.6306	25.9600
	FOPID6	12.0236	10.6232	66.3925	28.0700
1.5 (+50%)	FOPID1	0.9853	15.8986	4.9666	2.3000
	FOPID2	0.9847	15.8526	4.9462	2.3000
	FOPID3	21.0887	11.5764	91.7490	48.9700
	FOPID4	11.4625	15.6965	63.7201	28.4200
	FOPID5	10.1816	4.4812	35.8480	22.7300
	FOPID6	10.7334	10.5045	58.5451	25.0600

possessing iso-damping property and are functioning quite well compared to other fractional controllers.

8.3.2 Design of Fractional Order $PI^{\lambda}D^{\mu}D^{\mu 2}$ Controller for Automatic Voltage Regulator System

To demonstrate that the proposed method is valid to integer order processes and can handle various reference systems, we consider the

TABLE 8.7

Performance index (ISE) in case of setpoint tracking

k_p	FOPID1	FOPID2	FOPID3	FOPID4	FOPID5	FOPID6
1	0.6110	0.6103	14.6274	6.6540	5.2421	5.7233
0.9	0.6611	0.6597	15.7372	8.2469	5.7280	6.2314
0.7	0.7978	0.7960	18.8381	8.8685	8.1001	8.6290
0.5	1.0258	1.0233	24.2456	11.5756	8.5381	8.9900
1.1	0.5690	0.5678	13.7073	6.1575	4.8413	5.2990
1.3	0.5021	0.5010	12.2666	5.3710	4.2177	4.6287
1.5	0.4510	0.4501	11.1868	4.7742	3.7533	4.1216

TABLE 8.8

Performance index (ISE) in case of disturbance rejection

k_p	FOPID1	FOPID2	FOPID3	FOPID4	FOPID5	FOPID6
1	0.0246	0.0246	0.5843	0.2664	0.2089	0.2291
0.9	0.0266	0.0266	0.6284	0.2901	0.2282	0.2494
0.7	0.0321	0.0320	0.7517	0.3549	0.2823	0.3053
0.5	0.0412	0.0411	0.9668	0.4632	0.3780	0.3998
1.1	0.0230	0.0229	0.5476	0.2465	0.1931	0.2122
1.3	0.0203	0.0202	0.4902	0.2150	0.1683	0.1853
1.5	0.0182	0.0182	0.4472	0.1912	0.1499	0.1651

TABLE 8.9

Step response characteristics of various controllers for variations in τ_1, τ_2, τ_3

Change in τ_1, τ_2, τ_3	Controller	t_r	OS	t_s	t_p
–10%	FOPID1	1.2335	15.8770	6.2064	2.8800
	FOPID2	1.2329	15.8079	6.1824	2.8775
	FOPID3	28.7268	8.6140	118.2295	63.3800
	FOPID4	14.2492	16.7466	76.1763	34.0500
	FOPID5	13.4170	3.6522	44.1233	30.0900
	FOPID6	13.6139	10.7552	74.7097	31.7800
–30%	FOPID1	1.0220	15.8958	5.1498	2.3900
	FOPID2	1.0213	15.8465	5.1289	2.3855
	FOPID3	22.0480	11.2334	95.5205	51.0500
	FOPID4	11.8812	15.8287	65.4757	28.4100
	FOPID5	10.6458	4.3440	38.1127	23.7800
	FOPID6	11.1571	10.5447	61.8038	26.0500

(*Continued*)

TABLE 8.9 (Cont.)

Change in τ_1, τ_2, τ_3	Controller	t_r	OS	t_s	t_p
–50%	FOPID1	0.7945	15.9077	4.0114	1.8600
	FOPID2	0.7939	15.8831	3.9945	1.8533
	FOPID3	16.2206	13.8252	72.1217	38.4500
	FOPID4	8.2857	14.6113	53.3636	22.2200
	FOPID5	8.8402	5.3402	28.1376	18.4700
	FOPID6	8.5372	10.2742	48.6648	18.9300
+10%	FOPID1	1.4335	15.8555	8.2039	3.3500
	FOPID2	1.4331	15.7698	8.1778	3.3439
	FOPID3	33.3007	8.4761	138.5728	75.5200
	FOPID4	16.4516	18.4760	85.9496	38.2900
	FOPID5	16.1671	3.1336	50.2704	36.45900
	FOPID6	15.9496	10.9253	86.7029	38.2100
+30%	FOPID1	1.6246	15.8332	8.1557	3.7900
	FOPID2	1.6244	15.7331	8.1282	3.7869
	FOPID3	38.7941	8.6187	156.7783	88.5500
	FOPID4	18.5257	18.0755	95.0449	44.2300
	FOPID5	18.9032	2.7265	55.5343	42.8800
	FOPID6	18.1894	11.0664	98.9837	42.4200
+50%	FOPID1	1.8086	15.8107	8.0704	4.2200
	FOPID2	1.8086	15.6977	8.0420	4.2166
	FOPID3	44.2217	6.9424	175.0176	98.5000
	FOPID4	20.4961	18.5802	103.6118	48.9100
	FOPID5	21.6291	2.3968	58.7289	48.3800
	FOPID6	20.3504	11.1853	108.6836	48.4300

automatic voltage regulator system (AVR) shown in Figure 8.13, which is used to maintain the terminal voltage of a synchronous generator in power generation systems at a desired level.

TABLE 8.10

Performance evaluation (ISE) of various controllers for setpoint tracking

Change in τ_1, τ_2, τ_3	FOPID1	FOPID2	FOPID3	FOPID4	FOPID5	FOPID6
–10%	0.5647	0.6597	13.6146	6.1061	4.8010	5.2561
–30%	0.4678	0.7960	11.5401	4.9673	3.9050	4.2880
–50%	0.3635	1.0233	8.3795	3.7568	2.9804	3.2626
+10%	0.6562	0.5678	15.6269	8.1895	5.6794	6.1811
+30%	0.7435	0.5010	18.5931	8.2279	6.5466	8.0718
+50%	0.8274	0.4501	18.5244	8.2282	8.4062	8.9340

TABLE 8.11

Performance evaluation (ISE) of various controllers for disturbance rejection

Change in τ_1, τ_2, τ_3	FOPID1	FOPID2	FOPID3	FOPID4	FOPID5	FOPID6
–10%	0.0228	0.0266	0.5439	0.2444	0.1915	0.2104
–30%	0.0189	0.0320	0.4613	0.1989	0.1559	0.1717
–50%	0.0147	0.0411	0.3751	0.1505	0.1192	0.1307
+10%	0.0264	0.0229	0.6240	0.2878	0.2262	0.2474
+30%	0.0299	0.0202	0.7022	0.3293	0.2605	0.2831
+50%	0.033	0.0182	0.7790	0.3693	0.2943	0.3175

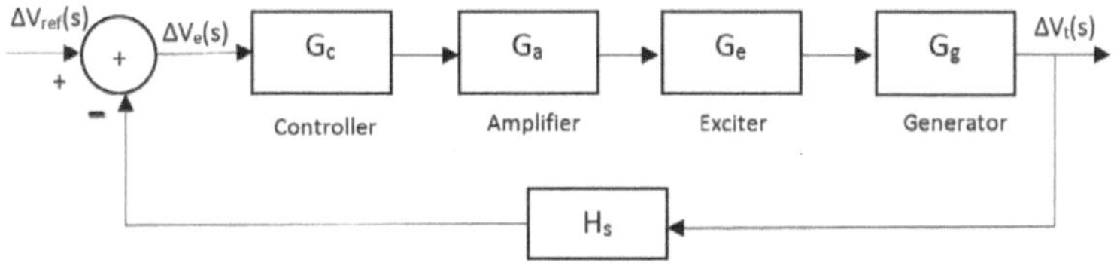

FIGURE 8.13
AVR system

The transfer functions of amplifier, exciter, generator, and sensor (Hs) are considered, as given here:

$$G_a(s) = \frac{k_a}{T_a s + 1}, \; G_e(s) = \frac{k_e}{T_e s + 1},$$
$$G_g(s) = \frac{k_g}{T_g s + 1}, \; H_s(s) = \frac{k_s}{T_s s + 1}, \tag{8.23}$$

where $k_a = 10$, $T_a = 0.1$, $k_e = 1$, $T_e = 0.4$, $k_g = 1$, $T_g = 1$.

We now choose a fractional version of the controller PIDD^2 proposed in [21]:

$$G_c(s) = K_C + \frac{K_I}{s^\lambda} + K_d s^\mu + K_{d_2} s^{\mu_2}. \tag{8.24}$$

The explicit form of reference system is

$$Gclr(s) = \frac{1}{0.04s + 1}. \tag{8.25}$$

Compared to the standard fractional PID controller designed in the preceding problem, the fractional $PI^\lambda D^\mu D^{\mu_2}$ has two additional unknowns (K_{d_2}, μ_2) that may present difficulties while tuning it in order to make the AVR system act exactly as the reference system in Equation (8.25).

The initial data for LJ is $N_P = 150$, $N_T = 70$, $R = 10$, $\varphi = 10^{-6}$, $\eta = 0.95$, $a = [0, 0, 0, 0, 0.01, 0.01, 0.01]$, $b = [5, 5, 5, 5, 2, 2, 3]$ and for PSO is $S = 50$, $Maxitr = 100$, $c_1 = 1.49$, $c_2 = 1.49$, $w = 1.1$, $b = [5, 5, 5, 5, 2, 2, 3]$, $a = [0, 0, 0, 0, 0.01, 0.01, 0.01]$.

First, the actual optimization problem is solved by minimizing the objective function in Equation (8.5) with PSO algorithm (Program 8.5). PSO requires 38.4533 seconds to reach minimum of 0.0224. Even LJ (Program 8.6) exhibits poor performance with minimum of 1.143219 which is far away from the global optimum. For the first run, LJ needs CPU time of 115.6363 seconds. To achieve better solutions, LJ's performance is enhanced by taking the optimum values of $PI^\lambda D^\mu D^{\mu_2}$ controller parameters obtained in the previous (first) run as the initial guess for the next (second) run. From the second run onward, the region size vector r^1 is computed as $K^{(0)}/2$. After the fourth run, the value of J is obtained as 0.015634. Proceeding in this fashion for five more runs, we get a minima of 0.013248. Because there is no notable improvement in results even after 15 runs, further runs are not performed. Using the same initial data and the step size of 0.1, PSO is applied to triangular strip operational matrix-based

TABLE 8.12

Controller parameters and performance indices for AVR system

	Controller parameters										
Algorithm-Controller	K_C	K_I	K_d	K_{d_2}	μ	λ	μ_2	**OS**	t_r	t_s	**ISE**
AC1	0.1179	4.4402	0.0582	1.5006	2.1397	0.4158	0.9693	0	0.0918	0.1541	0.0187
AC2	2.778	1.852	0.999	0.074	1.1980	1.3183	-	0	0.0929	0.1635	0.0233
AC3	1.6264	0.2956	0.3226	-	1.2081	1.8373	-	5.4124	0.1567	2.6848	0.0787
AC4	1.6986	0.1797	0.3122	-	1.1122	1.0624	-	5.7732	0.1579	33.518	0.0798
AC5	1.0537	0.4418	0.2510	-	1.0838	1.0275	-	3.8524	0.2191	0.5372	0.1075
AC6	0.9315	0.4776	0.2536		1.1296	1.0248		2.8362	0.2297	0.8949	0.1125
AC7	1.6947	0.8849	0.3964		0.8851	0.9384		8.26	0.1298	0.3395	0.0778
AC8	0.5073	0.5724	0.2550					2.0929	0.2745	0.39973	0.1554

AC1=LJ-$PI^{\lambda}D^{\mu}D^{\mu_2}$, AC2=PSO-$PIDD^2$ [22], AC3=PSO-FOPID (β=1), AC4=PSO-FOPID (β=1.5), AC5=CAS-FOPID (β=1), AC6=CAS-FOPID (β=1.5), AC7=GA-FOPID (β=1), AC8=GA-FOPID (β=1)

H_∞-optimal control problem (Program 8.7). Comparing with its previous performance, PSO bestows slightly better results ($\tilde{J} = 0.0028$ and elapsed time is 25.4031 seconds) but the solution accuracy is still not acceptable. The same procedure applied for LJ in case of minimizing Equation (8.5) is repeated here (Program 8.8) for finding the optimal values of K_C, K_I, K_d, K_{d_2}, λ, μ and μ_2 through the minimization of Equation (8.13). After 15 runs, $\tilde{J}$ is minimized to 0.000336. On average, CPU time of 28.9813 seconds is taken by LJ to complete each run. In Table 8.12, the optimum values of controller parameters are given.

The designed fractional $PI^\lambda D^\mu D^{\mu_2}$ controller is

$$LJ - PI^\lambda D^\mu D^{\mu_2} = 0.1179 + \frac{4.4402}{s^{0.4158}} + 0.0582s^{2.1397} + 1.5006s^{0.9693}. \tag{8.26}$$

It is evident from Figure 8.14 that the proposed method can be used to tune more complex controllers.

In [22], the authors have designed FOPID controller for the AVR system using gradient-free global optimization methods; Genetic algorithm (GA-FOPID), PSO (PSO-FOPID), and chaotic ant swarm (CAS-FOPID). They have employed the following equation as the objective function to tune FOPID,

$$f = \left(1 - e^{-\beta}\right)\left(M_P + E_{ss}\right) + e^{-\beta}\left(t_s - t_r\right), \tag{8.27}$$

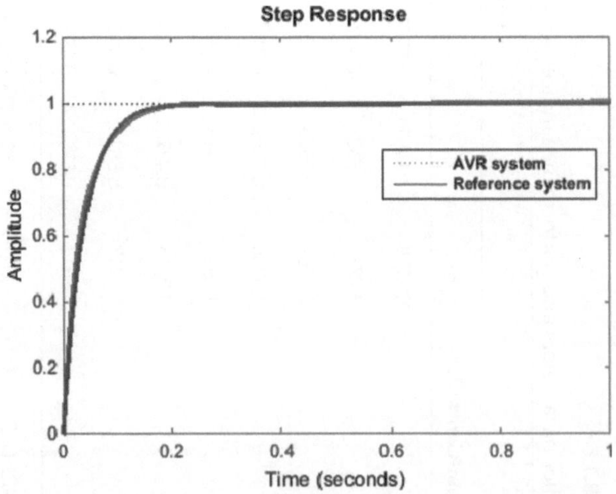

FIGURE 8.14
Comparison of step response of AVR system and Reference system

where M_P is the overshot, E_{ss} is the steady state error and β can be 1 or 1.5.

The optimum values of FOPID controller parameters obtained via GA, PSO, and CAS (in [22]) for the two cases; $\beta = 1$ and $\beta = 1.5$ are also given in Table 8.12. It is noticed from Figure 8.15 and Table 8.12 that the designed $PI^{\lambda}D^{\mu}D^{\mu_2}$controller demonstrates the best time domain performance. To check the robustness of $PI^{\lambda}D^{\mu}D^{\mu_2}$, each process parameter is varied to ±60% from their nominal values. The performance indices; rise time, settling time, overshoot, peak time, and peak value are computed and tabulated in Table 8.13. The results confirm that the fractional $PI^{\lambda}D^{\mu}D^{\mu_2}$ controller can meet the control objectives such as tracking command signal, and so on when it is applied to real AVR plant.

8.3.3 Design of Fractional Order PI$^{\lambda}$ Controller, Fractional PD$^{\mu}$ Controller, Fractional Order PI$^{\lambda}$D$^{\mu}$ Controller with Fractional Order Filter, and Series Form of Fractional Order PI$^{\lambda}$D$^{\mu}$ Controller

In this subsection, we aim to prove that the proposed method can be used to design fractional PI, fractional PD, fractional PID with fractional order filter and series form of fractional PID. The process models considered here are given in Table 8.14, among which G_p^I is a hypothetical model. The control law selected for each process model to track the dynamics of respective reference system is also presented in Table 8.14. Like heating furnace control system design, the statistical analysis (Table 8.15) is performed. The results made clear that

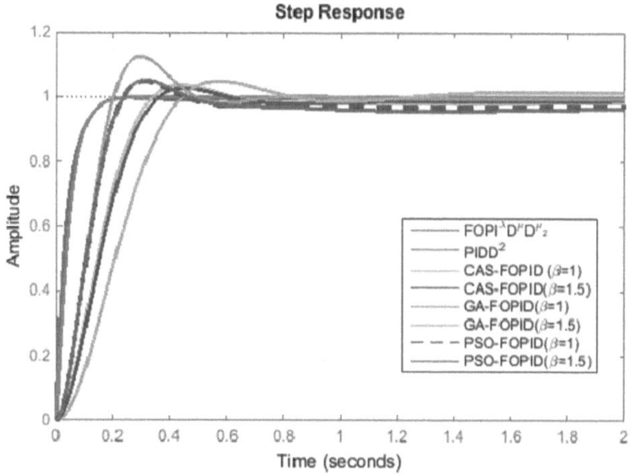

FIGURE 8.15
Closed loop response of AVR system to a unit step change in setpoint

TABLE 8.13

Robustness analysis using $FOPI^{\lambda}D^{\mu}D^{\mu_2}$

Perf. index	k_a –60%	k_a +60%	T_a –60%	T_a +60%	k_e –60%	k_e +60%	T_e –60%	T_e +60%	k_g –60%	k_g +60%	T_g –60%	T_g +60%	Nominal value
t_r	0.2476	0.0468	0.1196	0.1071	0.2476	0.0468	0.0214	0.1324	0.2476	0.0468	0.0195	0.1434	0.0918
t_s	0.5468	0.0952	0.3794	0.3673	0.5468	0.0952	0.5738	0.8956	0.5468	0.0952	0.4801	1.1092	0.1541
OS	1.3205	0.5121	0.5204	5.8430	1.3205	0.5121	0	2.9926	1.3205	0.5121	0	2.2448	0
P_u	1.0100	1.0043	1.0039	1.0571	1.0100	1.0043	0.9972	1.0286	1.0100	1.0043	0.9929	1.0211	0.9983
t_p	1.1259	0.1712	0.9755	0.2341	1.1259	0.1712	1.1157	0.3168	1.1259	0.1712	1.4303	0.9154	0.2333

TABLE 8.14

Test process models and respective reference system and controller

Process model	Reference system	Controller design
$G_p^I = \frac{1.883}{0.5s^{1.5} + 1.15s}$	$Gclr_1(s) = \frac{3.582}{45.975s^{1.5} + 3.582}$	$G_{c1} = K_C + \frac{K_I}{s^\lambda}$
$G_p^{II} = \frac{1}{0.4s^{2.4} + s}$	$Gclr_2(s) = \frac{1.397}{1.671s + 1.397}$	$G_{c2} = K_C + K_d s^\mu$
$G_p^{III} = \frac{400}{s^2 + 50s}$	$Gclr_3(s) = \frac{1}{s^{1.12} + 1}$	G_{C3}, G_{C4}

$$G_{C3} = K_C + \frac{K_I}{s^\lambda} + \frac{K_d s^\mu}{1+0.01K_d s^\mu},$$

$$G_{C4} = K_C\left(\frac{K_I s^\lambda + 1}{K_I s^\lambda}\right)\left(\frac{K_d s^\mu + 1}{0.01K_d s + 1}\right).$$

when the original H_∞-optimal control problem is simple (i.e., tuning G_{C2} to minimize $G_{clr2}(s)$), finding its optimal solution is certainly quite easy but as the controller structure becomes more complex (G_{C3}, G_{C4} or $PI^\lambda D^\mu D^{\mu_2}$ in Subsection 8.3.2), achieving feasible (practical) solutions in acceptable CPU time gets tougher. In such situations, formulating an approximation to original problem and obtaining fairly good solutions (at low CPU time) with the help of triangular strip operational matrices is proved here as a promising approach. Table 8.16 and Figure 8.16 reveal another important point that the proposed method is applicable for a range of reference systems and control laws.

8.4 MATLAB Codes for Simulation Examples

Program 8.1

```
clc
clear all
fun = @Obj_Fun;nvars = 5;lb = [0 0 0 0.001 0.001];ub =
  [10^4 10^4 10^5 1 1];
tic
options = optimoptions('particleswarm','SwarmSize',
  50,'MaxIter',100);
options.HybridFcn = @fmincon;
[x,fval,exitflag] = particleswarm(fun,nvars,lb,ub,options);
```

TABLE 8.15

Comparison of optimal solutions of J and $\tilde{J}$

		Original H_∞-optimization				TSOM-based H_∞-optimization			
Process model	**Method**	**Mean J**	**Std.**	**Best J**	**CT(s)**	**Mean J**	**Std.**	**Best J**	**CT(s)**
G_p^I	PSO	3.83e-05	5.94e-05	3.302e-07	44.038	8.081e-09	3.27e-09	3.850e-09	3.436
	LJ	0.001	0.0014	2.66e-05	108.24	1.742e-13	3.462e-13	1.472e-15	20.4943
G_p^{II}	PSO	1.319e-08	1.25e-08	3.385e-09	8.834	4.27e-09	1.58e-09	1.304e-09	2.108
	LJ	1.04e-16	2.73e-17	8.77e-17	71.124	3.373e-16	1.396e-17	3.27e-16	11.812
$G_p^{III}(G_{c3})$	PSO	0.024	0.035	0.0038	12.844	3.314e-06	1.309e-06	2.068e-06	4.6507
	LJ	0.019	0.0145	0.003	168.79	2.920e-04	3.590e-04	8.743e-06	40.5603
$G_p^{III}(G_{c4})$	PSO	0.008	0.002	0.004	35.095	1.344e-05	4.596e-06	8.097e-06	8.138
	LJ	0.0091	0.0045	0.0028	351.377	8.029e-05	4.457e-05	4.112e-05	82.051

Step size of 0.1 is used to minimize $\tilde{J}$. The initial data for PSO and LJ is taken from Subsection 8.3.1. For process models, G_p^I and G_p^{II}, the lower and upper bound on optimization variables are considered as $a = [0, 0, 0]$, $b = [10, 10, 2]$.

TABLE 8.16

Optimal values of controller parameters obtained from minimization of $\tilde{J}$

Process model	Method	K_C	K_I	K_d	λ	μ
G_p^I	PSO	0.02068	0.04758	–	0.4999	–
	LJ	0.02068	0.04758	–	0.500	–
G_p^{II}	PSO	0.8360	–	0.3344	–	1.3999
	LJ	0.83602	–	0.33441	–	1.4000
$G_p^{III}(G_{c3})$	PSO	0.0014	0.1235	0.00247	0.1230	0.8820
	LJ	0.02379	0.1017	0.0020	0.1450	0.92913
$G_p^{III}(G_{c4})$	PSO	0.0022	0.01806	0.01975	0.12185	1.0056
	LJ	0.0028	0.0237	0.0201	0.12309	0.9999

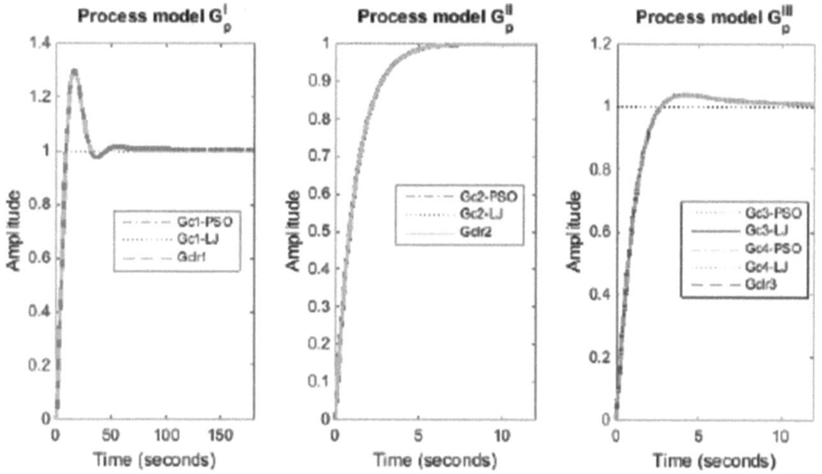

FIGURE 8.16
Closed loop step response of process models in Table 8.14

```
toc
%%%%%%%%%%%%%%%%%%%%%%%%%%%%%%%%%%%%%%%%%%%%%%%%%%%%%%
%%%%%%%%%%%%%%%%%%%%
function f= Obj_Fun(x)
Kp=x(1);KI=x(2);Kd=x(3);lambda=x(4);mu=x(5);
```

```
gamma=4/3;alpha1=1.31;alpha2=0.97;tau1=14994;
tau2=6008.5;tau3=1.69;
NumCoefs=[tau1,tau2,tau3,Kp,Kd,KI,-Kd,-Kp,-KI,-Kd,
  -Kp,-KI];
NumOrders=[alpha1+lambda,alpha2+lambda,lambda,lambda,mu+
  lambda,0,gamma+mu+lambda,lambda+gamma,gamma,mu+
  lambda,lambda,0];
DenCoefs=[tau1,tau2,tau3,Kp,Kd,KI,tau1,tau2,tau3,Kp,
  Kd,KI];
DenOrders=[alpha1+lambda+gamma,alpha2+lambda+gamma,
  lambda+gamma,lambda+gamma,mu+lambda+gamma,gamma,alpha1+
  lambda,alpha2+lambda,lambda,lambda,mu+lambda,0];
G=fotf( NumCoefs,NumOrders,DenCoefs,DenOrders,0);[A]=
  bode1(G);f=max(A);
end
```

Program 8.2

```
clc
clear all
N=5;    % no.of dimensions
LIM=20;  % value for every 20 itrs.
NIT=80; % no.of itrs.
NPTS=40; % no.of random points
RED=0.95; % region reduction factor
REN=0.7;N0=0;NP=50;LW=LIM;tol=10^-3;tol2=10^-6;h=0.1;
P=1/h;XS=zeros(1,N);XP=zeros(1,N);X=zeros(1,N);
  alw=[0 0 0 0.001 0.001];
buw=[10^4 10^4 10^5 1 1]Parm=0.5*(alw+buw);REGi=0.5*
  (buw-alw);
for k=1:1:N
  XP(k)=Parm(k);XS(k)=XP(k);REG(k)=REGi(k);
end
REG1=zeros(NP,N);REG1(1,:)=REG;TEST=zeros(1,NP);C1=[];
Test=10^20;
fprintf('%s\n','—————————————————————————————————————');
fprintf(' %s\t   \t%s\t   \t%s\n','itno','NFE','C');
fprintf('%s\n','—————————————————————————————————————');
tic
for w=1:1:NP
  Xold=XP;
for i=1:1:NIT
  for j=1:1:NPTS
     for k=1:1:1
    RN1=rand(1,1)-0.5;RN2=rand(1,1)-0.5;RN3=rand(1,1)-
      0.5;
    RN4=rand(1,1)-0.5;RN5=rand(1,1)-0.5;RANN=[RN1;RN2;
      RN3;RN4;RN5];
```

```
    RAN=diag(RANN);
    if j==1
      RAN=0;
    end
    X(k,:)=XP(k,:)+REG(k,:)*RAN;
    if X(1)<0
      break;
    end
    if X(4)<=0.001
      break;
    end
    if X(4)>1
      break;
    end
    if X(5)<=0.001
      break;
    end
    if X(5)>1
      break;
    end
    C= Obj_Fun(X);
    if C>Test
      break;
    else
      Test=C;C1=[C1;C];N0=N0+1;FM=C;XS(k,:)=X(k,:);
    end
    end
  end  % end of all random points
  if LW<LIM
    LW=LW+1;
  else
    LW=0;LW=LW+1;
  end
    for kkk=1:1:N
    REG(kkk)=REG(kkk)*RED;XP(kkk)=XS(kkk);
    end
end   % end of single pass
TEST(w)=Test;Xnew=XS;fprintf('%f\t  %f\t  %f\n',w,N0,FM);
reg=abs(Xnew-Xold);
for ki=1:1:N
  if reg(ki)<tol
    reg(ki)=tol;
  end
end
REG=reg;
  if w>5
     diff1=abs(TEST(w-2)-TEST(w-1));diff2=abs(TEST(w-1)-
TEST(w));
    diff=max(diff1,diff2);
```

```
  if diff<tol2
    tol=tol/2;tol2=tol2/2;
  end
  end
  if tol<10^-15
    break;
  end
end
toc
```

Program 8.3

```
clc
clear allfun = @objfn_HeatingFurnace;nvars = 5;
lb = [0 0 0 0.001 0.001];ub = [10^4 10^4 10^5 1 1];
tic
options = optimoptions('particleswarm','SwarmSize',150,
  'MaxIter',100);
options.HybridFcn = @fmincon;
[x,fval,exitflag] = particleswarm(fun,nvars,lb,ub,options);
toc
%%%%%%%%%%%%%%%%%%%%%%%%%%%%%%%%%%%%%%%%%%%%%%%
%%%%%%%%%%%%%%%%%%
function f= objfn_HeatingFurnace(x)
h=0.5;P=1/h;alpha1=1.31;alpha2=0.97;a=14994;b=6008.5;
  cc=1.69;X=x;
I=eye(P+1,P+1);lambda=X(4);mu=X(5);BN_lambda=
  ban(lambda,P+1,h);
BN_lambda_mu=ban(lambda+mu,P+1,h); %
BN_lambda_alpha1=ban(lambda+alpha1,P+1,h);
BN_lambda_alpha2=ban(lambda+alpha2,P+1,h);
  BN_ref_1=ban(4/3,P+1,h);
BN_num=X(1)*BN_lambda+X(2)*I+X(3)*BN_lambda_mu;
BN_den=a*BN_lambda_alpha1+b*BN_lambda_alpha2+(I*cc+
  X(1))*BN_lambda+X(2)*I+X(3)*BN_lambda_mu;BNR_num=
  (1^(4/3))*I;BNR_den=BN_ref_1+(1^(4/3))*I;
EN=norm(((BNR_num*(BN_den))-(BN_num*(BNR_den)))/
  (BN_den*BNR_den),2);f=EN;
end
```

Program 8.4

```
clc
clear all
N=5;     % no.of dimensions
LIM=20;  % value for every 20 itrs.
NIT=80;  % no.of itrs.
NPTS=40; % no.of random points
RED=0.95; % region reduction factor
```

```
REN=0.7;N0=0;NP=50;LW=LIM;tol=10^-3;tol2=10^-6;h=0.5;
  P=1/h;
XS=zeros(1,N);XP=zeros(1,N);X=zeros(1,N);
  alw=[0 0 0 0.01 0.01];
buw=[10^4 10^4 10^5 1 1];Parm=0.5*(alw+buw);REGi=0.5*
  (buw-alw);
for k=1:1:N
  XP(k)=Parm(k);XS(k)=XP(k);REG(k)=REGi(k);
end
REG1=zeros(NP,N);REG1(1,:)=REG;TEST=zeros(1,NP);C1=[];
Test=10^200;
fprintf('%s\n','——————————————————————————————————————');
fprintf(' %s\t   \t%s\t    \t%s\n','itno','NFE','C');
fprintf('%s\n','——————————————————————————————————————');
tic
for w=1:1:NP
  Xold=XP;
for i=1:1:NIT
  for j=1:1:NPTS
    for k=1:1:1
     RN1=rand(1,1)-0.5;RN2=rand(1,1)-0.5;RN3=rand
       (1,1)-0.5
     RN4=rand(1,1)-0.5;RN5=rand(1,1)-0.5;RANN=[RN1;
       RN2;RN3;RN4;RN5];
    RAN=diag(RANN);
    if j==1
      RAN=0;
    end
    X(k,:)=XP(k,:)+REG(k,:)*RAN;
    if X(1)<0
      break;
    end
    if X(4)<=0.01
      break;
    end
    if X(4)>1
      break;
    end
    if X(5)<=0.01
      break;
    end
    if X(5)>1
      break;
    end
    alpha1=1.31;alpha2=0.97;a=14994;b=6008.5;cc=1.69;I=
       eye(P+1,P+1);
     lambda=X(4);mu=X(5);BN_lambda=ban(lambda,P+1,h);  %
lambda
    BN_lambda_mu=ban(lambda+mu,P+1,h); %
```

```
    BN_lambda_alpha1=ban(lambda+alpha1,P+1,h);
    BN_lambda_alpha2=ban(lambda+alpha2,P+1,h);
    BN_ref_1=ban(4/3,P+1,h);
    BN_num=X(1)*BN_lambda+X(2)*I+X(3)*BN_lambda_mu;
BN_den=a*BN_lambda_alpha1+b*BN_lambda_alpha2+(I*cc+X
(1))*BN_lambda+X(2)*I+...X(3)*BN_lambda_mu;
    BNR_num=(1^(4/3))*I;BNR_den=BN_ref_1+(1^(4/3))*I;
    C=norm(((BNR_num*(BN_den))-(BN_num*(BNR_den)))/
     (BN_den*BNR_den),inf);
    if C>Test
      break;
    else
      Test=C;C1=[C1;C];N0=N0+1;FM=C;XS(k,:)=X(k,:);
    end
    end
  end  % end of all random points
  if LW<LIM
    LW=LW+1;
  else
    LW=0;
    LW=LW+1;
  end
    for kkk=1:1:N
    REG(kkk)=REG(kkk)*RED;XP(kkk)=XS(kkk);
    end
end   % end of single pass
TEST(w)=Test;Xnew=XS;fprintf('%f\t  %f\t  %f\n',w,N0,FM);
reg=abs(Xnew-Xold);
for ki=1:1:N
  if reg(ki)<tol
    reg(ki)=tol;
  end
end
REG=reg;
  if w>5
     diff1=abs(TEST(w-2)-TEST(w-1));diff2=abs(TEST(w-1)-
TEST(w));
    diff=max(diff1,diff2);
  if diff<tol2
    tol=tol/2;tol2=tol2/2;
  end
  end
  if tol<10^-15
    break;
  end
end
toc
```

Program 8.5

```
clc
clear all
fun = @hinfFOPIDD2;nvars =7;
lb = [0 0 0 0 0.01 0.01 0.01];ub = [5,5,5,3,2,2,3];
tic
options = optimoptions('particleswarm','SwarmSize',50);
options.HybridFcn = @fmincon;
[x,fval,exitflag] = particleswarm(fun,nvars,lb,ub,options);
toc
%%%%%%%%%%%%%%%%%%%%%%%%%%%%%%%%%%%%%%%%%%%%%%%%%%%%%%%%%%%
%%%%%%%
function f=hinfFOPIDD2(x)
Kp=x(1);KI=x(2);Kd=x(3);Kd2=x(4);lambda=x(5);mu=x(6);
  mu2=x(7);
a=0.00454;b=0.0555;c=0.151;d=0.1;
NumCoefs=[0.01*Kp,0.01*KI,0.01*Kd,0.01*Kd2,Kp,KI,
  Kd,Kd2];
NumOrders=[lambda+1,1,1+mu+lambda,1+mu2+lambda,
  lambda,0,mu+lambda,mu2+lambda];
DenCoefs=[0.01*a,0.01*b,0.01*c,0.01*d,a,b,c,d,Kp,KI,
  Kd,Kd2];
DenOrders=[4+lambda,3+lambda,2+lambda,1+lambda,3+
  lambda,2+lambda,1+lambda,...lambda,lambda,0,mu+lambda,
  mu2+lambda];
G=fotf( NumCoefs,NumOrders,DenCoefs,DenOrders,0);
Gclr=fotf([0,1],[0,0],[0.04 1],[1,0],0);
Gnorm=minus(Gclr,G);[A]=bode1(Gnorm);f=max(A);
end
```

Program 8.6

```
clc
clear all
N=7;    % no.of dimensions
LIM=20;  % value for every 20 itrs.
NIT=70;  % no.of itrs.
NPTS=10;  % no.of random points
RED=0.95; % region reduction factor
REN=0.7;N0=0;NP=150;LW=LIM;tol=10^-3;tol2=10^-6;h=0.25;
P=1/h;XS=zeros(1,N);
XP=zeros(1,N);X=zeros(1,N);alw=[0  0  0  0  0.01  0.01  0.01];
buw=[5,5,5,5,2,2,3];
Parm=0.5*(alw+buw);REGi=0.5*(buw-alw);
for k=1:1:N
   XP(k)=Parm(k);XS(k)=XP(k);REG(k)=REGi(k);
end
```

```
REG1=zeros(NP,N);REG1(1,:)=REG;TEST=zeros(1,NP);C1=[];
Test=10^20;
fprintf('%s\n',————————————————————————————————————');
fprintf(' %s\t   \t%s\t    \t%s\n','itno','NFE','C');
fprintf('%s\n','————————————————————————————————————');
tic
for w=1:1:NP
  Xold=XP;
for i=1:1:NIT
  for j=1:1:NPTS
    for k=1:1:1
     RN1=rand(1,1)-0.5;RN2=rand(1,1)-0.5;RN3=rand(1,1)-
       0.5;
     RN4=rand(1,1)-0.5;RN5=rand(1,1)-0.5;RN6=rand(1,1)-
       0.5;
   RN7=rand(1,1)-0.5;
    RANN=[RN1;RN2;RN3;RN4;RN5;RN6;RN7];RAN=diag(RANN);
    if j==1
      RAN=0;
    end
    X(k,:)=XP(k,:)+REG(k,:)*RAN;
    if X(1)<0
      break;
    end
     if X(2)<0
      break;
     end
    if X(3)<0
      break;
    end
    if X(4)<0
      break;
    end
     if X(5)<0.01
      break;
     end
    if X(6)<0.01
      break;
    end
    if X(7)<0.01
      break;
    end
    C=hinfFOPIDD2(X);
    if C>Test
      break;
    else
      Test=C;C1=[C1;C];N0=N0+1;FM=C;XS(k,:)=X(k,:);
    end
    end
```

```
    end  % end of all random points
    if LW<LIM
      LW=LW+1;
    else
      LW=0;
      LW=LW+1;
    end
      for kkk=1:1:N
      REG(kkk)=REG(kkk)*RED;XP(kkk)=XS(kkk);
      end
end   % end of single pass
TEST(w)=Test;Xnew=XS;fprintf('%f\t  %f\t  %f\n',w,N0,FM);
reg=abs(Xnew-Xold);
for ki=1:1:N
    if reg(ki)<tol
      reg(ki)=tol;
    end
end
REG=reg;
    if w>5
       diff1=abs(TEST(w-2)-TEST(w-1));diff2=abs(TEST(w-1)-
TEST(w));
      diff=max(diff1,diff2);
    if diff<tol2
      tol=tol/2;tol2=tol2/2;
    end
    end
    if tol<10^-15
      break;
    end
end
CPU(ii)=tocF(ii)=Test;XX(ii,:)=XS;
end
```

Program 8.7

```
clc
clear all
fun = @objfn_AVR_FOPIDD2;nvars =7;
lb = [0 0 0 0 0.001 0.001 0.001];ub =[5,5,5,5,2,2,3];
tic
options = optimoptions('particleswarm','SwarmSize',50,'
  SocialAdjustment',1.49);
options.HybridFcn = @fmincon;
[x,fval,exitflag] = particleswarm(fun,nvars,lb,ub,options);
toc
%%%%%%%%%%%%%%%%%%%%%%%%%%%%%%%%%%%%%%%%%%%%%%%
%%%%%%%
function f=objfn_AVR_FOPIDD2(x)
```

```
h=0.1;P=1/h;a=0.00454;b=0.0555;c=0.151;d=0.1;X=x;
  Kp=X(1);KI=X(2);Kd=X(3);
Kd2=X(4);lambda=X(5);mu=X(6);mu2=X(7);I=eye(P+1,P+1);
BN_1=ban(lambda,P+1,h);BN_2=ban(mu,P+1,h);BN_3=ban(mu2,
P+1,h);BN_4=ban(3,P+1,h); BN_5=ban(2,P+1,h); BN_6=ban(1,
P+1,h); BN_GC=Kp*I+((KI*I)/(BN_1))+Kd*BN_2+Kd2*BN_3;
BN_Gp=(1*I)/(a*BN_4+b*BN_5+c*BN_6+d*I);BN_Hs=I/(0.01*
  BN_6+I);
BN_GOP=BN_Gp*BN_GC;BN_GOP1=I+BN_Gp*BN_GC*BN_Hs;
BN_Gcls=(BN_GOP)*(BN_GOP1^-1);BN_ref1=ban(1,P+1,h);
BN_ref2=ban(2,P+1,h);BNR_num=1*I;BNR_den=0.04*
  BN_ref1+1*I;f1=BNR_num*BN_GOP1;
f2=BNR_den*BN_GOP;f3=f1-f2;
f4=BNR_den*BN_GOP1;f5=f3*(f4^-1);EN=norm(f5,2);f=EN;
end
```

Program 8.8

```
clc
clear all
N=7;    % no.of dimensions
LIM=20;  % value for every 20 itrs.
NIT=70; % no.of itrs.
NPTS=10; % no.of random points
RED=0.95; % region reduction factor
REN=0.7;N0=0;NP=150;LW=LIM;tol=10^-3;tol2=10^-6;h=0.1;
P=1/h;XS=zeros(1,N);
XP=zeros(1,N);X=zeros(1,N);alw=[0 0 0 0 0.01 0.01 0.01];
buw=[5,5,5,5,2,2,3]; Parm=buw;REGi=Parm/2;
for k=1:1:N
  XP(k)=Parm(k);XS(k)=XP(k);REG(k)=REGi(k);
end
REG1=zeros(NP,N);REG1(1,:)=REG;TEST=zeros(1,NP);C1=[];
Test=10^200;
fprintf('%s\n','————————————————————————————————————');
fprintf(' %s\t   \t%s\t    \t%s\n','itno','NFE','C');
fprintf('%s\n','————————————————————————————————————');
tic
for w=1:1:NP
  Xold=XP;
for i=1:1:NIT
   for j=1:1:NPTS
      for k=1:1:1
      RN1=rand(1,1)-0.5;RN2=rand(1,1)-0.5;RN3=rand(1,1)-
        0.5;
      RN4=rand(1,1)-0.5;RN5=rand(1,1)-0.5;RN6=rand(1,1)-
        0.5;
      RN7=rand(1,1)-0.5;RANN=[RN1;RN2;RN3;RN4;RN5;RN6;
        RN7];
```

```
RAN=diag(RANN);
if j==1
  RAN=0;
end
X(k,:)=XP(k,:)+REG(k,:)*RAN;
if X(1)<0
  break;
end
 if X(2)<0
  break;
 end
if X(3)<0
  break;
end
if X(4)<0
  break;
end
 if X(5)<0.01
  break;
end
if X(6)<0.01
  break;
end
 if X(7)<0.01
  break;
end
 if X(5)>buw(5)
  break;
end
if X(6)>buw(6)
  break;
end
 if X(7)>buw(7)
  break;
end
a=0.00454;b=0.0555;c=0.151;d=0.1;Kp=X(1);KI=X(2);
  Kd=X(3);Kd2=X(4);
lambda=X(5);mu=X(6);mu2=X(7);I=eye(P+1,P+1);BN_1=
  ban(lambda,P+1,h);
BN_2=ban(mu,P+1,h);BN_3=ban(mu2,P+1,h);BN_4=
  ban(3,P+1,h); %
BN_5=ban(2,P+1,h);BN_6=ban(1,P+1,h);
BN_GC=Kp*I+((KI*I)/(BN_1))+Kd*BN_2+Kd2*BN_3;
BN_Gp=(1*I)/(a*BN_4+b*BN_5+c*BN_6+d*I);BN_Hs=
  I/(0.01*BN_6+I);
BN_GOP=BN_Gp*BN_GC;BN_GOP1=I+BN_Gp*BN_GC*BN_Hs;
BN_Gcls=(BN_GOP)*(BN_GOP1^-1);BN_ref1=ban(1,P+1,h);
```

```
      BN_ref2=ban(2,P+1,h);BNR_num=1*I;BNR_den=0.040*
        BN_ref1+1*I;
        f1=BNR_num*BN_GOP1;f2=BNR_den*BN_GOP;f3=f1-f2;
          f4=BNR_den*BN_GOP1;
        ff5=f3*(f4^-1);C=norm(f5,2);
      if C>Test
        break;
      else
        Test=C;C1=[C1;C];N0=N0+1;FM=C;XS(k,:)=X(k,:);
      end
      end
   end  % end of all random points
   if LW<LIM
      LW=LW+1;
   else
      LW=0;LW=LW+1;
   end
      for kkk=1:1:N
      REG(kkk)=REG(kkk)*RED;XP(kkk)=XS(kkk);
      end
end   % end of single pass
TEST(w)=Test;
Xnew=XS;
fprintf('%f\t %f\t %f\n',w,N0,FM);
reg=abs(Xnew-Xold);
for ki=1:1:N
   if reg(ki)<tol
      reg(ki)=tol;
   end
end
REG=reg;
   if w>5
      diff1=abs(TEST(w-2)-TEST(w-1));diff2=abs(TEST(w-1)-
          fTEST(w));
      diff=max(diff1,diff2);
   if diff<tol2
      tol=tol/2;tol2=tol2/2;
   end
   end
   if tol<10^-15
      break;
   end
end
toc
```

8.5 Summary of Chapter Deliverables

We close this chapter by drawing the following significant conclusions.

- This chapter proposes a simple tuning technique aimed to produce a robust noninteger order PID controller exhibiting iso-damping property during the reparameterization of a plant.
- Subsections 8.3.2 and 8.3.3 disclose that the proposed robust control design method is not limited to Bode's ideal transfer function and is valid to any kind of integer and fractional order reference systems. The reference system design belongs to the instinct of the user and guided by the transient response of the system chosen.
- The proposed method is versatile enough to tune different versions of fractional PID, fractional PI, fractional PD, and an even more complicated controller, fractional $FOPIDD^2$. Anticipating the potential application of fractional PID controller in typical integer and noninteger order real processes like heat and mass diffusion process, the proposed method deserves immense significance.
- We presume that the method is also applicable for a linear time invariant integer as well as a fractional order delay processes.
- The adaptation and improvisation of fractional calculus for the solution of a real life problem needs a practical mathematical interface between the real problem and its real time solution. Triangular strip operational matrix as proposed by Podlubny [23] from the perspective of Mathematics (for the solution of fractional differential and partial differential equation) finds its elegant application in the presently proposed method of control system design. This very findings actually encourages that sort of development of techniques helpful in exploiting the power and the beauty of fractional calculus in engineering solution of real problems. As is evident from the present research endeavor that given a computation time, the solution of $H\infty$-optimal control problem using triangular strip operational matrix is much better than its original solution.
- The optimization methods adapted in the present work, particle swarm and Luus-Jaakola methods, render their varying individual performances for different problems chosen. It is ultimately a choice between accuracy and computation time to be accommodated.

References

[1] M. Manfred, Z. Evanghelos (1989). *Robust Process Control*. Englewood Cliffs, NJ: Prentice Hall.

[2] H.W. Bode (1945). *Network Analysis and Feedback Amplifier Design*. New York: Van Nostrand.

[3] A. Oustaloup (1991). *La Commade CRONE: Commade Robuste d'Ordre Non Entier*. Paris: Hermes.

[4] I. Podlubny (1999). Fractional-order systems and PID controllers. *IEEE T. Automat. Contr.*, vol. 44, pp. 208–214.

[5] C. Zhao, D. Xue, Y.Q. Chen (2005). A fractional order PID tuning algorithm for a class of fractional order plants. In: *Proceedings of the IEEE International Conference on Mechatronics & Automation*, Niagara Falls, Canada, pp. 216–221.

[6] I. Petras (2012). Tuning and implementation methods for fractional-order controllers. *Fract. Calc. Appl. Anal.*, vol. 15(2), pp. 282–303.

[7] R. Caponetto, G. Maione, A. Pisano, M.R. Rapaic, E. Usai (2013). Analysis and shaping of the self-sustained oscillations in relay controlled fractional-order systems. *Fract. Calc. Appl. Anal.*, vol. 16(1), pp. 93–108.

[8] V. Feliu-Batlle, R. Rivas-Perez, L. Sanchez-Rodriguez, M.A. Ruiz-Torija (2009). Robust fractional order PI controller implemented on a laboratory hydraulic canal. *J. Hydraul. Eng.*, vol. 135, pp. 271–282.

[9] Y. Luo, Y.Q. Chen (2009). Fractional-order proportional derivative controller for robust motion control: Tuning procedure and validation. In *Proc. of American Control Conference*, Hyatt Regency Riverfront, St. Louis, MO, pp. 1412–1417.

[10] Y. Luo, Y.Q. Chen (2012). Stabilizing and robust fractional order PI controller synthesis for first order plus time delay systems. *Automatica*, vol. 48, pp. 2159–2168.

[11] C.A. Monje, Y.Q. Chen, B.M. Vinagre, D. Xue, V. Feliu (2010). *Fractional-order Systems and Controls: Fundamentals and Applications*. London: Springer-Verlag.

[12] C.A. Monje, A.J. Calderon, B.M. Vinagre, Y. Chen, V. Feliu (2004). On fractional controllers: Some tuning rules for robustness to plant uncertainties. *Nonlinear Dynam.*, vol. 38, pp. 369–381.

[13] M. Talebpour, Y.M. Roshan, S. Mohseni (2009). Developing robust FOPID controllers based on fuzzy set point weighting algorithm. *Fract. Calc. Appl. Anal.*, vol. 12, pp. 373–390.

[14] M. Zamani, M. Karimi-Ghartemani, N. Sadati (2013). FOPID controller design for robust performance using particl swarm optimization. *Fract. Calc. Appl. Anal.*, vol. 10, pp. 169–188.

[15] P. Lanusse, J. Sabatier (2011). PLC implementation of a CRONE controller. *Fract. Calc. Appl. Anal.*, vol. 14(4), pp. 505–522.

[16] M.K. Bouafoura, N.B. Braiek (2010) $PI^{\lambda}D^{\mu}$ controller design for integer and fractional plants using piecewise orthogonal functions. *Commun. Nonlinear Sci. Numer. Simulat.*, vol. 15, pp. 1267–1278.

[17] B. Liao, R. Luus (2005). Comparison of the Luus-Jaakola optimization procedure and the genetic algorithm. *Eng. Optimiz.*, vol. 37, pp. 381–396.

[18] J. Robinson, Y. Rahmat-Samii (2004). Particle swarm optimization in electromagnetics. *IEEE T. Antenn. Propag.*, vol. 52, pp. 397–408.

[19] M. Tabatabaei, M. Haeri (2011). Design of fractional order proportional integral derivative controller based on moment matching and characteristic ratio assignment method. *Proc. IMechE., Part I: J. Syst. Cont. Eng.*, vol. 225, pp. 1040–1053.
[20] F. Merrikh-Bayat, M. Karimi-Ghartemani (2010). Method for designing PID stabilizers for minimum-phase fractional-order systems. *IET Control Theory Appl.*, vol. 4, pp. 61–70.
[21] M.A. Sahib (2015). A novel optimal PID plus second order derivative controller for AVR system. *Eng. Sci. Technol.*, vol. 18, pp. 194–206.
[22] Y. Tang, M. Cui, C. Hua, L. Li, Y. Yang (2012). Optimum design of fractional order $PI^{\lambda}D^{\mu}$ controller for AVR system using chaotic ant swarm. *Expert. Syst. Appl.*, vol. 39, pp. 6887–6896.
[23] I. Podlubny (2000). Matrix approach to discrete fractional calculus. *Fract. Calc. Appl. Anal.*, vol. 3, pp. 359–386.

9

Rational Integer Order System Approximation for Irrational Fractional Order Systems

The dynamical systems demonstrating the fractional order dynamics and requiring the fractional order differentiator and/or integrator to be mathematically modeled are called fractional order systems. The fractional order systems are naturally encountered in process control, certain types of electrical noise, electrochemistry, signal processing, image analysis and processing, biomedical engineering, electrode-electrolyte polarization, interface polarization, cardiac behavior, dielectric relaxation, spectral densities of music, smart mechatronic systems, and so on. The analog realization (practical implementation) of fractional order systems plays an important role in their real-world applications. The analogue realization of fractional order systems is generally carried out by employing the special electrical element called fractance (an electrical element with fractional order or noninteger order impedance). The fractance is, in some instances, constructed by designing the electrical circuit consisting of infinite resistors and capacitors. It is very problematic to implement an electric circuit with an infinite number of elements. The designed circuits are thus truncated, that is, approximated by integer-order filters. The use of fractional capacitors in building the fractance device is the second approach. Biswas et al. [1] and Jesus and Machado [2] proposed the use of some electrolyte processes, and the use of materials possessing fractal structure for fabrication of fractional capacitors [1, 2]. However, the proposed technologies do not seem quickly implementable and do not have wide practical applications. Therefore, utilization of integer-order filters in place of fractance is still the reliable way of hardware implementation of fractional order element. The finite dimensional integer-order filters are easy to handle and having characteristics close enough to that of original infinite dimensional fractional-order elements. The corresponding transfer functions for fractional-order systems resulting from applying continuous or discrete transforms are obviously irrational; therefore, FOSs have not been studied in detail in theory and practice. Such irrational transfer functions pose difficulties in practical implementation, causing the rise of a need for rational (or integer order) transfer functions for their (irrational) identification, analysis, and simulation, which in turn leads to the emergence of new research areas. What we mean by this need is that when simulations are to be performed or models are to be identified or controllers are to be implemented, the irrational transfer functions are generally replaced by approximated rational transfer functions. Researchers working in the applied fields mentioned earlier have paid close attention to this task, and have made efforts toward the formulation of reliable techniques yielding appropriate rational transfer functions for their counterparts. Oustaloup et al. [3] developed an approximation algorithm called a Crone approximation, which is widely used where a frequency band of interest is considered within which the frequency domain responses should be fit by a bank of integer-order filters to the fractional-order derivative. The resulting rational transfer function does not fit the actual irrational one around the frequency band boundaries. This boundary fitting problem was overcome by the approximation method developed by Xue et al. [4]. Although Xue et al.'s approximation method overcame the weak point of the Crone approximation, the rational transfer function resulting from this method is of very high order, which is unfavorable for practical implementation. In the year 2005, the first and third author of [4] proposed suboptimum H_2 rational approximation method, which takes the rational transfer function obtained by a Crone approximation technique as input, to produce a low-order integer order transfer function [5]. Carlson and Halijak [6] presented a third-order Newton process (this algorithm is known as a Carlson approximation algorithm) based on the predistortion of the algebraic expression $f(x) = x^n - a = 0$ for approximating $(1/s)^{1/n}$ (a general fractional capacitor) for any integer n. Das et al. [7] applied a Carlson algorithm to a fractional-order proportional derivative controller ($G_{FO[PD]}=(K_p+K_d s)^q$, q is a noninteger number) for its integer-order approximation. They found that higher-order Carlson approximations maintain a constant phase only around q=0.5, and when the fractional order (q) moves away from 0.5 (i.e., toward zero or one), the higher Carlson approximations do not exhibit the constant phase curve. On the contrary, the first-order approximation maintains exactly $(q\pi/2)$ radian of phase around frequency 1 rad/s for each q in the range [0, 1]. Thus, the first-order Carlson approximation is sufficient for hardware implementation of fractional order proportional derivative controller.

The Crone approximation technique, Carlson approximation technique, and the modified version of the Crone technique devised in [4] were originally proposed for finding rational approximation for a single fractional order differentiator. If they are used to approximate an irrational transfer function in other forms, the resulting rational approximations are generally of high order, which we shall discuss in Subsection 9.2. For approximating a fractal system or fractional power pole, Charef et al. [8] presented a singularity function method (this method is widely known as the Charef approximation technique) consisting of cascaded branches of a number of pole-zero (negative real) pairs; they demonstrated the application of their method to multiple fractional system. The Charef approximation technique uses magnitude data of the fractional power pole to produce rational approximation. Consequently, the magnitude Bode plot of the rational approximation precisely overlaps the magnitude Bode plot of the fractional system, but the phase Bode plot is not in compliance with that of its counterpart. Meng and Xue [9] proposed a new approximation scheme, which is an extension of the Charef approximation technique, for approximating fractional-order systems with a fractional power pole. Their method relaxes the restrictions on the location of pole-zero pairs of the rational approximation, which provides more flexibility and space for approximate systems to imitate the dynamics of the original system as close as possible, and was based on optimization processes that allow the rational approximation to represent the fractional-order system in a frequency domain, that is, both the magnitude and phase angle curve of the irrational as well as the rational are more or less the same. Matsuda and Fujii [10] developed an approximation technique that provides a rational transfer function for s^{α} (α is a noninteger) by computing gain at logarithmically spaced frequencies. Expanding the function $(1+x)^{\alpha}$with the help of continued fraction expansion and replacing x by $s-1$ in the resulting expression produces a rational transfer function for fractional-order differentiator (s^{α}) [11].

An attempt was made in [12] by extending the application of block pulse orthogonal functions to find a rational approximation to an infinite dimensional FOS. The FOS considered in [12] is given in the following equation:

$$G(s) = \frac{a_n s^{\alpha_n} + a_{n-1} s^{\alpha_{n-1}} + \cdots + a_1 s^{\alpha_1} + a_0 s^{\alpha 0}}{b_m s^{\beta_m} + b_{m-1} s^{\beta_{m-1}} + \cdots + b_1 s^{\beta_1} + b_0 s^{\beta_0}}, \tag{9.1}$$

where α_i, β_j ($i \in [0, n], j \in [0, m]$) are noninteger numbers.

Their method is based on the piecewise constant basis function. Therefore, an accurate rational or integer order approximation needs very small step sizes, leading to high CPU usage. Maiti and Konar [13] used a biologically inspired stochastic optimization technique, particle swarm optimization (PSO), which does not require the gradient information of the objective function and is not sensitive to an initial guess, to approximate FOS (Equation (9.1)) by an integer-order system in a larger time interval, $t \in [0, 10]$. The results obtained by them are not satisfactory because the rational approximation cannot track or mimic the dynamics of the original FOS in the time interval selected. To increase the convergence rate of conventional PSO, an adaptive PSO was developed and applied to the original PSO. However, adaptive PSO speeds up the convergence rate, reduces the iteration steps of calculating the parameters of the rational approximation (for a single fractional differentiator s^{α}), and reflects the convergence of the swarm better, it causes the particles to jump into a local optimum. A chaotic optimization strategy was adopted to prevent particles from getting stuck in the local optimum [14]. A comparative study of Oustaloup approximation (Crone), Carlson approximation, Matsuda approximation, the method proposed by Xue et al. [4], and the squared magnitude function [15] for rational approximation of fractional-order integrator is presented in [16]. Driven by the aspiration to approximate the characteristic impedance function of lossy RLGC (R stands for resistance, L for inductance, G for conductance, C for capacitance) transmission line by a circuit of lumped elements, Wing et al. [17] devised a systematic method based on linear fractional transformations derived from generalized cascaded two-port circuit theory to find a rational transfer function that approximates the square root of a rational transfer function ($f^{\alpha}(s) \cong (N(s)/D(s))^{0.5}$, where $N(s)$ and $D(s)$ are Hurwitz polynomials, over the domain of interval $\Omega = \{s = iw : w_{\min} \leq w \leq w_{\max}\}$). Each one of the approximation methods discussed here has its own merits and drawbacks. It is true that no method is capable enough to yield low order rational approximation which fits both frequency response and time behavior. Therefore, it is difficult to say that one of them is the best. Efforts are continuously made toward the analog realization of fractance or fractional-order differentiator or fractional-order integrator. Research in this field constantly goes on. The numerical method presented in this chapter is such an attempt.

9.1 The Proposed Integer-Order Approximation Method

The numerical method for approximating an irrational transfer function (or FOS) by a rational transfer function (or an integer-order system) is devised by using the triangular orthogonal functions.

Without loss of generality, the following general form of linear time invariant irrational transfer function is considered:

$$G(s) = \frac{Y(s)}{U(s)} = \frac{a_n s^{\alpha_n} + a_{n-1} s^{\alpha_{n-1}} + \cdots\cdots + a_1 s^{\alpha_1} + a_0 s^{\alpha_0}}{b_m s^{\beta_m} + b_{m-1} s^{\beta_{m-1}} + \cdots\cdots + b_1 s^{\beta_1} + b_0 s^{\beta_0}}, \tag{9.2}$$

where $\alpha_0 < \alpha_1 < \cdots < \alpha_n \leq k_1$($k_1$ is an integer), $\beta_0 < \beta_1 < \beta_2 \cdots < \beta_m \leq k_2$($k_2$ is an integer), $\alpha_i (i = 0, 1, 2, \ldots\ldots, n)$ and $\beta_j (j = 0, 1, 2, \ldots\ldots, m)$ are fractional orders (noninteger numbers), $a_i (i = 0, 1, 2, \ldots\ldots, n)$, and $b_j (j = 0, 1, 2, \ldots\ldots, m)$ are arbitrary constants, n and m are integers.

Equation (9.2) is rewritten as

$$\frac{Y(s)}{U(s)} = \frac{a_n s^{\alpha_n-\beta_m} + a_{n-1}s^{\alpha_{n-1}-\beta_m} + \cdots\cdots + a_1 s^{\alpha_1-\beta_m} + a_0 s^{\alpha_0-\beta_m}}{b_m + b_{m-1}s^{\beta_{m-1}-\beta_m} + \cdots\cdots + b_1 s^{\beta_1-\beta_m} + b_0 s^{\beta_0-\beta_m}}. \tag{9.3}$$

This equation is rearranged as follows:

$$b_m Y(s) + b_{m-1}\frac{Y(s)}{s^{\beta_m-\beta_{m-1}}} + \cdots + b_1\frac{Y(s)}{s^{\beta_m-\beta_1}} + b_0\frac{Y(s)}{s^{\beta_m-\beta_0}} = a_n\frac{U(s)}{s^{\beta_m-\alpha_n}} + A, \tag{9.4}$$

where $A = a_{n-1}\frac{U(s)}{s^{\beta_m-\alpha_{n-1}}} + \cdots + a_1\frac{U(s)}{s^{\beta_m-\alpha_1}} + a_0\frac{U(s)}{s^{\beta_m-\alpha_0}}$.

Using the definition of Laplace transform of Riemann-Liouville fractional order integral, the following can be obtained:

$$\begin{aligned} &b_m L(y(t)) + b_{m-1}L\left({}_0J_t^{\beta_m-\beta_{m-1}}y(t)\right) + \cdots\cdots + b_0 L\left({}_0J_t^{\beta_m-\beta_0}y(t)\right) \\ &\quad = a_n L\left({}_0J_t^{\beta_m-\alpha_n}u(t)\right) + B, \end{aligned} \tag{9.5}$$

where
$B = a_{n-1}L\left({}_0J_t^{\beta_m-\alpha_{n-1}}u(t)\right) + \cdots + a_1 L\left({}_0J_t^{\beta_m-\alpha_1}u(t)\right) + a_0 L\left({}_0J_t^{\beta_m-\alpha_0}u(t)\right)$.

The inverse Laplace transform is performed on Equation (9.5), and the following is acquired:

$$b_m y(t) + b_{m-1}{}_0J_t^{\beta_m-\beta_{m-1}}y(t) + \cdots\cdots + b_0{}_0J_t^{\beta_m-\beta_0}y(t) = a_n{}_0J_t^{\beta_m-\alpha_n}u(t) + C, \tag{9.6}$$

where $C = a_{n-1}{}_0J_t^{\beta_m-\alpha_{n-1}}u(t) + \cdots + a_1{}_0J_t^{\beta_m-\alpha_1}u(t) + a_0{}_0J_t^{\beta_m-\alpha_0}u(t)$.

The output and input signal are approximated in the TF domain as given in the next equation.

$$y(t) \cong C^T T1_m(t) + D^T T2_m(t),\ u(t) \cong C_0^T T1_m(t) + D_0^T T2_m(t). \tag{9.7}$$

The TF approximation for fractional order integral of output and input signal are derived here:

$${}_0J_t^{\alpha}y(t) \cong (C^T P_1^{\alpha} + D^T P_3^{\alpha})T1_m(t) + (C^T P_2^{\alpha} + D^T P_4^{\alpha})T2_m(t), \tag{9.8}$$

$${}_0J_t^{\alpha}u(t) \cong (C_0^T P_1^{\alpha} + D_0^T P_3^{\alpha})T1_m(t) + (C_0^T P_2^{\alpha} + D_0^T P_4^{\alpha})T2_m(t). \tag{9.9}$$

Equation (9.6) can be transformed into a set of algebraic equations as per Equation (9.10):

$$\begin{aligned}&b_m\left(C^T T1_m(t)+D^T T2_m(t)\right)\\&+b_{m-1}\left(\left(C^T P_1^{\gamma_1}+D^T P_3^{\gamma_1}\right)T1_m(t)+\left(C^T P_2^{\gamma_1}+D^T P_4^{\gamma_1}\right)T2_m(t)\right)+\\&\cdots\cdots+b_0\left(\left(C^T P_1^{\gamma_2}+D^T P_3^{\gamma_2}\right)T1_m(t)+\left(C^T P_2^{\gamma_2}+D^T P_4^{\gamma_2}\right)T2_m(t)\right)=AB,\end{aligned}\tag{9.10}$$

where

$$\begin{aligned}AB&=a_n\left(C_0^T P_1^{\gamma_3}+D_0^T P_3^{\gamma_3}\right)T1_m(t)+a_n\left(C_0^T P_2^{\gamma_3}+D_0^T P_4^{\gamma_3}\right)T2_m(t)+\\&a_{n-1}\left(C_0^T P_1^{\gamma_4}+D_0^T P_3^{\gamma_4}\right)T1_m(t)+a_{n-1}\left(C_0^T P_2^{\gamma_4}+D_0^T P_4^{\gamma_4}\right)T2_m(t)+\cdots+\\&a_1\left(C_0^T P_1^{\gamma_5}+D_0^T P_3^{\gamma_5}\right)T1_m(t)+a_1\left(C_0^T P_2^{\gamma_5}+D_0^T P_4^{\gamma_5}\right)T2_m(t)+\\&a_0\left(C_0^T P_1^{\gamma_6}+D_0^T P_3^{\gamma_6}\right)T1_m(t)+a_0\left(C_0^T P_2^{\gamma_6}+D_0^T P_4^{\gamma_6}\right)T2_m(t),\end{aligned}$$

$$\gamma_1=\beta_m-\beta_{m-1},\ \gamma_2=\beta_m-\beta_0,\ \gamma_3=\beta_m-\alpha_n,\ \gamma_4=\beta_m-\alpha_{n-1},\ \gamma_5=\beta_m-\alpha_1,\gamma_6=\beta_m-\alpha_0.$$

The following system of algebraic equations can be found by comparing the coefficients of LHTF and RHTF:

$$\begin{aligned}&b_m C^T+b_{m-1}\left(C^T P_1^{\gamma_1}+D^T P_3^{\gamma_1}\right)+\cdots\cdots+b_0\left(C^T P_1^{\gamma_2}+D^T P_3^{\gamma_2}\right)=a_n\left(C_0^T P_1^{\gamma_3}+D_0^T P_3^{\gamma_3}\right)\\&+a_{n-1}\left(C_0^T P_1^{\gamma_4}+D_0^T P_3^{\gamma_4}\right)+\cdots+a_1\left(C_0^T P_1^{\gamma_5}+D_0^T P_3^{\gamma_5}\right)+a_0\left(C_0^T P_1^{\gamma_6}+D_0^T P_3^{\gamma_6}\right),\end{aligned}\tag{9.11}$$

$$\begin{aligned}&b_m D^T+b_{m-1}\left(C^T P_2^{\gamma_1}+D^T P_4^{\gamma_1}\right)+\cdots\cdots+b_0\left(C^T P_2^{\gamma_2}+D^T P_4^{\gamma_2}\right)=a_n\left(C_0^T P_2^{\gamma_3}+D_0^T P_4^{\gamma_3}\right)\\&+a_{n-1}\left(C_0^T P_2^{\gamma_4}+D_0^T P_4^{\gamma_4}\right)+\cdots+a_1\left(C_0^T P_2^{\gamma_5}+D_0^T P_4^{\gamma_5}\right)+a_0\left(C_0^T P_2^{\gamma_6}+D_0^T P_4^{\gamma_6}\right).\end{aligned}\tag{9.12}$$

The unknown coefficient vectors C^T and D^T can be computed from Equations (9.11) and (9.12), hence the response of the original FOS to input signal $u(t)$ is approximated in the TF domain.

The finite dimensional rational transfer function which is used in place of the infinite dimensional FOS in Equation (9.1) is defined as

$$\tilde{G}(s)=\frac{\tilde{Y}(s)}{U(s)}=\frac{c_p s^p+c_{p-1}s^{p-1}+\cdots\cdots+c_1 s+c_0}{d_q s^q+d_{q-1}s^{q-1}+\cdots\cdots+d_1 s+d_0},\tag{9.13}$$

where p and q are fixed or known integers, c_i $(i=0,1,2,\ldots\ldots,p)$ and $d_j(j=0,1,2,\ldots\ldots,q)$ are unknown coefficients to be determined.

An equivalent form of Equation (9.13) is provided here:

$$d_q\tilde{Y}(s) + d_{q-1}\frac{\tilde{Y}(s)}{s^{q-(q-1)}} + \cdots + d_1\frac{\tilde{Y}(s)}{s^{q-1}} + d_0\frac{\tilde{Y}(s)}{s^q} = c_p\frac{U(s)}{s^{q-p}} + A1, \tag{9.14}$$

where $A1 = c_{p-1}\frac{U(s)}{s^{q-(p-1)}} + \cdots + c_1\frac{U(s)}{s^{q-1}} + c_0\frac{U(s)}{s^q}$.

In time domain, the following fractional-order integral equation is obtained after the inverse Laplace transform is carried out:

$$d_q\tilde{y}(t) + d_{q-1\,0}J_t^{q-(q-1)}\tilde{y}(t) + \cdots\cdots + d_{1\,0}J_t^{q-1}\tilde{y}(t) + d_{0\,0}J_t^{q}\tilde{y}(t) = c_{p\,0}J_t^{q-p}u(t) + C, \tag{9.15}$$

where $C = c_{p-1\,0}J_t^{q-(p-1)}u(t) + \cdots + c_{1\,0}J_t^{q-1}u(t) + c_{0\,0}J_t^{q}u(t)$.

Similar to the determination of the approximation of FOS' response to the input signal via TFs, the response of rational approximation to the same input signal is also estimated in Equation (9.16) by following the same procedure as described in Equation (9.10) through (9.12).

The TF approximation for the output signal $\tilde{y}(t)$ is

$$\tilde{y}(t) \cong \tilde{C}^T T1_m(t) + \tilde{D}^T T2_m(t). \tag{9.16}$$

If the given FOS is simple and its output signal can be analytically found out, it is not necessarily converted to a set of algebraic equations as we did in Equation (9.7) through (9.12). However, it is well known that most linear time-invariant as well as time-variant fractional-order systems do not have closed form solutions (the solutions in terms of known function); thus, the response of a given FOS to a particular input signal is required to be approximated. The reason for employing TFs for computing an FOS response is that TFs convert fractional order differential or integral equations into an easily solvable set of linear algebraic equations.

The objective of this chapter is to find the rational transfer function that reasonably approximates the given irrational transfer function. The objective is mathematically defined in the next equation:

$$J = \min_{\theta}\left\|y(t) - \tilde{y}(t)\right\|_\infty \cong \min_{\theta}\left\|\left(C^T T1_m(t) + D^T T2_m(t)\right) - \left(\tilde{C}^T T1_m(t) + \tilde{D}^T T2_m(t)\right)\right\|, \tag{9.17}$$

where $\theta = \left[d_q, d_{q-1}, \ldots\ldots, d_1, d_0, c_p, c_{p-1}, \ldots\ldots, c_1, c_0\right]$.

It is assumed that the integer differential orders in the rational transfer function are known and the coefficients of numerator polynomial

$(c_p, c_{p-1}, \ldots\ldots, c_1, c_0)$ and denominator polynomial $(d_q, d_{q-1}, \ldots\ldots, d_1)$ are the unknowns. The nature inspired stochastic optimization algorithm; particle swarm optimization algorithm is adopted here to find the optimal θ which gives the minimum value of the objective function in Equation (9.17).

9.2 Simulation Example

The following fractional order system is considered to examine whether the prime goal of this chapter is fulfilled or not. It is also shown that the proposed numerical method can stand among some of the rational approximation techniques reported in the literature.

The fractional order system is

$$Y(s) = \frac{1}{0.8s^{2.2} + 0.5s^{0.9} + 1} U(s). \tag{9.18}$$

The structure of the rational transfer function is selected as

$$Y(s) = \frac{1}{a_3s^3 + a_2s^2 + a_1s + a_0} U(s). \tag{9.19}$$

The unknowns to be found out are $\theta = [a_0, a_1, a_2, a_3]$. The time interval $t \in [0, 5]$ is considered and is divided into 250 subintervals using the constant step size of 0.02. The lower and upper bounds on the optimization variables are chosen as $lb = [0.001, 0.001, 0.001, 0.001]$ and $ub = [2, 2, 2, 2]$. The parameters (swarm size (S), maximum iterations (MaxItr), self-adjustment factor (C1), social-adjustment factor (C2)) of PSO are considered here:

$$\text{S=30, MaxItr=50, C1=2, C2=2.} \tag{9.20}$$

The response of the given FOS (Equation (9.18)) to a unit step input is computed in the TF domain. The source code written in MATLAB® for implementation of the proposed numerical method on the given FOS is given in Program 9.1.

The objective function is minimized to 0.00727650, and the rational transfer function obtained by TFs based numerical method is

$$Y_{TF}(s) = \frac{1}{0.08434s^3 + 0.82344s^2 + 0.32564s + 1.0765} U(s). \tag{9.21}$$

The computational time required by the proposed method is 87.62 seconds.

In addition to the proposed numerical method, for comparison, the rational approximation techniques; Crone, Carlson, Charef, Matsuda, and

CFE are also employed to get the integer-order system for the given FOS. The fractional-order differentiators; $s^{2.2}$ and $s^{0.9}$ are individually approximated by the rational transfer function by using each technique. The obtained rational approximation is used in place of the respective fractional order differentiator and the resulting transfer function is the finite dimensional transfer function approximating the entire FOS in Equation (9.18). The parameters used in Crone, Carlson, Charef, Matsuda, and CFE are provided in Table 9.1. Crone, Carlson, and Matsuda rational approximation are obtained by using the source code developed (noninteger toolbox) in MATLAB, which is freely available for download at http://web.ist.utl.pt/duarte.valerio/ninteger/ninteger.htm. The MATLAB codes for Charef as well as CFE approximation are provided in Programs 9.2 and 9.3.

n: number of zeros and poles, w: frequency band.

The rational approximation by the Crone technique is

$$G_1(s) = \frac{N_1(s)}{D_1(s)}, \tag{9.22}$$

where

$$\begin{aligned} N_1(s) = {} & 0.0005012s^{30}+1.371s^{29}+1462s^{28}+7.922e05s^{27}+2.415e08s^{26}+4.345e10s^{25} \\ & +4.741e12s^{24}+3.186e14s^{23}+1.331e16s^{22}+3.475e17s^{21}+5.693e18s^{20}+5.861e19s^{19} \\ & +3.799e20s^{18}+1.551e21s^{17}+3.991e21s^{16}+6.475e21s^{15}+6.624e21s^{14}+4.271e21s^{13} \\ & +1.736e21s^{12}+4.446e20s^{11}+7.167e19s^{10}+7.261e18s^{9}+4.615e17s^{8}+1.834e16s^{7} \\ & +4.528e14s^{6}+6.886e12s^{5}+6.351e10s^{4}+3.458e08s^{3}+1.059e06s^{2}+1649s+1, \end{aligned}$$

$$\begin{aligned} D_1(s) = {} & 0.001596s^{32}+4.059s^{31}+3981s^{30}+1.979e06s^{29}+5.518e08s^{28}+9.077e10s^{27} \\ & +9.051e12s^{26}+5.559e14s^{25}+2.124e16s^{24}+5.083e17s^{23}+7.66e18s^{22}+7.315e19s^{21} \\ & +4.467e20s^{20}+1.774e21s^{19}+4.715e21s^{18}+8.801e21s^{17}+1.218e22s^{16} \\ & +1.281e22s^{15}+1e22s^{14}+5.5e21s^{13}+2.035e21s^{12}+4.922e20s^{11}+7.657e19s^{10} \\ & +7.583e18s^{9}+4.749e17s^{8}+1.868e16s^{7}+4.585e14s^{6}+6.943e12s^{5}+6.385e10s^{4} \\ & +3.471e08s^{3}+1.062e06s^{2}+1651s+1.001. \end{aligned}$$

TABLE 9.1

Values of parameters

Approximation technique	Parameters
Crone	$n = 15$, $w = [0.001, 1000]$
Carlson	$n = 2$, $w = [0.001, 1000]$
Charef	$n = 15$, $w_{\max} = 10^9$, $y = 1$, $p_T = 1$
Matsuda	$n = 10$, $w = [0.001, 1000]$
CFE	$n = 5$

The finite dimensional transfer function via Carlson technique is

$$G_2(s) = \frac{N_2(s)}{D_2(s)}, \tag{9.23}$$

where

$$\begin{aligned} N_2(s) = {} & s^{18}+69.89s^{17}+1787s^{16}+2.339e04s^{15}+1.871e05s^{14}+1.008e06s^{13} \\ & +3.878e06s^{12}+1.103e07s^{11}+2.374e07s^{10}+3.914e07s^{9}+4.97e07s^{8}+4.852e07s^{7} \\ & +3.609e07s^{6}+2.006e07s^{5}+8.082e06s^{4}+2.239e06s^{3}+3.906e05s^{2}+3.676e04s+1384, \end{aligned}$$

$$\begin{aligned} D_2(s) = {} & 1.8s^{20}+112.9s^{19}+2509s^{18}+2.966e04s^{17}+2.211e05s^{16}+1.142e06s^{15}+4.328e06s^{14} \\ & +1.255e07s^{13}+2.87e07s^{12}+5.294e07s^{11}+7.991e07s^{10}+9.934e07s^{9} \\ & +1.014e08s^{8}+8.394e07s^{7}+5.522e07s^{6}+2.803e07s^{5}+1.053e07s^{4}+2.754e06s^{3} \\ & +4.568e05s^{2}+4.085e04s+1461. \end{aligned}$$

The rational transfer function by Charef technique is

$$G_3(s) = \frac{N_3(s)}{D_3(s)}, \tag{9.24}$$

where

$$\begin{aligned} N_3(s) = {} & 1.097e130s^{30}+4.859e146s^{29}+1.546e162s^{28}+3.786e176s^{27}+7.178e189s^{26} \\ & +1.054e202s^{25}+1.197e213s^{24}+1.054e223s^{23}+7.178e231s^{22}+3.786e239s^{21}+1.546e246s^{20} \\ & +4.89e251s^{19}+1.197e256s^{18}+2.256e259s^{17}+3.097e261s^{16}+7.827e261s^{15}-1.511e263s^{14} \\ & +6.237e263s^{13}-1.418e264s^{12}+2.126e264s^{11}-2.272e264s^{10}+1.798e264s^{9}-1.076e264s^{8} \\ & +4.915e263s^{7}-1.713e263s^{6}+4.516e262s^{5}-8.831e261s^{4}+1.238e261s^{3}-1.173e260s^{2} \\ & +6.688e258s-1.727e257, \end{aligned}$$

$$\begin{aligned} D_3(s) = {} & 1.838e129s^{31}+8.269e145s^{30}+2.686e161s^{29}+6.754e175s^{28}+1.323e189s^{27}+ \\ & 2.023e201s^{26}+2.418e212s^{25}+2.263e222s^{24}+1.661e231s^{23}+9.57e238s^{22}+4.336e245s^{21}+ \\ & 1.546e251s^{20}+4.335e255s^{19}+9.537e258s^{18}+1.587e261s^{17}+1.096e262s^{16}-9.057e262s^{15}+ \\ & 2.161e263s^{14}-1.528e263s^{13}-3.286e263s^{12}+1.041e264s^{11}-1.477e264s^{10}+1.363e264s^{9}- \\ & 8.991e263s^{8}+4.388e263s^{7}-1.604e263s^{6}+4.382e262s^{5}-8.806e261s^{4}+1.261e261s^{3}- \\ & 1.215e260s^{2}+7.024e258s-1.833e257. \end{aligned}$$

The Matsuda rational approximation is

$$G_4(s) = \frac{N_4(s)}{D_4(s)}, \tag{9.25}$$

where

$$N_4(s) = s^{10}+9662s^9+6.63e06s^8+5.898e08s^7+1.253e10s^6+5.375e10s^5+6.117e10s^4 +1.465e10s^3+9.141e08s^2+1.024e07s+2.453e04,$$

$$D_4(s) = 4.05s^{12}+3.773e04s^{11}+1.418e07s^{10}+1.019e09s^9+1.486e10s^8+5.634e10s^7 +7.788e10s^6+9.335e10s^5+7.065e10s^4+1.529e10s^3+9.23e08s^2+1.027e07s +2.454e04.$$

The integer-order transfer function attained by the Continued Fraction Expansion is

$$G_5(s) = \frac{N_5(s)}{D_5(s)}, \tag{9.26}$$

where

$$N_5(s) = 2.841s^{10}+752.9s^9-1.239e04s^8+1.383e06s^7+2.365e07s^6+1.277e08s^5 +2.877e08s^4+2.992e08s^3+1.435e08s^2+2.939e07s+1.938e06,$$

$$D_5(s) = 7631s^{10}+2.189e06s^9+2.92e07s^8+1.39e08s^7+3.266e08s^6+4.851e08s^5 +5.319e08s^4+3.944e08s^3+1.622e08s^2+3.075e07s+1.943e06.$$

The finite dimensional approximation by Xue's method [4] is given in the next equation.

$$G_6(s) = \frac{N_6(s)}{D_6(s)}, \tag{9.27}$$

where

$$\begin{aligned} N_6(s) = {} & (5.615 \times 10^4)s^{37}+(9.711 \times 10^8)s^{36}+(4.766 \times 10^{12})s^{35}+(1.139 \times 10^{16})s^{34} \\ & +(1.567 \times 10^{19})s^{33}+(1.33 \times 10^{22})s^{32}+(7.177 \times 10^{24})s^{31}+(2.476 \times 10^{27})s^{30} \\ & +(5.431 \times 10^{29})s^{29}+(7.49 \times 10^{31})s^{28}+(6.475 \times 10^{33})s^{27})+(3.529 \times 10^{35})s^{26} \\ & +(1.222 \times 10^{37})s^{25}+(2.709 \times 10^{38})s^{24}+(3.907 \times 10^{39})s^{23}+(3.717 \times 10^{40})s^{22} \\ & +(2.333 \times 10^{41})s^{21}+(9.523 \times 10^{41})s^{20}+(2.472 \times 10^{42})s^{19}+(4.023 \times 10^{42})s^{18} \\ & +(4.073 \times 10^{42})s^{17}+(2.536 \times 10^{42})s^{16}+(9.589 \times 10^{41})s^{15}+(2.203 \times 10^{41})s^{14} \\ & +(3.086 \times 10^{40})s^{13}+(2.623 \times 10^{39})s^{12}+(1.341 \times 10^{38})s^{11}+(4.131 \times 10^{36})s^{10} \\ & +(7.682 \times 10^{34})s^9+(8.551 \times 10^{32})s^8+(5.595 \times 10^{30})s^7+(2.126 \times 10^{28})s^6 \\ & +(4.586 \times 10^{25})s^5+(5.29 \times 10^{22})s^4+(2.858 \times 10^{19})s^3+(5.918 \times 10^{15})s^2 \\ & +(4.319 \times 10^{11})s+(5.132 \times 10^6), \end{aligned}$$

$$\begin{aligned}D_6(s)= {} & 3658\times s^{38}+(6.283\times 10\text{^}7)s^{37}+(3.035\times 10^{11})s^{36}+(7.131\times 10\text{^}14)s^{35}\\ & +(9.65\times 10^{17})s^{34})+(8.068\times 10^{20})(s^{33})+(4.292\times 10^{23})s^{32}\\ & +(1.466\times 10^{26})s^{31})+(3.205\times 10^{28})s^{30}+(4.455\times 10^{30})s^{29}\\ & +(3.951\times 10^{32})(s^{28})+(2.265\times 10^{34})s^{27}+(8.547\times 10^{35})s^{26}\\ & +(2.17\times 10^{37})s^{25}+(3.805\times 10^{38})s^{24}+(4.689\times 10^{39})(s^{23})\\ & +(4.065\times 10^{40})s^{22})+(2.433\times 10^{41})s^{21}+(9.717\times 10^{41})s^{20}\\ & +(2.499\times 10^{42})s^{19}+(4.053\times 10^{42})(s^{18})+(4.095\times 10^{42})s^{17}\\ & +(2.548\times 10^{42})s^{16})+(9.632\times 10^{41})s^{15})+(2.212\times 10^{41})s^{14}\\ & +(3.098\times 10^{40})(s^{13})+(2.633\times 10^{39})s^{12})+(1.346\times 10^{38})s^{11}\\ & +(4.147\times 10^{36})s^{10}+(7.711\times 10^{34})s^{9}+(8.583\times 10^{32})(s^{8})\\ & +(5.616\times 10^{30})s^{7}+(2.133\times 10^{28})s^{6}+(4.603\times 10^{25})s^{5}\\ & +(5.309\times 10^{22})s^{4}+(2.868\times 10^{19})(s^{3})+(5.938\times 10^{15})s^{2}\\ & +(4.331\times 10^{11})s+5.132\times 10^{6}.\end{aligned}$$

The authors in [12] and [13] obtained the following approximations via block pulse functions and PSO, respectively:

$$Y_{BPF}(s)=\frac{1}{0.1685s^3+0.6780s^2+0.4528s+1.0279}U(s), \tag{9.28}$$

$$Y_{PSO}(s)=\frac{1}{0.1772s^3+0.7329s^2+0.4463s+1.0265}U(s). \tag{9.29}$$

The response of the TF rational approximation and the other rational approximations provided above to different input excitation signals such as step signal ($u_1(t)$), pseudo random binary signal ($u_2(t)$), sinusoidal wave signal ($u_3(t)$), square wave signal ($u_4(t)$), Sawtooth wave signal ($u_5(t)$), and pulse signal ($u_6(t)$) are determined and compared with the response of the original FOS in Figure 9.1 through Figure 9.9. The maximum or infinite norm of the error between the response of each rational approximation and the actual response to every input signal is calculated and tabulated in Table 9.2.

The following can be noted from Figure 9.1 through Figure 9.9 and Table 9.2.

- Charef approximation as well Xue approximation cannot be regarded as substitutes for the original FOS, as they fail to mimic the dynamics of FOS to all input excitation signals considered in this chapter. The

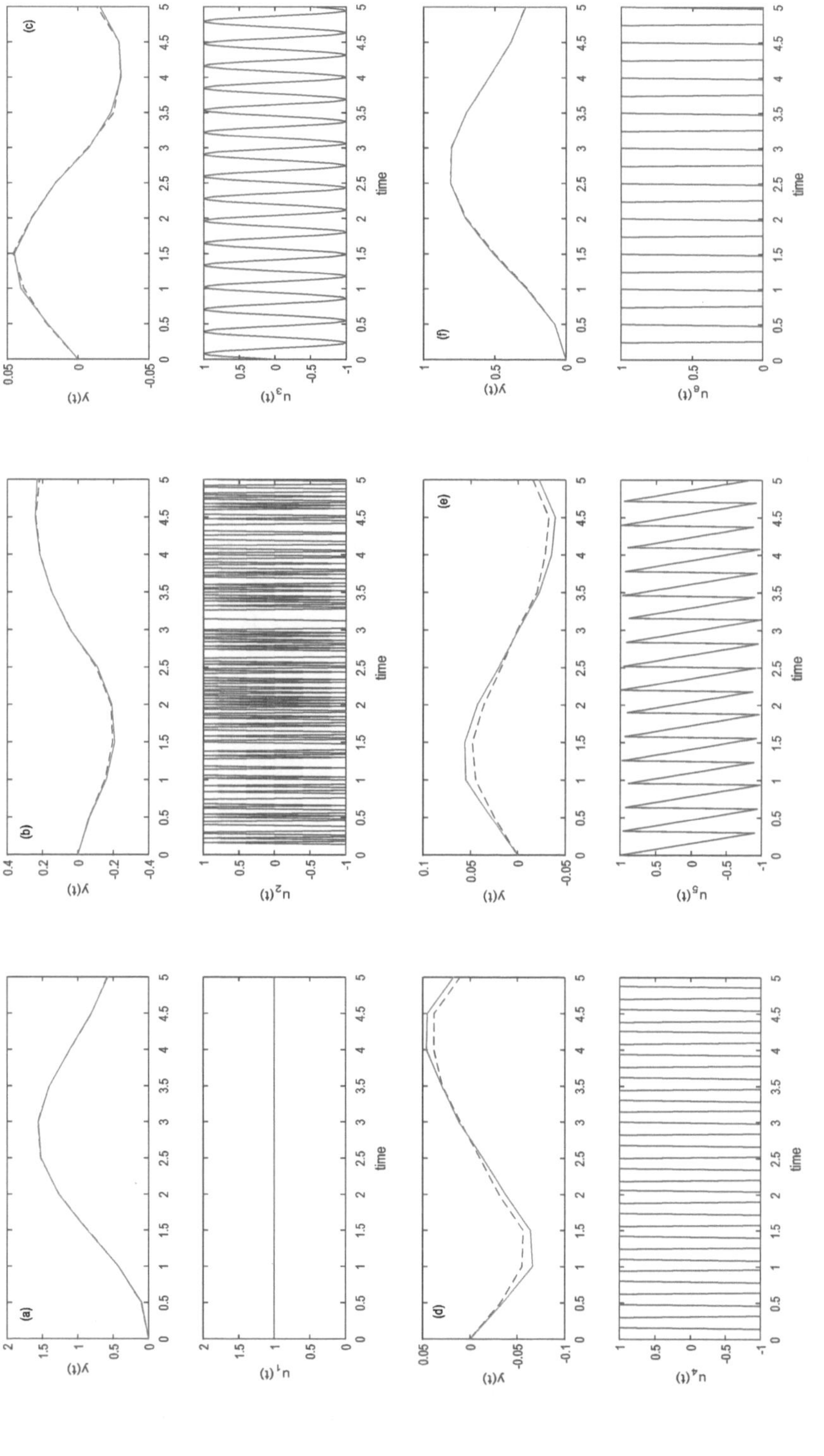

FIGURE 9.1
Response of TF rational approximation (solid line) and original FOS (dashed line) to various input signals.

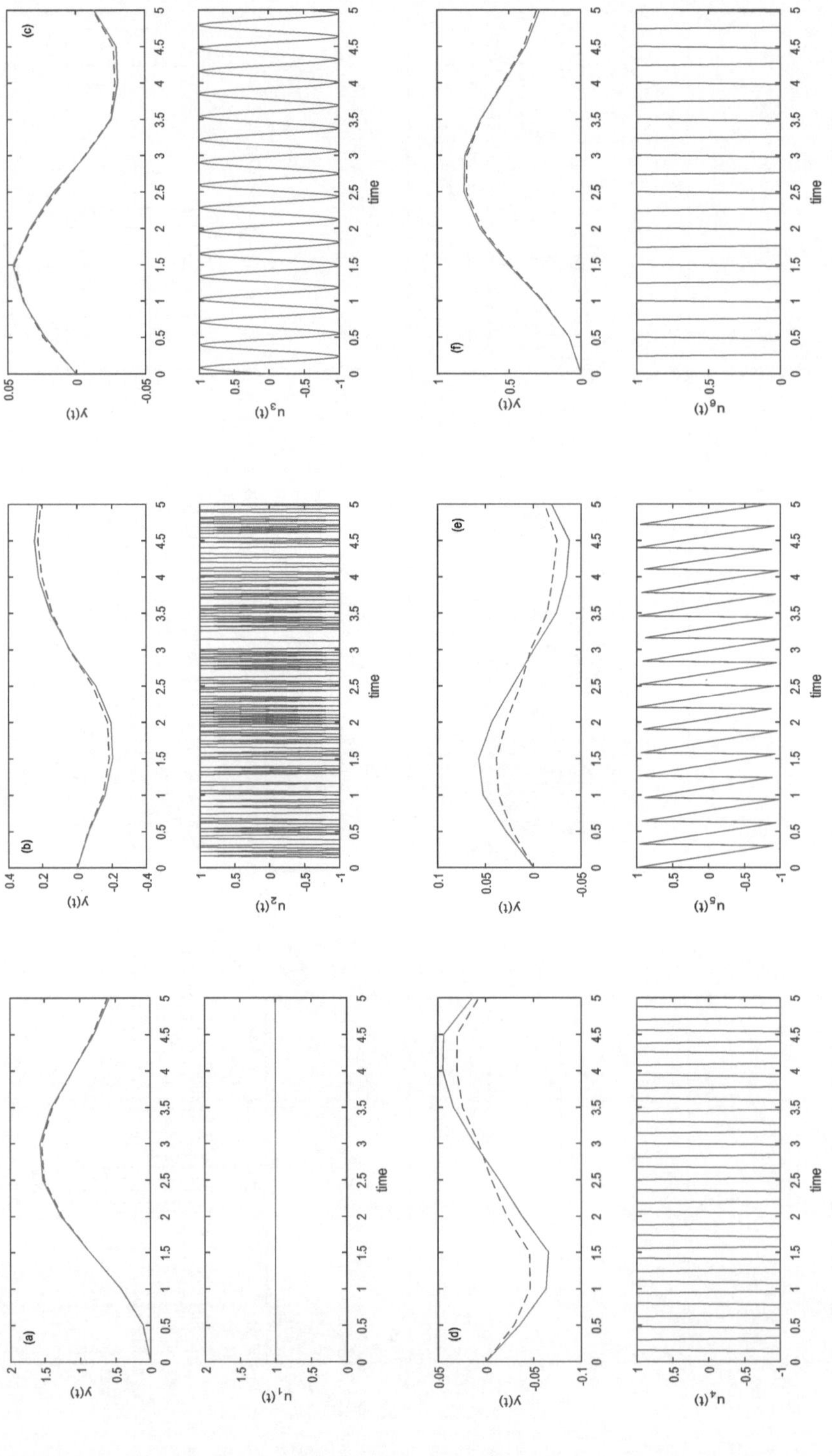

FIGURE 9.2
Response of Crone rational approximation (solid line) and original FOS (dashed line) to various input signals

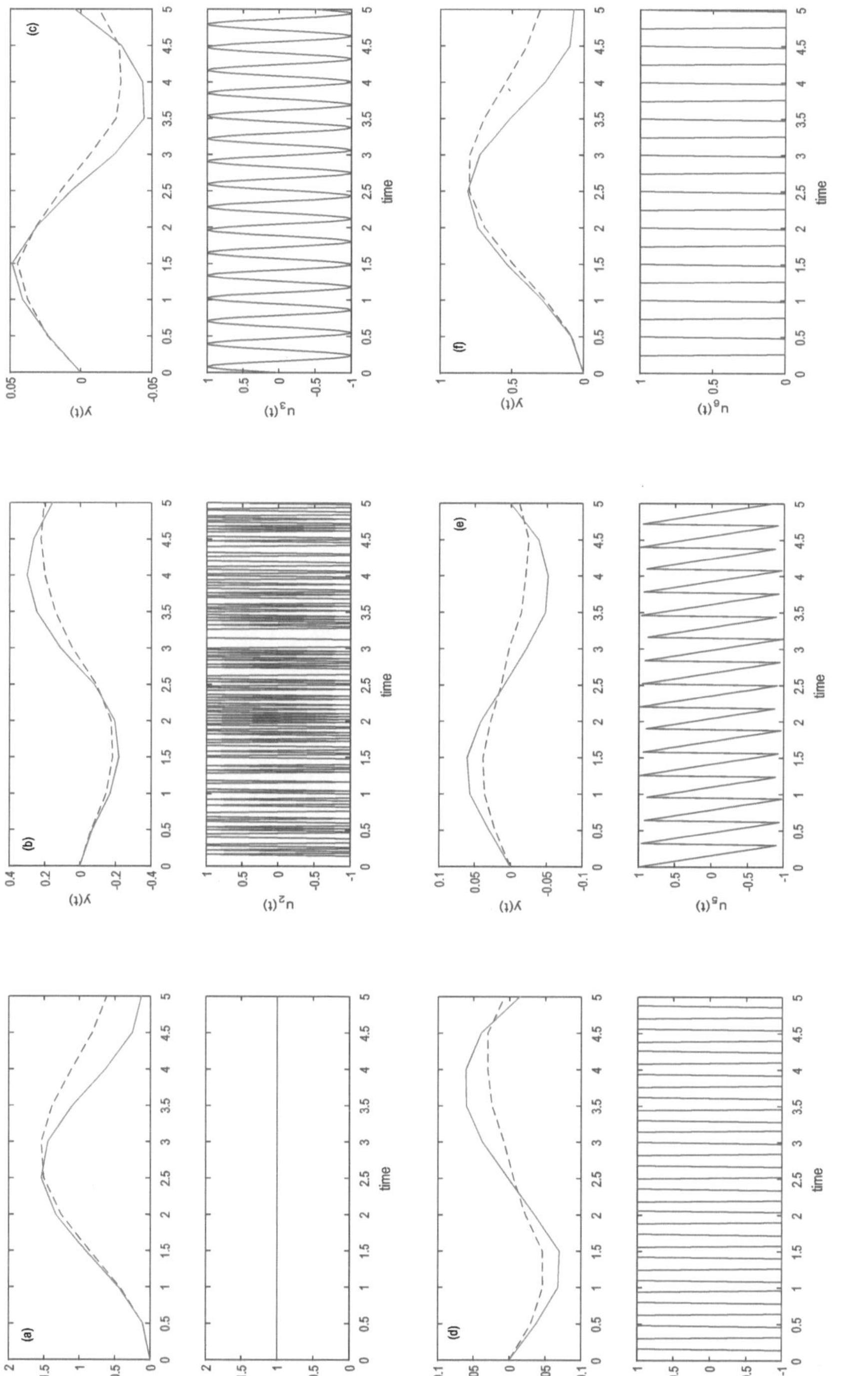

FIGURE 9.3
Response of Carlson rational approximation (solid line) and original FOS (dashed line) to various input signals

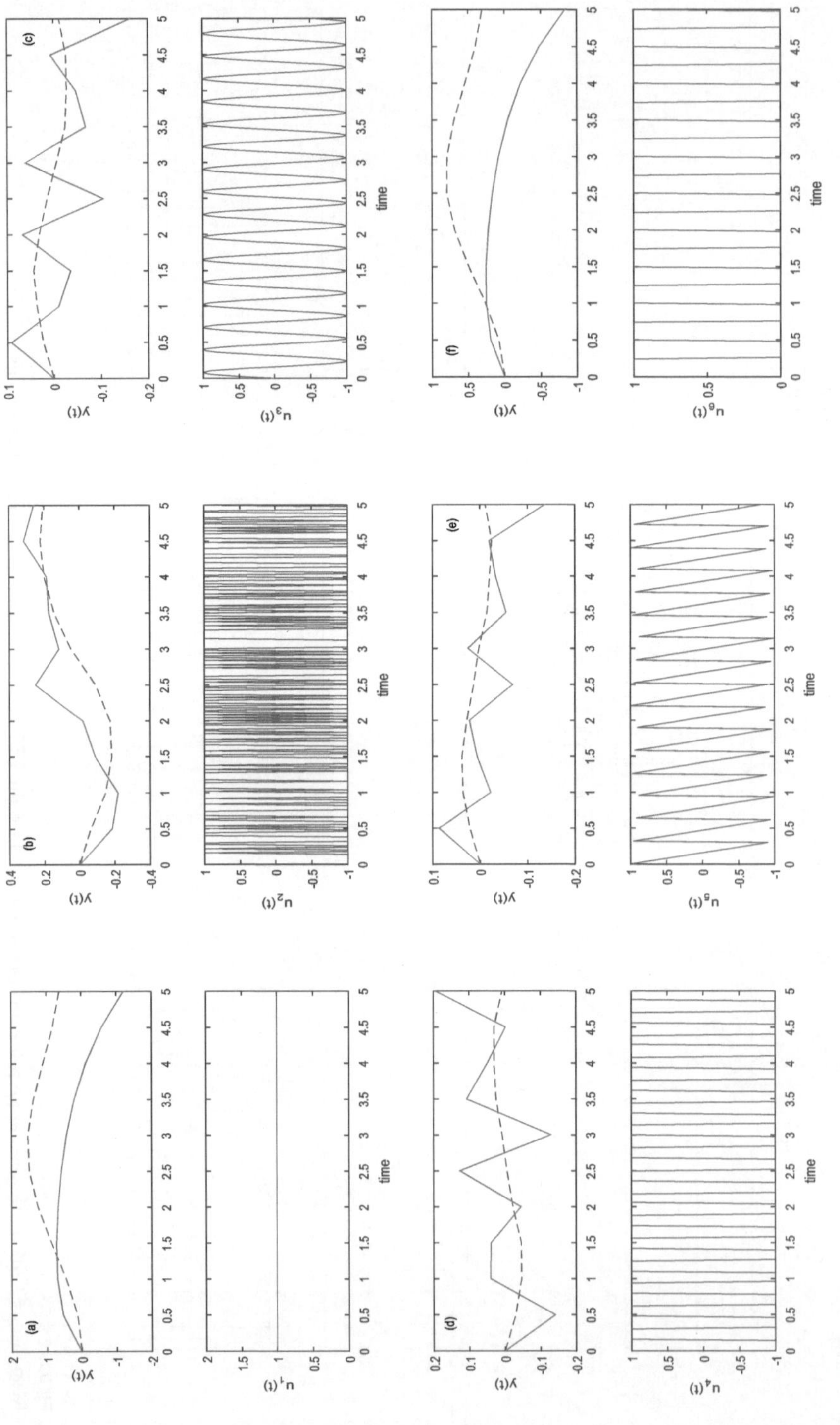

FIGURE 9.4
Response of Charef rational approximation (solid line) and original FOS (dashed line) to various input signals

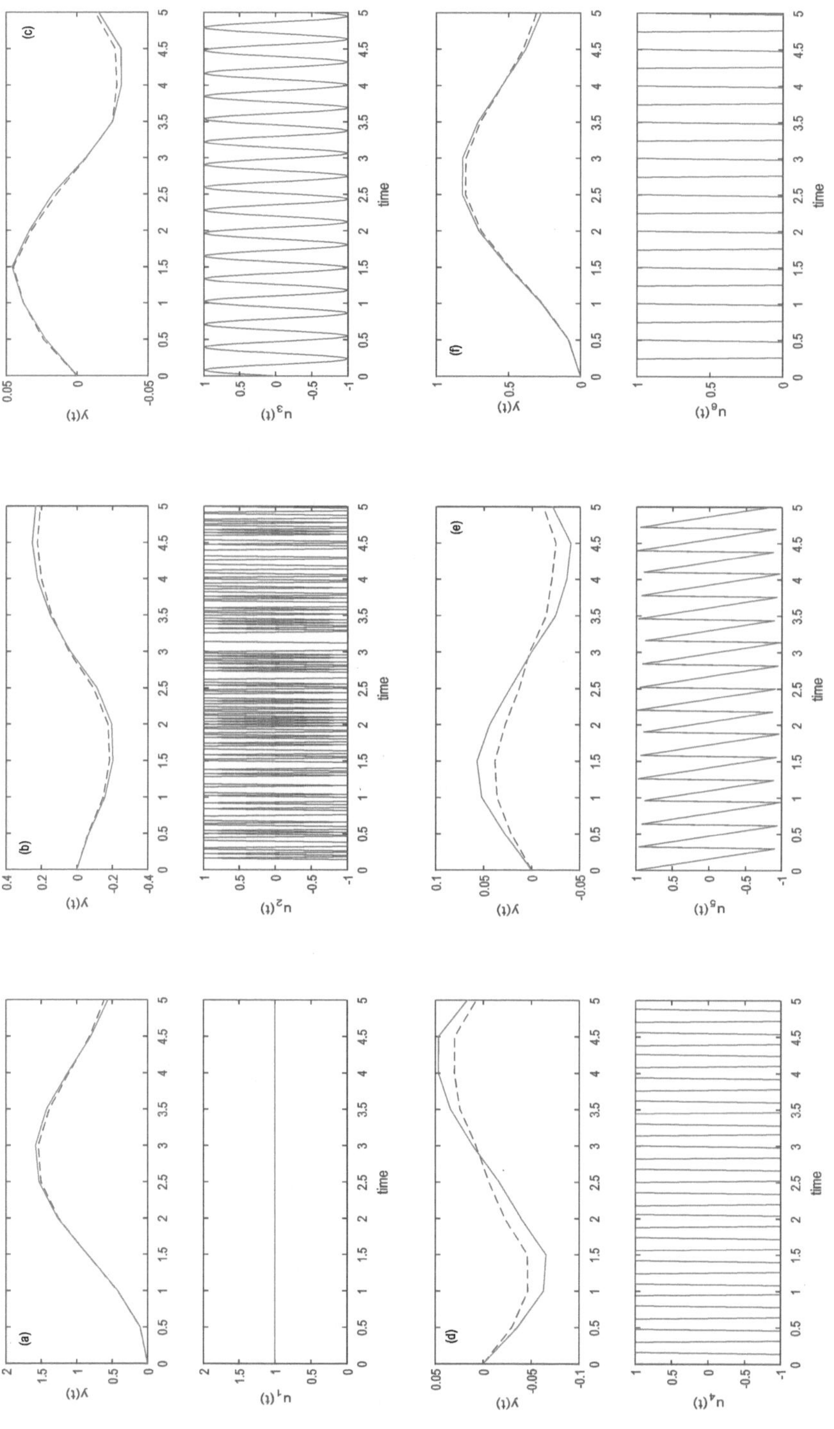

FIGURE 9.5
Response of Matsuda rational approximation (solid line) and original FOS (dashed line) to various input signals

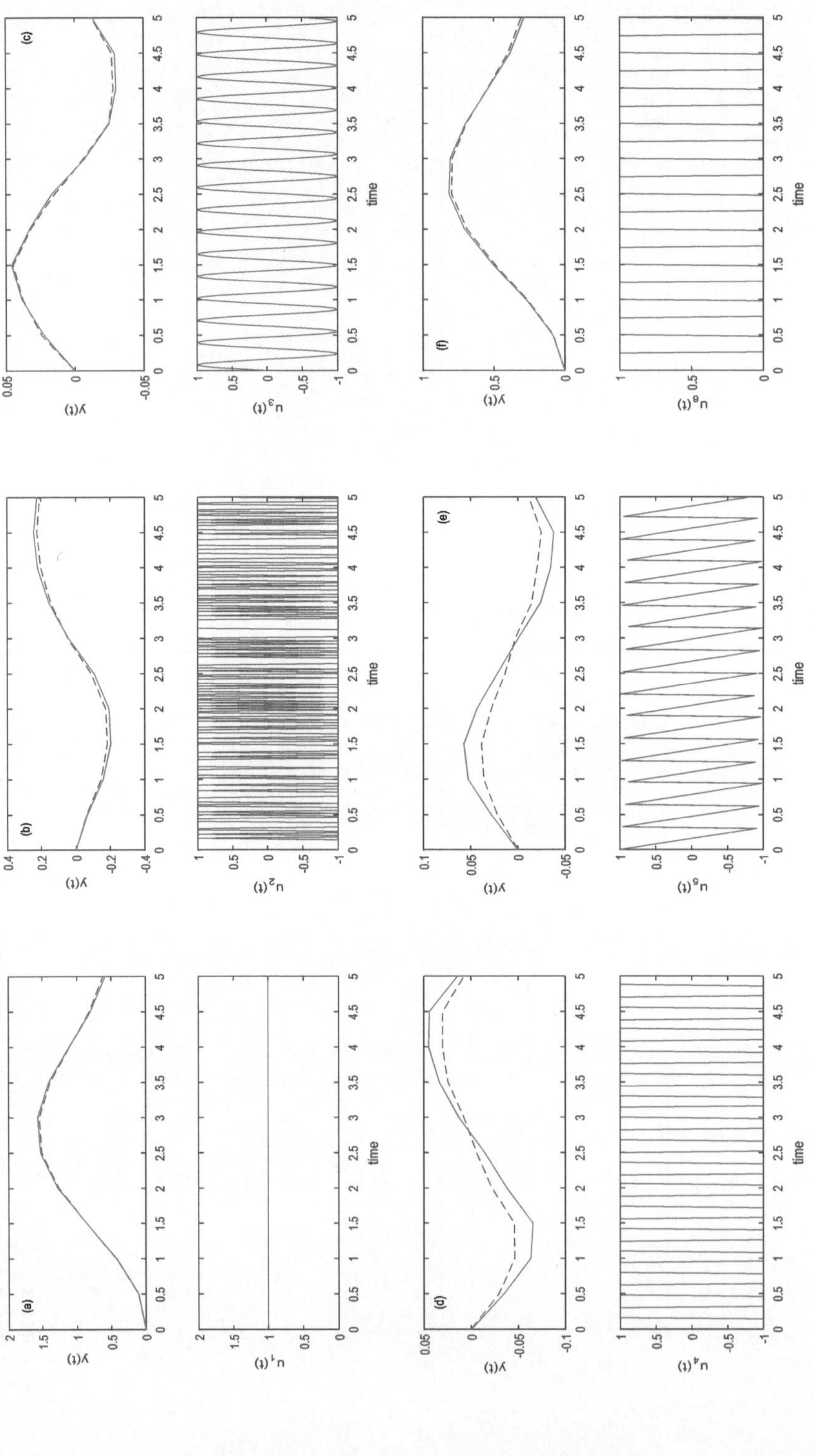

FIGURE 9.6
Response of CFE rational approximation (solid line) and original FOS (dashed line) to various input signals

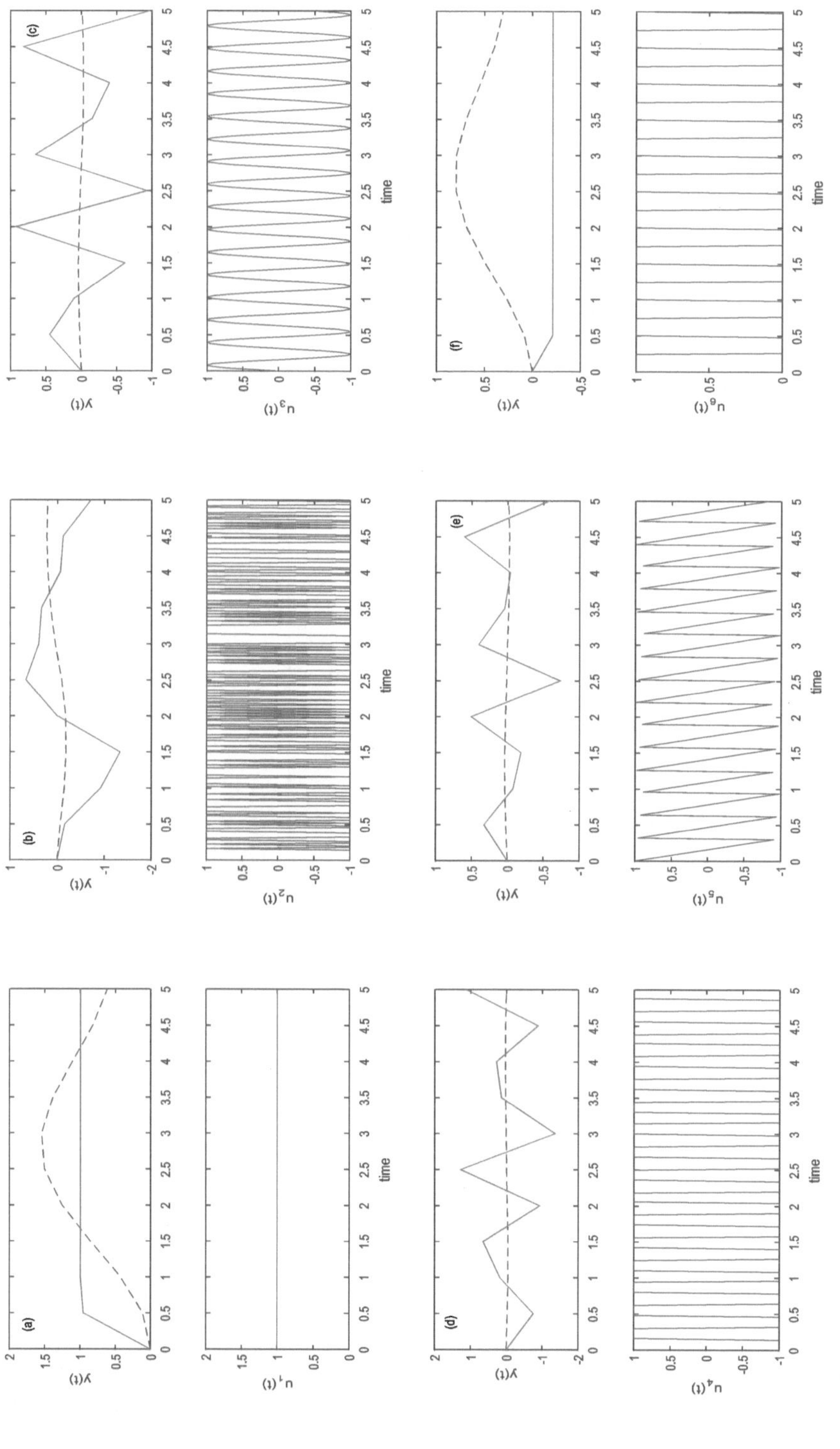

FIGURE 9.7
Response of Xue's rational approximation (solid line) and original FOS (dashed line) to various input signals

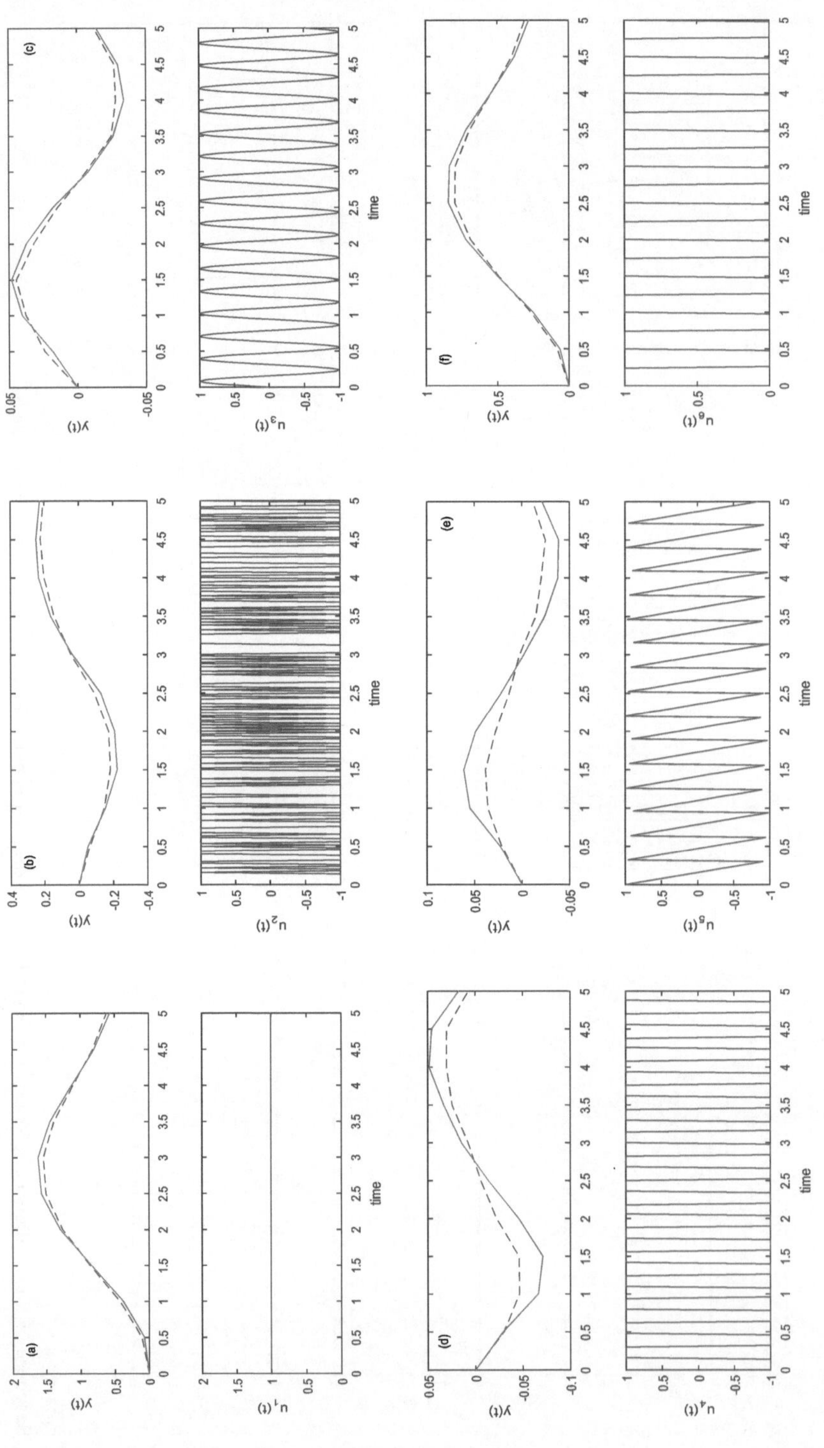

FIGURE 9.8
Response of PSO rational approximation (solid line) and original FOS (dashed line) to various input signals

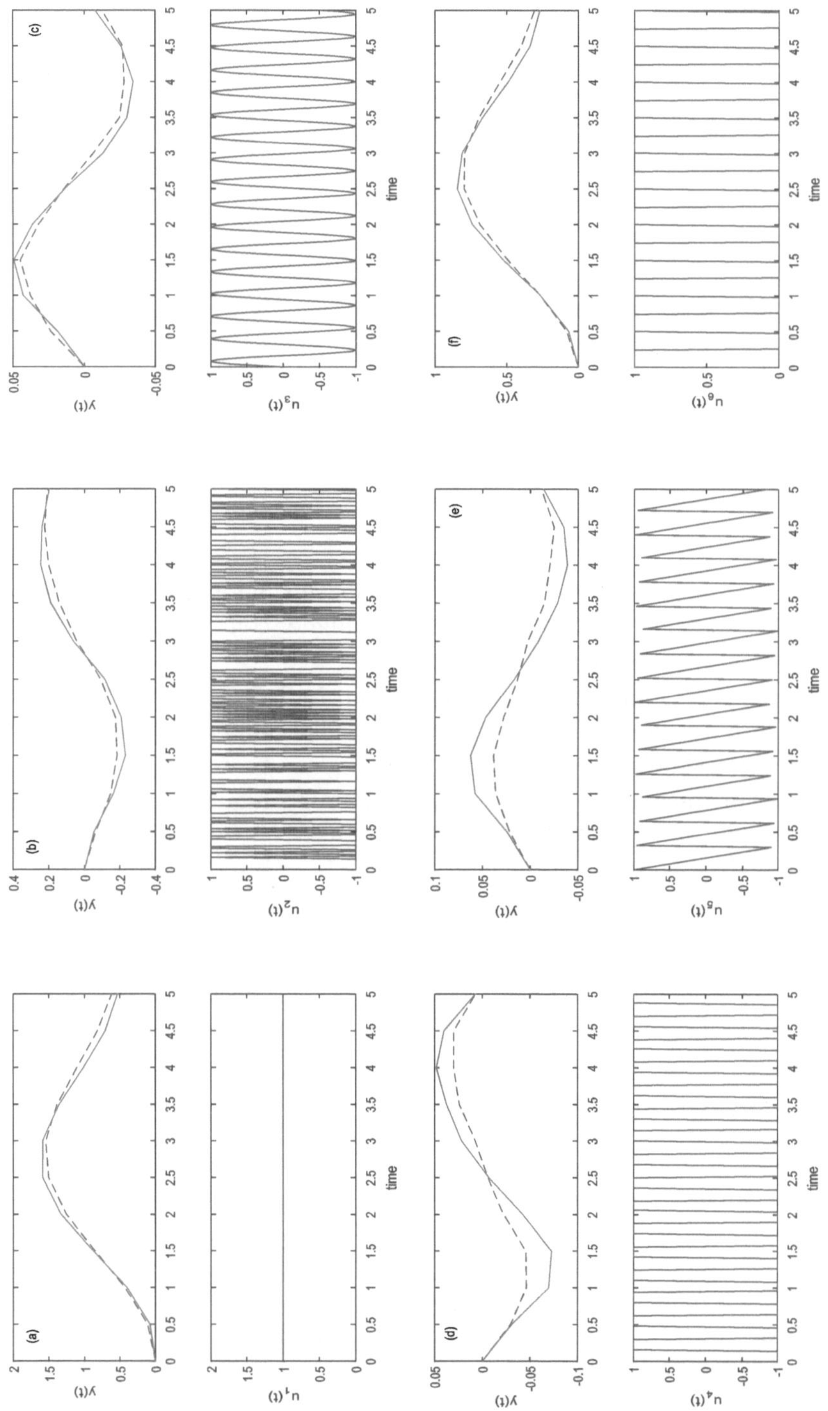

FIGURE 9.9
Response of BPF rational approximation (solid line) and original FOS (dashed line) to various input signals

TABLE 9.2
Maximum absolute errors

IES	TF	Crone	Carlson	Charef	Matsuda	CFE	Method in [12]	PSO	BPF
S1	**0.00727**	0.03155	0.56470	1.82491	0.05201	0.03064	1.24013	0.07986	0.11160
S2	**0.01581**	0.02156	0.11099	0.34977	0.02948	0.02192	1.38325	0.03904	0.05027
S3	0.00318	**0.00266**	0.02043	0.15012	0.00469	0.00298	1.06355	0.00700	0.00879
S4	**0.01207**	0.02110	0.03791	0.23225	0.02123	0.02208	1.48040	0.02943	0.03109
S5	**0.01062**	0.01960	0.03517	0.14868	0.01966	0.02039	1.00552	0.02474	0.02713
S6	**0.01027**	0.02115	0.30213	1.14019	0.03275	0.02102	1.24013	0.04713	0.06788

IES: input excitation signal, S1: step signal, S2: pseudo random binary signal, S3: sinusoidal signal, S4: square wave signal, S5: sawtooth wave signal, S6: pulse signal

order of rational transfer functions (order of Charef approximation is 31 and of Xue approximation is 38) are higher compared to the order of the remaining approximations.

- The Crone rational transfer function can replace the given FOS, but its order is still higher compared to the rational approximation obtained via the proposed method.
- A better integer-order system by Carlson technique can be acquired by taking more number of zeros and poles. However, the larger n, the higher the order of the approximation.
- The TFs are formed from the mother function; BPFs and are piecewise polynomials of degree 1, which is greater than the degree of the piecewise constant polynomials, that is, BPFs. So the proposed numerical method always shows better performance than the BPF based numerical method, which is pointed out in all the preceding chapters.
- The response of the rational approximation by the Matsuda technique and the continued fraction expansion (CFE) method closely follow the response of FOS to step signal, pseudo random binary signal, sinusoidal wave signal, and pulse signal, but significantly deviates from the actual ones for the square and Sawtooth wave signal. Such variations can be suppressed by increasing n and the width of the frequency band.
- By increasing the swarm size and number of iterations, and taking different values for self-adjustment factor and social adjustment factor, the performance of PSO based numerical method formulated in [13] can be enhanced.

- The proposed TFs-based numerical method beats all techniques studied in this chapter. Similar to PSO- and BPFs-based methods, the order of the resulting rational transfer function can be controlled in the proposed method while accurately approximating the given FOS. This characteristic makes the method a better alternative to Crone, Carlson, Charef, Matsuda, and CFE approximation techniques.
- The proposed method can be extended to fractional order system with delay, provided that the delay term is properly estimated in the TF domain.

9.3 MATLAB Codes for Simulation Example

Program 9.1

```
clc
clear all
global Ytrue t0 T m h t u
t0=0;T=5;h=0.02;m=(T-t0)/h;t=[t0:h:T];
u=[ones(1,m+1)];
% unit step input
Ytrue=TrueOutput(t0,T,m,t,u);  % TF based output
fun = @Objective_Fun;nvars = 4;
lb = [0.001;0.001;0.001;0.001];ub = [2;2;2;2];
tic
options =
optimoptions('particleswarm','SwarmSize',30,
  'MaxIter',50,'SelfAdjustment',2,'SocialAdjustment',2);
options.HybridFcn = @fmincon;
[x,fval,exitflag] =particleswarm(fun,nvars,lb,ub,options);
toc
a0=x(1);a1=x(2);a2=x(3);a3=x(4);
%%%%%%%%%%%%%%%%%%%%%%%%%%%%%%%%%%%%%%%%%%%%%%%%%%
%%%%%%%%%%%%
function Ytrue=TrueOutput(t0,T,m,t,u)
a0t=1;a1t=0.5;a2t=0.8;beta1t=0.9;beta2t=2.2;alpha1t=be-
ta2t-beta1t;
alpha2t=beta2t;
C00=u;    % unit step input
C01t=C00(1:end-1);D01t=C00(2:end);I=eye(m,m);
[P1alph1t,P2alph1t]=TOF1(t0,T,m,alpha1t);
[P3alph1t,P4alph1t]=TOF12(t0,T,m,alpha1t);
[P1alph2t,P2alph2t]=TOF1(t0,T,m,alpha2t);
[P3alph2t,P4alph2t]=TOF12(t0,T,m,alpha2t);
A1t=a2t*I+a1t*P1alph1t+a0t*P1alph2t;B1t=a1t*P3alph1t
  +a0t*P3alph2t;
```

```
A2t=a1t*P2alph1t+a0t*P2alph2t;B2t=a2t*I+a1t*P4alph1t
  +a0t*P4alph2t;
Z1t=(C01t*P1alph2t+D01t*P3alph2t);Z2t=(C01t*P2alph2t+
  D01t*P4alph2t);
At=[A1t A2t;B1t B2t];Zt=[Z1t Z2t];
Xt=Zt/At;x1t=Xt(1:m);x2t=Xt(m+1:end);Ytrue=[x1t
x2t(end)];
end
%%%%%%%%%%%%%%%%%%%%%%%%%%%%%%%%%%%%%%%%%%%%%%%
%%%%%%%%%%%%%
function f=Objective_Fun(x)
global Ytrue t0 T m h t u
a0t=x(1);a1t=x(2);a2t=x(3);a3t=x(4);beta1t=1;beta2t=2;
  beta3t=3;
alpha1t=beta3t-beta2t;alpha2t=beta3t-beta1t;
  alpha3t=beta3t;
C00=u;C01t=C00(1:end-1);D01t=C00(2:end);I=eye(m,m);
[P1alph1t,P2alph1t]=TOF1(t0,T,m,alpha1t);
[P3alph1t,P4alph1t]=TOF12(t0,T,m,alpha1t);
[P1alph2t,P2alph2t]=TOF1(t0,T,m,alpha2t);
[P3alph2t,P4alph2t]=TOF12(t0,T,m,alpha2t);
[P1alph3t,P2alph3t]=TOF1(t0,T,m,alpha3t);
[P3alph3t,P4alph3t]=TOF12(t0,T,m,alpha3t);
A1t=a3t*I+a2t*P1alph1t+a1t*P1alph2t+a0t*P1alph3t;
B1t=a2t*P3alph1t+a1t*P3alph2t+a0t*P3alph3t;
A2t=a2t*P2alph1t+a1t*P2alph2t+a0t*P2alph3t;
B2t=a3t*I+a2t*P4alph1t+a1t*P4alph2t+a0t*P4alph3t;
Z1t=(C01t*P1alph3t+D01t*P3alph3t);Z2t=(C01t*P2alph3t
  +D01t*P4alph3t);
At=[A1t A2t;B1t B2t];Zt=[Z1t Z2t];
Xt=Zt/At;x1t=Xt(1:m);x2t=Xt(m+1:end);Y=[x1t   x2t(end)];
e=abs(Ytrue-Y);
f=norm(e,inf);
end
```

Program 9.2

```
function G1=Charef(alpha,pT,y,wmax,N)
% This code generates the integer order approximation for
the fractional power pole transfer function of the form H(s)
=1/(1+(s/pT))^alpha. It is required to have approximation
for s^alpha, so necessary changes are made in the code in
order to get the desired approximation.
alpha is the order of fractional differentiator, N is
desired number of zeros and poles, y is the error, wmax is
the maximum frequency,pT=1;
A1=y/(10-10*alpha);a=10^A1;A2=y/(10*alpha);b=10^A2;
A3=y/(10*alpha*(1-alpha));ab=10^A3;A41=y/(20*alpha);
  p0=pT*(10^A41);
A51=y/(10*(1-alpha));z0=p0*(10^A51);P(1)=p0;Z(1)=z0;
```

```
A4=log(wmax/p0);A5=log(ab);AA=A4/A5;
if isinteger(AA)==1
  AA=AA;
else
  AA=floor(AA);
end
N=(AA)+1;N=abs(N);
if nargin<5
  N=N;
elseif nargin==5
  N=N1;
end
disp(AA)
disp(N)
for i=1:1:N
  P(i+1)=p0*(ab)^i;
end
for i=1:1:N
  Z(i+1)=a*p0*(ab)^i;
end
s=tf('s');s=s-1;num1=1;den1=1;
for j=0:1:N-1
  num1=num1*(1+(s/Z(j+1)));
end
for k=0:1:N
  den1=den1*(1+(s/P(k+1)));
end
num=num1;den=den1;G=num/den;G1=1/G;
end
```

Program 9.3

```
function G=CFE(alpha)
% alpha is the order of fractional differentiator.
% This code provides only fifth order rational transfer func-
tion to the fractional order differentiator.
a=alpha;
P0=-(a^5)-15*(a^4)-85*(a^3)-225*(a^2)-274*a-120;
P1=5*(a^5)+45*(a^4)+5*(a^3)-1005*a*a-3250*a-3000;
P2=-10*(a^5)-30*(a^4)+410*(a^3)+1230*(a^2)-4000*a-
  12000;
P3=10*(a^5)-30*(a^4)-410*(a^3)+1230*(a^2)+4000*a-12000;
P4=-5*(a^5)+45*(a^4)-5*(a^3)-1005*(a^2)+3250*a-3000;
P5=(a^5)-15*(a^4)+85*(a^3)-225*(a^2)+274*a-120;
Q5=P0;Q4=P1;Q3=P2;Q2=P3;Q1=P4;Q0=P5;s=tf('s');
num=P0*(s^5)+P1*(s^4)+P2*(s^3)+P3*(s^2)+P4*s+P5;
den=Q0*(s^5)+Q1*(s^4)+Q2*(s^3)+Q3*(s^2)+Q4*s+Q5;
G=num/den;
end
```

References

[1] K. Biswas, S. Sen, P.K. Dutta (2006). Realization of a constant phase element and its performance study in a differentiator circuit. *IEEE T. Circ. Syst. II*, vol. 53(9), pp. 802–806.

[2] I.S. Jesus, J.A.T. Machado (2009). Development of fractional order capacitors based on electrolyte processes. *Nonlinear Dynamics*, vol. 56(1–2), pp. 45–55.

[3] A. Oustaloup, F. Levron, B. Mathieu, F.M. Nanot (2000). Frequency-band complex noninteger differentiator: Characterization and synthesis. *IEEE T. Circ. Syst. I*, vol. 47(1), pp.25–39.

[4] D. Xue, C. Zhao, Y.Q. Chen (2006). A modified approximation method of fractional order system. In: *Proceedings of the 2006 IEEE International Conference on Mechatronics and Automation*, pp.1043–1048.

[5] D. Xue, Y. Chen (2005). Sub-optimum H_2 rational approximations to fractional order linear systems. In: *ASME. International Design Engineering Technical Conferences and Computers and Information in Engineering Conference, Volume 6: 5th International Conference on Multibody Systems, Nonlinear Dynamics, and Control, Parts A, B, and C*, pp. 1527–1536.

[6] G.E. Carlson, C.A. Halijak (1964). Approximation of a fractional capacitors $(1/s)^{(1/n)}$ by a regular Newton process. *IEEE T Circ. Theory CT*, vol. 11(2), pp. 210–213.

[7] S. Das, S. Saha, A. Gupta, S. Das (2011). Analog realization of fractional order hybrid differentiators via Carlson's approach. In: *Proceedings of 2011 International Conference on Multimedia, Signal Processing and Communication Technologies*, pp. 60–63.

[8] A. Charef, H.H. Sun, Y.Y. Tsao, B. Onaral (1992). Fractal system as represented by singularity function. *IEEE T. Automat. Contr.*, vol. 37(9), pp.1465–1470.

[9] L. Meng, D. Xue (2012). A New Approximation Algorithm of Fractional Order System Models Based Optimization. *J. Dyn. Syst.*, vol. 134(4), pp. 044504-1–044504-7.

[10] K. Matsuda, H. Fujii (1993). $H\infty$-optimized wave-absorbing control: Analytical and experimental results. *J. Guid. Control. Dynam.*, vol. 16(6), pp.1146–1153.

[11] B.T. Krishna (2011). Studies on fractional order differentiators and integrators: A survey. *Signal Processing*, vol. 91, pp.386–426.

[12] C.-H. Wang, C.-Y. Chen (2012). Finding the integer order systems for fractional order systems via fractional operational matrices. In: *Proceedings of 2012 9th IEEE International Conference on Networking, Sensing and Control*, Beijing, pp. 267–270.

[13] D. Maiti, A. Konar (2008). Approximation of a fractional order system by an integer order model using particle swarm optimization technique. In: *IEEE Sponsored Conference on Computational Intelligence, Control and Computer Vision in Robotics & Automation*, India, pp. 149–152.

[14] Z. Gao, X. Liao (2012). Rational approximation for fractional-order system by particle swarm optimization. *Nonlinear Dyn.*, vol. 67, pp.1387–1395.

[15] M. Khanra, J. Pal, K. Biswas (2010). Rational approximation of fractional order SISO System using squared-magnitude function. In: *Mathematical Methods in Engineering International Symposium*, Coimbra, Portugal.

[16] M. Khanra, J. Pal, K. Biswas (2010). Rational approximation of fractional operator: A comparative study. In: *2010 International Conference on Power, Control and Embedded Systems (ICPCES)*, Allahabad, pp. 1–5.

[17] O. Wing, Y.-L. Jiang, Q.-J. Yu (1998). Rational approximation of irrational functions by linear fractional transformations. *IEEE T. Circ. Syst. I*, vol. 45(11), pp. 1216–1221.

10

Numerical Method for Solving Fractional-Order Optimal Control Problems

An optimal control problem can be defined as the minimization of a criterion function (or cost function or performance index) of the states and control inputs of the system over a set of admissible control functions. The system is subject to constrained dynamics (in form of a single or system of differential equations or integral equations) and equality or inequality constraints on state and control variables. The right end point (i.e., final time) of the interval, in which the optimal control problem is defined, is sometimes considered as a constraint. Optimal control problems emerge in a variety of fields [1, 2]. A fractional optimal control problem (FOCP) is an optimal control problem in which the cost function (performance index) and/or the differential (or integral) equations governing the dynamics of the system contain at least one fractional-order operator (fractional-order differentiator or integrator). An FOCP can be defined with respect to different definitions of fractional order operators. However, the most important definitions are the Riemann-Liouville and Caputo fractional derivatives. The Caputo fractional derivative is often chosen as it enables one to incorporate traditional initial conditions into the problem formulation in a straightforward manner. The integer-order or classical optimal control problems are a subclass of FOCPs. Theoretical aspects of FOCPs and advanced computational methods for solving FOCPs are discussed in [3–5]. The classical optimal control theory has been under development for years, whereas FOCP is a fresh area. FOCPs encountered in applied fields usually do not have closed-form solutions. Therefore, formulation of effective numerical schemes for solving practical fractional optimal control problems has become an active research area, and it is gaining increasing interest. The approaches for numerical solutions of optimal control problems are classified into two major groups: indirect methods and direct methods. The indirect methods are based on the Pontryagin maximum principle and require the numerical solution of boundary value problems that result from the necessary conditions of optimal control [6]. Direct optimization methods transcribe the (infinite-dimensional) continuous problem to a finite-dimensional nonlinear programming problem through some parametrization of the state and/or control variables. In direct methods, initial guesses are provided only for physically intuitive quantities such as states and possibly controls. Indirect schemes are based on optimizing, then discretizing, the main optimal control problem (OCP), whereas the direct methods are based on discretizing, then optimizing, the main OCP.

The aim of this chapter is to develop and employ a new numerical method that is based on the triangular functions for solving FOCPs. The stated objective is accomplished as described in the following steps:

1. The Pontryagin maximum principle is imposed on the FOCP under consideration; consequently, a fractional two-point boundary value problem is obtained.
2. The state and control variables are approximated in the triangular function domain.
3. Integrator and differentiator (integer or noninteger order) are replaced by their equivalent TF operational matrices.
4. After steps 2 and 3 are carried out, the fractional two-point boundary value problem becomes a system of algebraic equations, which is solved using an algebraic solver.
5. The solution of resultant system of algebraic equations is the approximate solution to the original FOCP.
6. If the error between the exact optimal solution and the approximated solution is not negligible, a larger number of subintervals is used. Steps 2 to 4 are repeated.

The proposed numerical scheme is used to solve various kinds of classical (integer) as well as fractional (noninteger) unconstrained optimal control problems. This chapter shows another important application of the triangular functions, and significantly contributes to the ongoing research.

10.1 The Proposed Numerical Method

The fractional order optimal control (FOCP) is to find the optimal control $u(t)$ and the state variable $x(t)$, which minimize the performance index:

$$\min_{u(t)} J = \int_a^b f(x(t), u(t), t)dt, \tag{10.1}$$

subject to the fractional order dynamic system

$$ {}_0^C D_t^\alpha x(t) = g(x(t), u(t), t),\ x^{(k)}(a) = p_k,\ k = 0, 1, 2, \ldots\ldots, \lfloor \alpha \rfloor - 1. \tag{10.2} $$

In this problem, p_ks are arbitrary constants, ${}_0^C D_t^\alpha$ is the Caputo fractional-order derivative, $x(t)$ is the state variable, $u(t)$ is the control variable, and f and g are given functions that are continuously differentiable with respect to all their arguments in the domain of definition. When α takes only integers, FOCP in Equations (10.1) and (10.2) become the integer-order optimal control problem.

The Hamiltonian function of the above FOCP is defined as

$$ H(x(t), u(t), \lambda(t), t) = f(x(t), u(t), t) + \lambda g(x(t), u(t), t), \tag{10.3} $$

where λ is the Lagrange multiplier, also known as a costate or adjoint variable.

The necessary conditions for optimality of FOCP are given here:

$$ {}_0^c D_t^\alpha x(t) = \frac{\partial H}{\partial \lambda(t)} = h_1(x(t), u(t), t), \tag{10.4} $$

$$ {}_a^c D_t^\alpha \lambda(t) = \frac{\partial H}{\partial x(t)} = h_2\big(x(t), u(t), \lambda(t), t\big), \tag{10.5} $$

$$ \frac{\partial H}{\partial u(t)} = 0 \Rightarrow u(t) = h_3(x(t), \lambda(t), t). \tag{10.6} $$

If Equation (10.6) is used in Equations (10.4) and (10.5), the following system of fractional-order differential equations with initial conditions can be attained:

$$ {}_0^c D_t^\alpha x(t) = h_1(x(t), h_3(x(t), \lambda(t), t), t) = h_4(x(t), \lambda(t), t), \tag{10.7} $$

$$ {}_a^c D_t^\alpha \lambda(t) = h_2(x(t), h_3(x(t), \lambda(t), t), \lambda(t), t) = h_5(x(t), \lambda(t), t), \tag{10.8} $$

$$ x^{(k)}(a) = p_k,\ \lambda^{(k)}(a) = q_k,\ k = 0, 1, 2, \ldots\ldots, \lfloor \alpha \rfloor - 1. \tag{10.9} $$

The state variable and the Lagrange multiplier are approximated using the triangular functions:

$$ x(t) \approx C_1^T T1_m(t) + D_2^T T2_m(t), \tag{10.10} $$

$$\lambda(t) = C_2^T T1_m(t) + D_2^T T2_m(t). \tag{10.11}$$

The TF approximations for the functions h_4 and h_5 are determined as

$$h_4(x(t), \lambda(t), t) \approx C_{10}^T T1_m(t) + D_{10}^T T2_m(t), \tag{10.12}$$

$$h_5(x(t), \lambda(t), t) \approx C_{20}^T T1_m(t) + D_{20}^T T2_m(t). \tag{10.13}$$

Equation (10.7) is converted into fractional-order Volterra integral equation, as provided in the next equation:

$$x(t) = \underbrace{\sum_{k=0}^{\lfloor\alpha\rfloor-1} x^{(k)}(0)\left(\frac{t^k}{k!}\right)}_{h_6(t)} + J^\alpha(h_4(x(t), \lambda(t), t)), \tag{10.14}$$

$$\lambda(t) = \underbrace{\sum_{k=0}^{\lfloor\alpha\rfloor-1} \lambda^{(k)}(0)\left(\frac{t^k}{k!}\right)}_{h_7(t)} + J^\alpha(h_5(x(t), \lambda(t), t)). \tag{10.15}$$

The functions $h_6(t)$ and $h_6(t)$ are approximated in the TF domain:

$$h_6(t) \approx C_{30}^T T1_m(t) + D_{30}^T T2_m(t), \tag{10.16}$$

$$h_7(t) \approx C_{40}^T T1_m(t) + D_{40}^T T2_m(t). \tag{10.17}$$

The system of fractional-order Volterra integral equations can be transformed into a system of algebraic equations, as shown in the following equations, by using Equations (10.12), (10.13), (10.16), and (10.17):

$$\begin{aligned} C_1^T T1_m(t) + D_1^T T2_m(t) &= C_{30}^T T1_m(t) + D_{30}^T T2_m(t) + J^\alpha(C_{10}^T T1_m(t) + D_{10}^T T_{2m}(t)) \\ &= C_{30}^T T1_m(t) + D_{30}^T T_{2m}(t) + (C_{10}^T P_1^\alpha + D_{10}^T P_3^\alpha) T1_m(t) + (C_{10}^T P_2^\alpha + D_{10}^T P_4^\alpha) T_{2m}(t), \end{aligned} \tag{10.18}$$

$$\begin{aligned} C_2^T T1_m(t) + D_2^T T2_m(t) &= C_{40}^T T1_m(t) + D_{40}^T T2_m(t) + J^\alpha(C_{20}^T T1_m(t) + D_{20}^T T2_m(t)) \\ &= C_{40}^T T1_m(t) + D_{40}^T T2_m(t) + (C_{20}^T P_1^\alpha + D_{20}^T P_3^\alpha) T1_m(t) + (C_{20}^T P_2^\alpha + D_{20}^T P_4^\alpha) T2_m(t). \end{aligned} \tag{10.19}$$

The coefficients of LHTF vector ($T1_m(t)$) and RHTF vector ($T2_m(t)$) are compared as given in the following equations:

$$C_1^T = C_{30}^T + \left(C_{10}^T P_1^\alpha + D_{10}^T P_3^\alpha\right), \tag{10.20}$$

$$D_1^T = D_{30}^T + \left(C_{10}^T P_2^\alpha + D_{10}^T P_4^\alpha\right), \tag{10.21}$$

$$C_2^T = C_{40}^T + +\left(C_{20}^T P_1^\alpha + D_{20}^T P_3^\alpha\right), \tag{10.22}$$

$$D_2^T = D_{40}^T + \left(C_{20}^T P_2^\alpha + D_{20}^T P_4^\alpha\right). \tag{10.23}$$

The TF approximation for $x(t)$ and $\lambda(t)$ can be obtained by the above system of algebraic equations.

The approximate solution for the original optimal control variable can be found by substituting the approximation of $x(t)$ and $\lambda(t)$ in Equation (10.6), hence, the optimal control $u(t)$ and $x(t)$ (which minimize the given objective function J) are numerically found out.

10.2 Simulation Examples

In this section, the proposed numerical method is implemented on different kinds of fractional order optimal control problems. Table 10.1 provides the step size and CPU usage needed by the proposed method in each case study.

TABLE 10.1

Step size and CPU times

Case study	Step size	CT (seconds)
10.1	1/1500	347.700938
10.2	1/1000	126.126337
10.3	1/1500	387.031503
10.4	1/1000	486.681786

CT: computational time

Case study 10.1: Optimal control of linear time invariant integer order system (IOS)

Consider the following time-invariant optimal control problem.

$$\min_{u(t)} J = \frac{1}{2}\int_0^1 (x^2(t) + u^2(t))dt, \tag{10.24}$$

subject to the system dynamics:

$$\frac{dx(t)}{dt} = -x(t) + u(t),\ x(0) = 1. \tag{10.25}$$

The exact solution of the given integer-order optimal control problem can easily be determined as given here:

$$\begin{aligned} x(t) &= \cosh(\sqrt{2}t) + w\sinh(\sqrt{2}t), \\ u(t) &= (1+\sqrt{2}w)\cosh(\sqrt{2}t) + (\sqrt{2}+w)\sinh(\sqrt{2}t), \end{aligned} \tag{10.26}$$

where $w = -\frac{\cosh(\sqrt{2})+\sqrt{2}\sinh(\sqrt{2})}{\sqrt{2}\cosh(\sqrt{2})+\sinh(\sqrt{2})}$.

The TF approximate state variable $\tilde{x}(t)$ and optimal control $\tilde{u}(t)$, which minimize the cost function J to 0.192909343901483, are subtracted from the original state and control variable. The obtained absolute error at each time point is compared (Table 10.2) with the results obtained by Legendre wavelets–based numerical scheme [7] and Bernoulli polynomials–based numerical method [8]. It is noticed that the proposed method yields more accurate approximate solution than the methods in [7] and [8].

Case study 10.2: Optimal control of linear time-varying fractional-order system (FOS)

Consider the following fractional-order optimal control

$$\min_{u(t)} J = \int_0^1 \left((x(t) - t^2)^2 + \left(u(t) + t^2 - \frac{20t^{9/10}}{9\Gamma(9/10)} \right) \right) dt, \tag{10.27}$$

subject to the following dynamical system with initial conditions

TABLE 10.2
Absolute errors for case study 10.1

	$\|u(t) - \tilde{u}(t)\|$			$\|x(t) - \tilde{x}(t)\|$		
t	**Method in Ref. [7]**	**Method in Ref. [8]**	**Proposed method**	**Method in Ref. [7]**	**Method in Ref. [8]**	**Proposed method**
0	NV	1.88e-06	7.5052e-10	NV	6.25e-06	5.6443e-09
0.1	2.3140e-05	7.08e-07	2.5167e-09	7.6389e-05	2.39e-06	9.5014e-09
0.2	2.9270e-05	5.20e-07	1.7245e-10	5.6050e-05	1.21e-06	1.6563e-08
0.3	4.5399e-06	4.98e-07	1.3453e-09	3.9829e-05	1.72e-06	2.1698e-08
0.4	5.1000e-06	2.49e-07	2.2037e-09	2.7510e-05	6.82e-06	2.5302e-08
0.5	0.00085952	5.81e-07	2.5413e-09	1.8309e-05	1.93e-06	2.7691e-08
0.6	1.8760e-05	4.97e-07	2.4753e-09	1.1680e-05	3.11e-07	2.9124e-08
0.7	2.4880e-05	5.93e-07	2.1057e-09	7.1600e-06	1.90e-06	2.9814e-08
0.8	3.9610e-05	2.41e-07	1.5194e-09	4.6499e-06	9.17e-07	2.9929e-08
0.9	3.9529e-05	7.61e-07	7.9379e-10	3.6699e-06	2.49e-06	2.9604e-08
1	0.00356862	NV	2.9080e-08	1.6199e-06	NV	1.8939e-13

NV: no value is reported at the particular value of t in the cited paper.

$$ {}_0^CD^{1.1}x(t) = t^2x(t) + u(t),\ x(0) = x^{(1)}(0) = 0. \tag{10.28}$$

For this problem, the following actual solution minimizes the performance index to 0.

$$x(t) = t^2,\ u(t) = \frac{20t^{9/10}}{9\Gamma\left(9/10\right)} - t^4. \tag{10.29}$$

The minimum value of J obtained by the proposed method is compared (Table 10.3) with the optimal cost function achieved by using spectral method based on shifted second-kind Chebyshev polynomials and Legendre-Gauss quadrature [9], the Legendre orthonormal polynomials-based numerical method [10], computational method based on hat functions [11], and numerical method based on Legendre polynomials and Gauss quadrature [12]. The absolute errors between the actual optimal solution and the TF approximate solution are provided in Table 10.4.

TABLE 10.3
Optimal cost function of case study 10.2

Method	Cost function J
Proposed method	4.85535e-14
Method in Ref. [9]	6.00953e-09
Method in Ref. [10]	5.81653e-09
Method in Ref. [11]	2.44e-09
Method in Ref. [12]	7.03371e-08

TABLE 10.4
Absolute errors for case study 10.2

t	$\lvert x(t) - \tilde{x}(t)\rvert$	$\lvert u(t) - \tilde{u}(t)\rvert$
0	3.0114e-18	1.9621e-07
0.1	1.6960e-07	1.7936e-07
0.2	1.8013e-07	1.6009e-07
0.3	1.8246e-07	1.4012e-07
0.4	1.8284e-07	1.1976e-07
0.5	1.8363e-07	9.9166e-08
0.6	1.8621e-07	7.8434e-08
0.7	1.9160e-07	5.7735e-08
0.8	2.0074e-07	3.7311e-08
0.9	2.1468e-07	1.7575e-08
1	2.3466e-07	4.4408e-16

Case study 10.3: Optimal control of nonlinear FOS

Find the control $u(t)$ that minimizes the nonlinear performance index

$$J(u(t)) = \int_0^1 \left(x^2(t) - 2t^{3/2}x(t) + u^2(t) - \frac{3}{4}\sqrt{\pi}e^{-t}u(t) + e^{-t+t^{3/2}}u(t) + t^3 + A \right) dt, \tag{10.30}$$

TABLE 10.5

Absolute errors for case study 10.3

t	$\|x(t) - \tilde{x}(t)\|$	$\|u(t) - \tilde{u}(t)\|$
0	7.1624e-15	3.6237e-13
0.1	1.6889e-14	3.5055e-13
0.2	2.5354e-14	3.4301e-13
0.3	2.2232e-14	3.3308e-13
0.4	2.1516e-13	3.2143e-13
0.5	7.3829e-14	2.9317e-13
0.6	1.6520e-13	2.3599e-13
0.7	2.0772e-13	1.4641e-13
0.8	5.1059e-13	1.4488e-14
0.9	4.1111e-13	6.8889e-14
1	4.6629e-15	1.4677e-13

subject to the following fractional-order dynamic system with initial conditions.

$$ {}_0^C D_t^{1.5} x(t) = e^{x(t)} + 2e^t u(t),\ x(0) = x^{(1)}(0) = 0. \tag{10.31}$$

The expression for A is given as $A = \left(\frac{9\pi}{64}\right)e^{-2t} - \frac{3\sqrt{\pi}}{8}e^{-2t+t^{3/2}} + \frac{1}{4}e^{-2t+2t^{3/2}}$ $+e^{2t}$.

The optimum value of the performance index for this problem is $J^* = 3.19453$, with the following exact solutions for state and control functions.

$$x(t) = t^{3/2},\ u(t) = \frac{1}{2}e^{-t}\left(e^{-t^{3/2}} + \frac{3\sqrt{\pi}}{4}\right). \tag{10.32}$$

As per the results shown in Table 10.5, the proposed method produces an approximate solution almost equal to the exact solution. The nonlinear performance index is minimized to 3.19452852272872, which is very close to the actual optimal J^*.

TABLE 10.6

Absolute errors for case study 10.4

	$\lvert x_1(t)-\tilde{x}_1(t)\rvert$		$\lvert x_2(t)-\tilde{x}_2(t)\rvert$		$\lvert u(t)-\tilde{u}(t)\rvert$	
t	**LMCM in Ref. [13]**	**Proposed method**	**LMCM in Ref. [13]**	**Proposed method**	**LMCM in Ref. [13]**	**Proposed method**
0.1	6.94706e-06	6.98498e-06	1.80223e-06	5.46170e-08	2.59777e-06	1.8180e-06
0.2	5.66267e-06	6.09315e-06	1.28185e-06	8.93967e-08	2.12128e-06	1.5404e-06
0.3	1.42083e-06	5.31176e-06	1.754e-06	1.09773e-07	8.64065e-07	1.2903e-06
0.4	5.48337e-07	4.62904e-06	2.10988e-06	1.19822e-07	5.08921e-07	1.0638e-06
0.5	3.53749e-06	4.03423e-06	6.63399e-07	1.22622e-07	1.34629e-06	8.5728e-07
0.6	5.8476e-06	3.51755e-06	2.86712e-06	1.20467e-07	1.96033e-06	6.6695e-07
0.7	3.90867e-06	3.07024e-06	1.5552e-06	1.15064e-07	1.31925e-06	4.8939e-07
0.8	1.1543e-07	2.68447e-06	1.5283e-06	1.07659e-07	9.16113e-06	3.2125e-07
0.9	1.84314e-06	2.35334e-06	1.28547e-06	9.91562e-08	1.00325e-07	1.5921e-07

Case study 10.4: Optimal control of two-dimensional IOS

Consider the integer-order optimal control problem:

$$\min J = \frac{1}{2}\int_0^1 \left(x_1^2(t) + x_2^2(t) + u^2(t)\right)dt, \tag{10.33}$$

subject to the dynamical system

$$\frac{dx_1(t)}{dt} = -x_1(t) + x_2(t) + u(t), \tag{10.34}$$

$$\frac{dx_2(t)}{dt} = -2x_2(t), \tag{10.35}$$

and the initial conditions $x_1(0) = x_2(0) = 1$.
For the above problem, the exact solution can be found as follows:

$$x_1(t) = -\frac{3}{2}e^{-2t} + 2.48164e^{-\sqrt{2}t} + 0.018352e^{\sqrt{2}t}, \tag{10.36}$$

$$x_2(t) = e^{-2t}, \tag{10.37}$$

$$u(t) = \frac{e^{-2t}}{2} - 1.02793e^{-\sqrt{2}t} + 0.0443056e^{\sqrt{2}t}. \tag{10.38}$$

It is seen from Table 10.6 that the performance of proposed method is similar to that of Legendre multiwavelet collocation method formulated in [13]. The optimal cost function found via the proposed method is J=0.431987437877140, which is equal to the optimal value J=0.431987 achieved in [14] by Epsilon-Ritz method.

10.3 MATLAB® Codes for Simulation Examples

This section provides the source codes developed in MATLAB for implementation of the proposed method in case studies 10.1 to 10.4.

Program 10.1

```
%%%%%%%%%%%%% MATLAB code for case study 10.1 %%%%%%%%%%%
function [J,EE1,EE2]=Casestudy_10_1
t0=0;T=1;aa=150;m=10*aa;h=(T-t0)/m;t=[t0:h:T];I=ones(1,
length(t));
x0=zeros(1,2*(m+1));
tic
C0=xt(1)*I(1:end-1);D0=xt(1)*I(1:end-1);C00=-ut(end)*I
  (1:end-1);
D00=-ut(end)*I(1:end-1);[P1alph,P2alph]=TOF1(t0,T,m,1);
[P3alph,P4alph]=TOF12(t0,T,m,1);P1alphr=zeros(m,m);
P2alphr=zeros(m,m);
P3alphr=zeros(m,m);P4alphr=zeros(m,m);
for j=1:m
  P1alphr(j,:)=-wrev(P1alph(m-j+1,:));P2alphr(j,:)=-wrev
(P2alph(m-j+1,:));
  P3alphr(j,:)=-wrev(P3alph(m-j+1,:));P4alphr(j,:)=-wrev
(P4alph(m-j+1,:));
end
[x,fval]=fsolve(@Problem_Fun,x0);
function f=Problem_Fun(x)
for i=1:1:length(t)
  CC1(i)=x(i);
end
f1=C00-ff0(1:end-1)-C2;C1=CC1(1:end-1);D1=CC1(2:end);
for ii=length(t)+1:1:2*length(t)
  CC2(ii-length(t))=x(ii);
end
```

```
C2=CC2(1:end-1);D2=CC2(2:end);CC3=CC1-CC2;C01=CC3(1:
  end-1);D01=CC3(2:end);
f01=((C01)*P1alphr+(D01)*P3alphr);f02=((C01)*P2alphr
  +(D01)*P4alphr);
ff0=[f02(1) f01];CC4=-CC1-CC2;C02=CC4(1:end-1);D02=CC4
  (2:end);
f2=D00-ff0(2:end)-D2;f3=C0+((C02)*P1alph+(D02)
  *P3alph)-C1;
f4=D0+((C02)*P2alph+(D02)*P4alph)-D1;f=[f1 f2 f3 f4];
end
toc
% TF solution
x=real(x);xtf=x(1:m+1);utf=x((m+1)+1:end);utf=-utf;
 % Exact Solution
beta=-
[cosh(sqrt(2))+sqrt(2)*sinh(sqrt(2))]/[sqrt(2)*cosh
(sqrt(2))+sinh(sqrt(2))];
xt=cosh(sqrt(2)*t)+beta*sinh(sqrt(2)*t);
ut=(1+sqrt(2)*beta)*cosh(sqrt(2)*t)+(sqrt(2)+beta)*sinh
  (sqrt(2)*t);
ee1=abs(xt-xtf);ee2=abs(ut-utf);norminf1=norm(ee1,inf);
normrms1=norm(ee1,2);
norminf2=norm(ee2,inf);normrms2=norm(ee2,2);
% minimum value of J
CC=0.5*(xtf.*xtf+utf.*utf);C01=CC(1:end-1);D01=CC
  (2:end);
C02=((C01)*P1alph+(D01)*P3alph);D02=((C01)*P2alph+(D01)
  *P4alph);
CC1=[C02 D02(end)];cc=CC1(end)-CC1(1);J=cc;
for i=1:11
  bb=(i-1)*aa+1;EE1(i)=ee1(bb);
end
EE1=EE1';
for i=1:11
  cc=(i-1)*aa+1;EE2(i)=ee2(cc);
end
EE2=EE2';
end
```

Program 10.2

```
%%%%%%%%%%%%% MATLAB code for case study 10.2 %%%%%%%%%%%
function [J,EE1,EE2]= Casestudy_10_2
t0=0;T=1;aa=100;m=10*aa;h=(T-t0)/m;t=[t0:h:T];I=ones(1,
  length(t));
x0=zeros(1,2*(m+1));
tic
C0=xt(1)*I(1:end-1);D0=xt(1)*I(1:end-1);C00=lambdat
  (end)*I(1:end-1);
```

```
D00=lambdat(end)*I(1:end-1);[P1alph,P2alph]=TOF1(t0,
  T,m,1.1);
[P3alph,P4alph]=TOF12(t0,T,m,1.1);P1alphr=zeros(m,m);
P2alphr=zeros(m,m);
P3alphr=zeros(m,m);P4alphr=zeros(m,m);
for j=1:m
  P1alphr(j,:)=-wrev(P1alph(m-j+1,:));P2alphr(j,:)=-wrev
(P2alph(m-j+1,:));
  P3alphr(j,:)=-wrev(P3alph(m-j+1,:));P4alphr(j,:)=-wrev
(P4alph(m-j+1,:));
end
[x,fval]=fsolve(@Problem_Fun,x0);
function f=Problem_Fun(x)
for i=1:1:length(t)
  CC1(i)=x(i);
end
C1=CC1(1:end-1);D1=CC1(2:end);
for ii=length(t)+1:1:2*length(t)
  CC2(ii-length(t))=x(ii);
end
C2=CC2(1:end-1);D2=CC2(2:end);CC3=2*CC1-2*t.*t+CC2.*t.
*t;C01=CC3(1:end-1);
D01=CC3(2:end);f01=((C01)*P1alphr+(D01)*P3alphr);
f02=((C01)*P2alphr+(D01)*P4alphr);ff0=[f02(1) f01];
AA=(40*(t.^(9/10)))/(9*gamma(9/10));CC4=t.*t.*CC1+0.5*
  (-2*(t.^4)+AA-CC2);
C02=CC4(1:end-1);D02=CC4(2:end);f1=C0+((C02)*P1alph
+(D02)*P3alph)-C1;
f2=D0+((C02)*P2alph+(D02)*P4alph)-D1;f3=C00-ff0(1:end-
  1)-C2;
f4=D00-ff0(2:end)-D2;f=[f1 f2 f3 f4];
end
toc
% TF solution
xtf=x(1:m+1);utf=x((m+1)+1:end);AA=(40*(t.^(9/10)))/
  (9*gamma(9/10));
utf=0.5*(-2*(t.^4)+AA-utf);
 % Exact Solution
xt=t.*t;ut=(20/(9*gamma(9/10)))*(t.^(9/10))-(t.^4);
% absolute error
ee1=abs(xt-xtf);ee2=abs(ut-utf);norminf1=norm(ee1,inf);
normrms1=norm(ee1,2);
norminf2=norm(ee2,inf);normrms2=norm(ee2,2);
% minimum value of J
[P1alph1,P2alph1]=TOF1(t0,T,m,1);[P3alph1,P4alph1]
  =TOF12(t0,T,m,1);
AA=(20*(t.^(9/10)))/(9*gamma(9/10));CC=((xtf-t.*t).^2)
  +([utf+(t.^4)-AA].^2);
```

```
C01=CC(1:end-1);D01=CC(2:end);C02=((C01)*P1alph1+(D01)
  *P3alph1);
D02=((C01)*P2alph1+(D01)*P4alph1);CC1=[C02    D02(end)];
cc=CC1(end)-CC1(1);
J=cc;
for i=1:11
  bb=(i-1)*aa+1;EE1(i)=ee1(bb);
end
EE1=EE1';
for i=1:11
  cc=(i-1)*aa+1;EE2(i)=ee2(cc);
end
EE2=EE2';
end
```

Program 10.3

```
%%%%%%%%%%%% MATLAB code for case study 10.3 %%%%%%%%%%%
function [J,EE1,EE2]= Casestudy_10_3
t0=0;T=1;aa=150;m=10*aa;h=(T-t0)/m;t=[t0:h:T];I=ones(1,
  length(t));
x0=zeros(1,2*(m+1));
tic
C0=xt(1)*I(1:end-1);D0=xt(1)*I(1:end-1);C00=lambdat
  (end)*I(1:end-1);
D00=lambdat(end)*I(1:end-1);[P1alph,P2alph]=TOF1(t0,
  T,m,1.5);
[P3alph,P4alph]=TOF12(t0,T,m,1.5);P1alphr=zeros(m,m);
P2alphr=zeros(m,m);
P3alphr=zeros(m,m);P4alphr=zeros(m,m);
for j=1:m
  P1alphr(j,:)=-wrev(P1alph(m-j+1,:));P2alphr(j,:)=-wrev
  (P2alph(m-j+1,:));
  P3alphr(j,:)=-wrev(P3alph(m-j+1,:));P4alphr(j,:)=-wrev
  (P4alph(m-j+1,:));
end
[x,fval]=fsolve(@Problem_Fun,x0);
function f=Problem_Fun(x)
for i=1:1:length(t)
  CC1(i)=x(i);
end
C1=CC1(1:end-1);D1=CC1(2:end);
for ii=length(t)+1:1:2*length(t)
  CC2(ii-length(t))=x(ii);
end
C2=CC2(1:end-1);D2=CC2(2:end);CC4=2*CC1-2*(t.^1.5)+exp
  (CC1).*CC2;
C01=CC4(1:end-1);D01=CC4(2:end);f01=((C01)*P1alphr
  +(D01)*P3alphr);
f02=((C01)*P2alphr+(D01)*P4alphr);ff0=[f02(1) f01];
```

```
CC3=exp(CC1)+exp(t).*[(3/4)*sqrt(pi)*exp(-t)-exp(-t+(t.
  ^1.5))-2*exp(t).*CC2];
C02=CC3(1:end-1);D02=CC3(2:end);f1=C0+((C02)*P1alph
  +(D02)*P3alph)-C1;
f2=D0+((C02)*P2alph+(D02)*P4alph)-D1;f3=C00-ff0(1:end-
  1)-C2;
f4=D00-ff0(2:end)-D2;f=[f1 f2 f3 f4];
end
toc
% TF solution
xtf=x(1:m+1);utf=x((m+1)+1:end);
utf=0.5*((3/4)*sqrt(pi)*exp(-t)-exp(-t+(t.^(1.5)))-
  2*exp(t).*utf);
 % Exact Solution
xt=t.^1.5;ut=0.5*exp(-t).*[-exp(t.^(1.5))+(3/4)
  *sqrt(pi)];
% absolute error
ee1=abs(xt-xtf);ee2=abs(ut-utf);norminf1=norm(ee1,inf);
normrms1=norm(ee1,2);
norminf2=norm(ee2,inf);normrms2=norm(ee2,2);
% minimum value of J
[P1alph1,P2alph1]=TOF1(t0,T,m,1);[P3alph1,P4alph1]
  =TOF12(t0,T,m,1);
CC=xtf.*xtf-2*(t.^(1.5)).*xtf+utf.*utf-(3/4)*sqrt(pi)
*exp(-t).*utf+exp(-t+(t.^(1.5))).*utf+(t.^3)+((9*pi)/
  64)*exp(-2*t)-(3/8)*sqrt(pi)*exp(-2*t+(t.^(1.5)))
  +0.25*exp(-2*t+2*(t.^(1.5)))+exp(2*t);
C01=CC(1:end-1);D01=CC(2:end);C02=((C01)*P1alph1+(D01)
*P3alph1);
D02=((C01)*P2alph1+(D01)*P4alph1);CC1=[C02 D02(end)];
cc=CC1(end)-CC1(1);J=cc;
for i=1:11
   bb=(i-1)*aa+1;EE1(i)=ee1(bb);
end
EE1=EE1';
for i=1:11
   cc=(i-1)*aa+1;EE2(i)=ee2(cc);
end
EE2=EE2';
end
```

Program 10.4

```
%%%%%%%%%%%%%% MATLAB code for case study 10.4 %%%%%%%%%%
function [J,EE1,EE2]= Casestudy_10_4
t0=0;T=1;aa=100;m=10*aa;h=(T-t0)/m;t=[t0:h:T];I=ones(1,
length(t));
x0=zeros(1,4*(m+1));
tic
```

```
C01=I(1:end-1);D01=I(1:end-1);C02=I(1:end-1);D02=I(1:
end-1);
C03=lambdat(end)*I(1:end-1);D03=lambdat(end)*I(1:end-
1);C04=0*I(1:end-1);
D04=0*I(1:end-1);[P1alph,P2alph]=TOF1(t0,T,m,1);
[P3alph,P4alph]=TOF12(t0,T,m,1);P1alphr=zeros(m,m);
P2alphr=zeros(m,m);
P3alphr=zeros(m,m);P4alphr=zeros(m,m);
for j=1:m
  P1alphr(j,:)=-wrev(P1alph(m-j+1,:));P2alphr(j,:)=-wrev
  (P2alph(m-j+1,:));
  P3alphr(j,:)=-wrev(P3alph(m-j+1,:));P4alphr(j,:)=-wrev
  (P4alph(m-j+1,:));
end
[x,fval]=fsolve(@Problem_Fun,x0);
function f=Problem_Fun(x)
for i=1:1:length(t)
  CC1(i)=x(i);
end
C1=CC1(1:end-1);D1=CC1(2:end);
for ii=length(t)+1:1:2*length(t)
  CC2(ii-length(t))=x(ii);
end
C2=CC2(1:end-1);D2=CC2(2:end);
for iii=2*length(t)+1:1:3*length(t)
  CC3(iii-2*length(t))=x(iii);
end
C3=CC3(1:end-1);D3=CC3(2:end);
for iiii=3*length(t)+1:1:4*length(t)
  CC4(iiii-3*length(t))=x(iiii);
end
C4=CC4(1:end-1);D4=CC4(2:end);x1t=CC1;x2t=CC2;lamb
  da1t=CC3;lambda2t=CC4;
AC1=-x1t+x2t-lambda1t;A01=AC1(1:end-1);A02=AC1(2:end);
f1=C01+((A01)*P1alph+(A02)*P3alph)-C1;f2=D01+((A01)
  *P2alph+(A02)*P4alph)-D1;
AC2=-2*x2t;A03=AC2(1:end-1);A04=AC2(2:end);
f3=C02+((A03)*P1alph+(A04)*P3alph)-C2;f4=D02+((A03)
  *P2alph+(A04)*P4alph)-D2;
AC3=-lambda1t+x1t;A05=AC3(1:end-1);A06=AC3(2:end);
f01=((A05)*P1alphr+(A06)*P3alphr);f02=((A05)*P2alphr
  +(A06)*P4alphr);
ff0=[f02(1)    f01];f5=C03-ff0(1:end-1)-C3;f6=D03-ff0(2:
end)-D3;
AC4=-2*lambda2t+lambda1t+x2t;A07=AC4(1:end-1);A08=AC4
(2:end);
f03=((A07)*P1alphr+(A08)*P3alphr);f04=((A07)*P2alphr
  +(A08)*P4alphr);
```

```
ff00=[f04(1) f03];f7=C04-ff00(1:end-1)-C4;f8=D04-ff00(2:
  end)-D4;
f=[f1 f2 f3 f4 f5 f6 f7 f8];
end
toc
% TF solution
xtf1=x(1:m+1);xtf2=x((m+1)+1:2*(m+1));lamd1t=x(2*(m+1)+
  1:3*(m+1));
utf=-lamd1t;
 % Exact Solution
xt1=-1.5*exp(-2*t)+2.48164*exp(-sqrt(2)*t)+0.018352*exp
  (sqrt(2)*t);
xt2=exp(-2*t);
ut=0.5*exp(-2*t)-1.02793*exp(-sqrt(2)*t)+0.0443056*exp
  (sqrt(2)*t);
% absolute error
ee1=abs(xt1-xtf1);ee2=abs(xt2-xtf2);ee3=abs(ut-utf);nor
  minf1=norm(ee1,inf);
normrms1=norm(ee1,2);norminf2=norm(ee2,inf);normrms2=-
norm(ee2,2);
norminf3=norm(ee3,inf);normrms3=norm(ee3,2);
% minimum value of J
[P1alph1,P2alph1]=TOF1(t0,T,m,1);[P3alph1,P4alph1]
  =TOF12(t0,T,m,1);
CC=0.5*(xtf1.*xtf1+xtf2.*xtf2+utf.*utf);C011=CC(1:end-
  1);D011=CC(2:end);
C022=((C011)*P1alph1+(D011)*P3alph1);D022=((C011)
  *P2alph1+(D011)*P4alph1);
CC1=[C022 D022(end)];cc=CC1(end)-CC1(1);J=cc;
  for i=1:11
   bb=(i-1)*aa+1;EE1(i)=ee1(bb);
end
EE1=EE1';
for i=1:11
   cc=(i-1)*aa+1;EE2(i)=ee2(cc);
end
EE2=EE2';
for i=1:11
   dd=(i-1)*aa+1;EE3(i)=ee3(dd);
end
EE3=EE3';
end
```

References

[1] J.T. Betts(2001). *Practical Methods for Optimal Control Using Nonlinear Programming*. Philadelphia: SIAM.

[2] J.T. Betts, S.O. Erb (2003). Optimal low thrust trajectories to the moon. *SIAM J. Appl. Dyn. Syst.*, vol. 2, pp.144–170.

[3] A.B. Malinowska, D.F.M. Torres (2012). *Introduction to the Fractional Calculus of Variations*. London: Imperial College Press.

[4] R. Almedia, S. Pooseh, D.F.M. Torres (2015). *Computational Methods in the Fractional Calculus of Variations*. London: Imperial College Press.

[5] A.B. Malinowska, T. Odzijewicz, D.F.M. Torres (2015). *Advanced Methods in the Fractional Calculus of Variations*. New York: Springer Briefs in Applied Sciences and Technology.

[6] L.S. Pontryagin, V. Boltyanskii, R. Gamkrelidze, E. Mischenko (1962). *The Mathematical Theory of Optimal Processes*. New York: John Wiley & Sons.

[7] M.H. Heydari, M.R. Hooshmandasl, F.M. Maalek Ghaini, C. Cattani (2016). Wavelets method for solving fractional optimal control problems. *Appl. Math. Comput.*, vol. 286, pp.139–154.

[8] E. Keshavarz, Y. Ordokhani, M. Razzaghi (2015). A numerical solution for fractional optimal control problems via Bernoulli polynomials. *J. Vib. Cont.*, vol. 22(18), pp.3889–3903.

[9] S. Nemati (2016). A spectral method based on the second kind Chebyshev polynomials for solving a class of fractional optimal control problems. *Sahand Commun. Math. Anal.*, vol. 4(1), pp.15–27.

[10] A.H. Bhrawy, E.H. Doha, D. Baleanu, S.S. Ezz-Eldien, M.A. Abdelkawy (2015). An accurate numerical technique for solving fractional optimal control problems. *Proc. Romanian Acad.*, vol. 16(1), pp.47–54.

[11] M.H. Heydari, M.R. Hooshmandasl, A. Shakiba, C. Cattani (2016). An efficient computational method based on the hat functions for solving fractional optimal control problems. *Tbilisi Math. J.*, vol. 9(1), pp.143–157.

[12] A. Lotfi, S.A. Yousefi, M. Dehghanb (2013). Numerical solution of a class of fractional optimal control problems via the Legendre orthonormal basis combined with the operational matrix and the Gauss quadrature rule. *J. Comput. Appl. Math.*, vol. 250, pp.143–160.

[13] S.A. Yousefi, A. Lotfi, M. Dehghan (2011). The use of a Legendre multiwavelet collocation method for solving the fractional optimal control problems. *J. Vib. Cont.*, vol. 17(13), pp.2059–2065.

[14] A. Lotfi, S.A. Yousefi (2014). Epsilon-Ritz method for solving a class of fractional constrained optimization problems. *J. Optimiz. Theory. App.*, vol. 163, pp.884–899.

Index

R

S

W

X